Frontiers in Earth Sciences

Series Editors

J. P. Brun, Clermont-Ferrand, France

Onno Oncken, Potsdam, Brandenburg, Germany

Helmut Weissert, Zürich, Switzerland

Wolf-Christian Dullo, Paleoceanography, Helmholtzzentrum für Ozeanforschrung | G, Kiel, Germany

More information about this series at http://www.springer.com/series/7066

Gilles Ramstein • Amaëlle Landais •
Nathaelle Bouttes • Pierre Sepulchre •
Aline Govin
Editors

Paleoclimatology

Volume 1

Editors
Gilles Ramstein
LSCE/IPSL, CEA-CNRS-UVSQ
Université Paris-Saclay
Gif-sur-Yvette, Ariège, France

Nathaelle Bouttes
LSCE/IPSL, CEA-CNRS-UVSQ
Université Paris-Saclay
Gif-sur-Yvette, Ariège, France

Aline Govin
LSCE/IPSL, CEA-CNRS-UVSQ
Université Paris-Saclay
Gif-sur-Yvette, Ariège, France

Amaëlle Landais
LSCE/IPSL, CEA-CNRS-UVSQ
Université Paris-Saclay
Gif-sur-Yvette, Ariège, France

Pierre Sepulchre
LSCE/IPSL, CEA-CNRS-UVSQ
Université Paris-Saclay
Gif-sur-Yvette, Ariège, France

ISSN 1863-4621   ISSN 1863-463X (electronic)
Frontiers in Earth Sciences
ISBN 978-3-030-24981-6   ISBN 978-3-030-24982-3 (eBook)
https://doi.org/10.1007/978-3-030-24982-3

Translated from the original French by Mary Minnock
© Springer Nature Switzerland AG 2021
This work is subject to copyright. All rights are reserved by the Publisher, whether the whole or part of the material is concerned, specifically the rights of translation, reprinting, reuse of illustrations, recitation, broadcasting, reproduction on microfilms or in any other physical way, and transmission or information storage and retrieval, electronic adaptation, computer software, or by similar or dissimilar methodology now known or hereafter developed.
The use of general descriptive names, registered names, trademarks, service marks, etc. in this publication does not imply, even in the absence of a specific statement, that such names are exempt from the relevant protective laws and regulations and therefore free for general use.
The publisher, the authors and the editors are safe to assume that the advice and information in this book are believed to be true and accurate at the date of publication. Neither the publisher nor the authors or the editors give a warranty, expressed or implied, with respect to the material contained herein or for any errors or omissions that may have been made. The publisher remains neutral with regard to jurisdictional claims in published maps and institutional affiliations.

This Springer imprint is published by the registered company Springer Nature Switzerland AG
The registered company address is: Gewerbestrasse 11, 6330 Cham, Switzerland

Foreword

A brief history of paleoclimates

Climate is undeniably a topical issue of utmost importance. It has been the focus of attention for several decades now, during which the study of ancient climates (paleoclimatology) has progressed and gained a solid reputation. Currently, this work has become fundamental to our understanding of how the climate system functions and to validate the models used to establish future projections. Thanks to the study of past climates, a database documenting a much greater diversity of climate changes than during recent centuries, has been created. This diversity makes it possible to test climate models in situations that are vastly different from those we have known over the last 150 years and, in some cases, for climates that are closer to those that await us in the future if we apply the conclusions of the Intergovernmental Panel for Climate Change.

The Earth's climate changes have always changed over time and will continue to change in the future. While we are all aware of the weather phenomena that condition our daily lives, few of us are aware of what climate really is. Climatology is the science that explores the great variability in meteorological conditions over time and space, throughout history. This word comes from the Greek word *klima* meaning inclination, in this case referring to that of the rays of the Sun. Since the dawn of our civilization, therefore, we have linked the variations in climate and in the energy received from the Sun in a relationship of cause and effect. For a long time, the term 'climate' was reserved to describe the characteristics of air temperature and precipitation particular to different parts of the globe. This description was based on meteorological measurements and their averages conducted over a few decades. It is only recently understood that climate also varies over much longer time scales and therefore concerns more than just the atmosphere. At present, specialists studying climate and its variations analyze all of the fluid and solid envelopes of the Earth. Along with the atmosphere, we associate the hydrosphere and the cryosphere which, together, represent the systems where water exists in solid (snow, glaciers, and ice sheets) and liquid form (rivers, lakes, and seas), the continents where plate tectonics and volcanic activity occur, and finally at the surface, the whole living world (biosphere) that influences nature, the properties of soil cover, and the biogeochemical cycles.

Climatology has evolved from being a descriptive discipline to become a multidisciplinary science involving five complex systems and their various interactions. It is therefore not surprising that the resulting climate studies vary on scales ranging from the season to millions of years. Although it is only in the last few decades that this science has exploded, the first discovery and study of climate change beyond the annual and decadal scales date back to the eighteenth century. It was at this time that the presence of erratic boulders in the mountainous landscape became associated for the first time with the massive extension of glaciers. In 1744, the Grenoble geographer Pierre Martel (1706–1767) reported that the inhabitants of the Chamonix Valley in the Alps attributed the dispersion of *roches moutonnées* to the glaciers themselves, which would have extended much further in the past. This was a revolutionary idea, because until then, most scientists still referred to the myth of the Biblical Flood to explain landscape structures. This was the case of Horace Benedicte de Chaussure (1740–1799) from Geneva, the French paleontologist, Georges Cuvier (1769–1832) and the Scottish geologist,

Charles Lyell (1797–1875), who continued to assume that these boulders were carried by the strength of strongly flowing waters. However, the location and nature of these boulders and other moraines led some scientists to admit that ice transport would provide a better explanation for the various observations. The Scottish naturalist, James Hutton (1726–1797), was the first to subscribe to this idea. Others followed his lead and detected the imprint of climatic changes in the fluctuations of the extent of the glaciers. These pioneers were the Swiss engineer, Ignace Venetz (1788–1859); the German forestry engineer, Albrecht Reinhart Benhardi (1797–1849); the Swiss geologist, Jean de Charpentier (1786–1855); and the German botanist, Karl Friedrich Schimper (1803–1867), who introduced the notion of ice ages. But it was the Danish-Norwegian geologist, Jens Esmark (1763–1839), who, in pursuing his analysis of glacier transport, proposed in 1824, for the first time, the notion that climate changes could be the cause and that these could have been instigated by variations of Earth's orbit.

It was the work of these pioneers that led the Swiss geologist, Louis Agassiz (1801–1873) to make the address to the Swiss Society of Natural Sciences of Neufchatel in 1837 entitled 'Upon glaciers, moraines and erratic blocks'. It was also at the beginning of the nineteenth century that the Frenchman Joseph Adhémar (1797–1862), not content with studying the polar ice caps, attempted to explain in his book, *Révolutions de la Mer, Déluges Périodiques* (1842), the pattern of ice ages stemming from the precession of the equinoxes. The astronomical theory of the paleoclimates was born and would be continued, thanks to the development of celestial mechanics, by the Frenchmen, Jean le Rond d'Alembert (1717–1783), Jean-Baptiste Joseph Delambre (1749–1822), Pierre-Simon Laplace (1749–1827), Louis Benjamin Francoeur (1773–1849), and Urban Le Verrier (1811–1877). In parallel, other advances were made with the first calculations of the long-term variations in the energy received from the Sun, variations due to the astronomical characteristics of the eccentricity of the Earth's orbit, the precession of the equinoxes, and the obliquity of the ecliptic. This was demonstrated by the work of John Frederick William Herschel (1792–1871), L.W. Meech (1821–1912), and Chr. Wiener (1826–1896), supported by the work of the mathematicians André-Marie Legendre (1751–1833) and Simon-Denis Poisson (1781–1840).

This sets the stage for James Croll (1821–1890) to develop a theory of ice ages based on the combined effect of the three astronomical parameters, a theory according to which winter in the northern hemisphere played a determining role. This theory was much appreciated by the naturalist, Charles Robert Darwin (1809–1882), and was taken up by the Scottish geologist brothers, Archibald (1835–1924) and James (1839–1914) Geikie, who introduced the notion of the interglacial. It is also the basis for the classification of alpine glaciations by Albrecht Penck (1858–1945) and Edward Brückner (1862–1927) and American glaciations by Thomas Chowder Chamberlin (1843–1928). However, geologists became increasingly dissatisfied with Croll's theory and many critics of it emerged. Many refuted the astronomical theory and preferred explanations that related to the Earth alone. The Scottish geologist, Charles Lyell (1797–1875), claimed that the geographical distribution of land and seas explained the alternation of hot and cold climates, while others turned to variations in the concentration of certain gases in the atmosphere. Hence, the French physicist, Joseph Fourier (1786–1830), expounded on the first notion of the theory of the greenhouse effect. He was followed by the Irish chemist, John Tyndall (1820–1893), to whom we owe the first experiments on the absorption of infrared radiation and the hypothesis of the fundamental role played by water vapor in the greenhouse effect. Later, the Italian, Luigi de Marchi (1857–1937) and the Swedish chemist, Svante Arrhenius (1859–1927) proposed, along with other scientists of their time, that the ice ages were caused by decreases in atmospheric carbon dioxide concentration. In 1895, Arrhenius suggested, in an article published by the Stockholm Physics Society, that a 40% reduction or increase in CO_2 concentration in the atmosphere could lead to feedback processes that would explain glacial advances or retreats.

A revival of the astronomical theory became, however, possible with advances in the calculation of astronomical elements by the American astronomer John Nelson Stockwell (1822–1920) and the Serbian astronomer Vojislava Protich Miskovitch (1892–1976) and of solar irradiation (1904) by the German mathematician, Ludwig Pilgrim (1879–1935). It was Joseph John Murphy (1827–1894), however, who, as early as 1869, proposed that cool summers of the northern hemisphere had instigated the ice ages. This original idea was taken up in 1921 by the German paleoclimatologist Rudolf Spitaler (1859–1946), but was popularized by the Serbian geophysicist engineer, Milutin Milankovich (1879–1958), mainly through his books *Mathematical Theory of Thermal Phenomena Produced by Solar Radiation* (1920) and *Kanon der Erdbestrahlung und seine Anwendung auf des Eizeitenproblem* (1941). Milankovitch was a contemporary of the German geophysicist Alfred Wegener (1880–1930) with whom he became acquainted through the Russsian-born climatologist Wladimir Köppen (1846–1940), Wegener's father-in-law (Thiede, 2017). The modern era of astronomical theory was born, even if there remained much criticism related to the lack of reliable paleoclimatic data and of a reliable timescale, both by geologists and meteorologists. It was not until the 1950s and 1960s that new techniques made it possible to date, measure, and interpret the climate records contained in marine sediments, in ice and on land. In 1955, the American, Cesare Emiliani (1922–1995), proposed a stratigraphy, which still applies today, based on the succession of minima and maxima of the oxygen-18 / oxygen-16 isotopic ratio measured in the foraminiferal shells found in sediments taken from the deep ocean. The interpretation of this isotopic ratio in terms of salinity was made by Jean-Claude Duplessy (1970), and in terms of temperature and volume of ice (1973) by Nicholas Shackleton (1937–2006) and Niels Opdyke (1933–2019). Mathematical tools made it possible to establish transfer functions to quantitatively interpret information collected in the oceans in 1974 by the American paleoceanographers John Imbrie (1925–2016) and Nilva Kipp (1925–1989),) and in tree rings (Harold Fritts, 1968). Efforts by the CLIMAP group (1976) resulted in the first seasonal climate chart of the Last Glacial Maximum and the pivotal article by James Hays, John Imbrie, and Nicholas Shackleton (1976). The arrival of big computers allowed the first climate simulations to be conducted using general circulation models (Fred Nelson Alyea, 1972), and further astronomical calculations led to the establishment of a high-precision time scale reference, as well as the determination of the daily and seasonal irradiation essential for climate modeling (André Berger, 1973 and Berger and Loutre, 1991).These calculations of the astronomical parameters were based on the 1974 and 1988 developments of the orbital elements by the French astronomers Pierre Bretagnon (1942–2002) and Jacques Laskar, respectively. These are valid over a few million years. The Laskar solution was extended over a few tens of millions of years by Laskar et al. (2011) and over the whole Mesozoic with the American paleobiologist Paul Olsen and colleagues (2019).

This evolution and the recent advances in paleoclimatology show the difficulties involved in tackling the study of the climate system. Overcoming these difficulties requires high-quality books to improve understanding and to update the range of disciplines involved. It is with this perspective in mind that this book was written. Written originally in French, it unquestionably fills a gap in the field of graduate and postgraduate third-level education that goes far beyond its description. It provides an overview of the state of knowledge on a number of key topics by outlining the information necessary to understand and appreciate the complexity of the disciplines discussed, making it a reference book on the subject. The first of the two volumes is devoted to the methods used to reconstruct ancient climates, the second to the behavior of the climate system in the past. Many of the thirty-one chapters are written by researchers from the *Laboratoire des Sciences du Climat et de l'Environnement* and associated research laboratories each focusing on his or her area of expertise, which ensures a reliable document founded on solid experience.

Understanding the evolving climate of the Earth and its many variations is not just an academic challenge. It is also fundamental in order to better understand the future climate and its possible impacts on the society of tomorrow. Jean-Claude Duplessy and Gilles Ramstein have achieved this huge feat by bringing together fifty or so of the most highly reputed researchers in the field.

The book they have written is a whole, providing both the necessary bases on the reconstruction techniques of ancient climates, their chronological framework, and the functioning of the climate system in the past based on observations and models. This book will allow all those who want to know more, to explore this science, which, although difficult, is hugely exciting. It will also give them the essential information to establish an objective idea of the climate and its past and future variations.

You may find most of the references and pioneering studies mentioned in this preface in BERGER A. 2012. A brief history of the astronomical theories of paleoclimates. In: "Climate change at the eve of the second decade of the century. Inferences from paleoclimates and regional aspects". Proceedings of Milankovitch 130th Anniversary Symposium, A. Berger, F. Mesinger, D. Sijacki (eds). 107–129. Springer-Verlag/Wien.DOI 10.1007/978-3-7091-0973-1.

André Berger
Emeritus Professor at the Catholic University of Louvain
Louvain la Neuve, Belgium

Introduction

For a long time, geology books devoted only a few lines to the history of past climates of our planet, mostly to establish the deposition framework for the sediments that geologists found on the continents, the only area of enquiry available to them. Scientists soon realized that the copious coal deposits of England, Belgium, Northern France, Germany, and Poland resulted from the fossilization of abundant vegetation facilitated by a warm and humid equatorial climate that reigned over Western Europe, some 350 million years ago (an illustrated insert in Chap. 2 volume I provides a diagram of continental drift since 540 Ma). Fifty million years later, the sediments of these same regions, red sandstone, poor in fossils and associated with evaporites testify to the replacement of forests by desert areas, dotted with occasional highly saline lakes, similar to what we currently find in Saharan Africa. Humidity gave way to intense aridity and we had no idea why. It was not until the discovery of plate tectonics that we realized that Europe had slowly drifted toward the tropics. This transformation of the face of the Earth due to tectonics is illustrated through 16 maps in Chap. 2 volume I.

The discovery of glaciations was a revelation for the geologists of the nineteenth century. A major polemic broke out at the Swiss Society of Natural Sciences in Neuchâtel when, in 1837, its president Louis Agassiz presented his explanation, incredible at the time, for the presence of gigantic boulders that dot the Jura mountains. He daringly claimed that these erratic boulders were not the remnants of the Biblical Flood, but rather enormous rocks transported over long distances by gigantic glaciers which used to cover the high latitudes of our hemisphere.

The controversy died down quickly, when European and American geologists discovered traces of glaciers all over the Northern Hemisphere, just as Agassiz imagined. In Europe, as in North America, mapping of the terminal moraines left behind by glaciers when they melted showed proof of the presence of gigantic ice caps in a past that seemed distant. especially since there was no idea how to date them.

As the idea of the Biblical Flood fell out of favor, a new theory, based on astronomical phenomena, soon appeared. Scientists like Joseph Adhémar and James Croll realized that there were small, quasi-periodic variations over time in the movement of the Earth around the Sun and suggested that associated mechanisms could periodically cause glacial advances and retreats. Finally, it was Milutin Milankovitch, a professor in Belgrade, who would lay the foundations for a complete mathematical theory of glaciations, the legitimacy of which was proven when paleoceanographers found the frequencies of orbital parameters reflected in the isotopic analysis of marine cores. We now know that the last one of these glacial periods culminated only 20,000 years ago and was preceded by many others.

The great contribution by Milankovitch was to plant a new idea within the scientific community: Ancient climates are not only of immense curiosity to geologists; they obey the same physical laws as those governing the current climate.

This intellectual revolution has had far-reaching consequences and has profoundly altered the approach to the study of ancient climates making paleoclimatology a science with many links to geology, geochemistry, oceanography, glaciology as well as the approach to the physical and dynamic dimensions of the climate. The first part of this book describes the physical, chemical, and biological phenomena that govern the functioning of the climate system and shows how it is possible to reconstruct the variations in the past at all timescales.

This is the work of paleoclimatologists. As soon as the means became available to them in the second half of the twentieth century, they undertook to track down all traces of climate change so as to establish a planetary vision. This led them to develop new methods of sampling continental sediments, marine sediments in the context of major oceanographic campaigns, and ice cores by carrying out large-scale drilling campaigns of mountain glaciers and the ice sheets of Greenland and of Antarctica. The level of resources that needs to be mobilized is such that the drilling campaigns of polar ice and of marine sediments from all the world's oceans could only be carried out in an international cooperative framework which makes it possible to coordinate the efforts of the various teams.

This scientific investment has produced an abundant harvest of samples containing records of past climates. On the continents, lake sediments; peat bogs; concretions in caves; and fossil tree rings have provided many indicators of environmental conditions, especially of the behavior of vegetation and the atmosphere. In the ocean, samples have been taken from all of the large basins and cores are able to trace the history of the last tens of millions of years. Finally, the large drillings in the ice sheets have provided information not only on polar temperatures, but also on the composition of the atmosphere (dust and the concentrations of greenhouse gases, such as carbon dioxide and methane).

Unfortunately, nature has no paleothermometer or paleopluviometer, and therefore, there is no direct indicator of the changes in temperature or precipitation: Everything has had to be built from scratch, not only to reconstruct the climates, but also to date them. Extracting a reconstruction of the evolution of the climate from these samples has necessitated considerable developments using the most innovative methods from the fields of geochemistry, biology, and physics. Firstly, it was essential to establish a timeframe to know which period was covered by each sample. Many methods were developed, and they are the subject of the second part of this book. Radioactive decay, which is governed by strict physical laws, plays a vital role. It has made it possible to obtain timescales converted into calendar years, and it has provided clarification on stratigraphic geology. Other more stratigraphic approaches have been implemented: identification of characteristic events that need to be dated elsewhere; counting of annual layers; or modeling of ice flow. It has thus been possible to establish a chronological framework, and paleoclimatologists are now trying to make it common to all data via an on-going effort to make multiple correlations between the various recordings. Few climatologists rely on one indicator. The confidence that they have in reconstructing a climate change at a given time is obtained by intersecting reconstructions from independent indicators but also by confronting them with results from models. Methods of reconstructing the evolution of the different components of the climate system from geological indicators then had to be developed. These are extremely varied, and their description constitutes the main and third part of volume I. Many use the latest developments in paleomagnetism, geochemistry, and statistical methods to empirically link the distribution of fossil plants and animals with environmental parameters, primarily air and water temperature. Reconstructions achieved in this way have now reached a level of reliability such that, for certain periods, not only qualitative variations (in terms of hot/cold, dry/wet) can be obtained, but even quantitative ones with the associated uncertainties also quantified. This is the level of climate reconstruction necessary to allow comparison with climate models.

The use of climate models also gained momentum during the second half of the twentieth century. First established to simulate atmospheric circulation, they have progressed by integrating more and more efficiently the physics, processes, and parameterization of the radiative budget and the hydrological cycle, in particular, by incorporating satellite data. However, the atmosphere only represents the rapid component of the climate system.

The late 1990s dramatically demonstrated the need to link atmospheric models to global patterns of the ocean and vegetation to reconstruct climate change. Indeed, teams from the GISS in the USA and from Météo-France bolstered by their atmospheric models that had succeeded in reconstructing the current climate, independently tried to use the disruption of the radiative budget calculated by Milankovitch to simulate the last entry into glaciation 115,000 years ago. In both cases, it was a total fiasco. The changes induced by the variation of the orbital parameters in these models were far too small to generate perennial snow. The components and feedbacks related to the ocean and terrestrial vegetation needed to be included. Developing a model that couples all three of these components is what modelers have been striving to achieve over the last 20 years, and these are the models that now contribute to the international IPCC effort.

Today, the so-called Earth system models that incorporate aspects from the atmosphere–ocean–terrestrial and marine biosphere, chemistry, and ice caps are used to explore the climate of the future and the climates of the past. Spatially, they are increasingly precise, they involve a very large number of processes and are run on the largest computers in the world. But, the flip side of this complexity is that they can only explore a limited number of trajectories because of the considerable computing time they require. Also, from the beginning, climate modelers armed themselves with a whole range of models. From behemoths like the 'general circulation models' to conceptual models, with models of intermediate complexity in-between. From this toolbox, depending on the questions raised by the paleoclimatic data, they choose the most appropriate tool or they develop it if it does not exist. With the simplest models, they can explore the possible parameter variations and, by comparing them with the data, try to establish the most plausible scenario. All of these modeling strategies are described in detail in volume II, which constitutes the last and fourth part of this book.

This investigative approach at each step of the research work, dating, reconstruction, modeling, and the back and forth between these stages allows us to develop and refine the scenarios to understand the evolution of the past climates of the Earth. We are certain that this approach also allows us, by improving our understanding of the phenomena that govern the climate of our planet and through continuous improvement of the models, to better predict future climate change. This comparison between models and data, which makes it possible to validate numerical simulations of the more or less distant past, is an essential step toward the development of climate projections for the centuries to come, which will, in any case, involve an unprecedented transition.

Acknowledgement

This book would never have been possible without the very efficient help of Guigone Camus and Sarah Amram. We are grateful to Mary Minnock for the translation of each chapter. We also thank the LSCE for its financial support and Nabil Khelifi for his support from the origin to the end. Last but not least, thank you to all our colleagues—more than 50—who patiently contributed to this book.

Preface

Before taking this journey together into the Earth's paleoclimates, it is important to know what we will be facing. This exploration will bring us into the heart of the 'Earth system': a tangle of interwoven components with very different characteristics and response times, a system in constant interaction.

The first volume is dedicated (Chaps. 1 and 2) to an introduction to climate of the Earth. Chapters 3–9 focus on different time measurement and datation technics. The most important part of this first volume deals with the reconstructions of different climatic parameters from the three major reservoirs (ocean, continent, cryosphere, Chaps. 10–21). The second volume is devoted to modeling the Earth system to better understand and simulate its evolution (Chaps. 1–9). Last but not least, the final chapter (Chap. 10) describes the future climate of the Earth projection from next century to millennia.

The first part of this book (Chaps. 1 and 2) will equip the reader with a 'climate kit' before delving into the study of paleoclimates. This quick overview shows the great diversity in the systems involved. From the microphysics of the clouds that can be seen evolving over our heads by the minute to the huge ice caps that take nearly 100,000 years to reach their peak, the spatiotemporal differences are dizzying (Chap. 1). Yet, it is the same 'Earth system' that, throughout the ages, undergoes various disturbances that we will address. Chapter 2 takes us on a journey through the geological history of our planet. The distribution of continents, oceans, and reliefs changes how energy and heat are transported at the Earth's surface by the ocean and the atmosphere.

The study of paleoclimates requires an understanding of two indispensable concepts in order to describe the past climates of the Earth.

The first is the concept of time. Measuring time is fundamental to our research, and an understanding of the diversity of temporalities particular to paleoclimatic records is essential. The second part (Chaps. 3 to 9) of this book is devoted to the question of the measurement of time. Different techniques may be implemented depending on the timescales considered in Chap. 3. Thus, although carbon-14 (Chap. 4) provides us with reliable measurements going back to 30,000–40,000 years ago, other radioactive disequilibria (Chaps. 5 and 6) need to be used to access longer timescales. But it is not only the radioactivity-based methods that inform us of the age of sediments; the use of magnetism (Chap. 7) is also a valuable way of placing events occurring on the geological timescale into the context of climate. On shorter timescales, the use of tree rings is also a valuable method (Chap. 8). Ice core dating techniques will also be outlined (Chap. 9). This gamut of different methods shows how researchers have succeeded in developing 'paleo-chronometers' which are essential to locate climate archives within a temporal context, but also to establish the connections of cause and effect between the different components of the Earth system during periods of climatic changes.

The second concept is that of climate reconstruction. Indeed, in the same way that there is no single chronometer that allows us to go back in time, there is not one paleothermometer, pluviometer, or anemometer. Just as it was necessary to invent paleo-chronometers based on physical or biological grounds in order to attribute an age and an estimate of its uncertainty to archives, the relevant climatic indicators had to be invented to quantify the variations in temperature, hydrological cycle, and deepwater current. The third part of this book (Chaps. 10

–21) is devoted to the slow and complex work of reconstruction by applying this whole range of indicators. Thus, we can reconstruct the climate of the major components of the climate system: the atmosphere, the ocean, the cryosphere, and the biosphere. But we can also take advantage of the specificities of temperate or tropical lakes, of caves and their concretions (speleothems), of tree rings and even, more recently, of harvesting dates (Chap. 17). How can paleo-winds or, to put it in more scientific jargon, the variations in atmospheric dynamics be reconstructed? Based on the isotopic composition of precipitation (Chap. 10) or of the loess (Chap. 13), not only can the evolution of the surface and deep ocean be reconstructed, but also the geometry and dynamics of large water masses (Chap. 21). For land surfaces, palynology and dendroclimatology enable us to retrace the evolution of vegetation and climate, respectively (Chaps. 12 and 16). Finally, the cores taken from the ice caps of both hemispheres make it possible to reconstruct the polar climate (Chap. 11).

In addition to these two main concepts, we also need to understand how fluctuations in the hydrology of the tropics have caused variations in lakes (Chaps. 18 and 19) and glaciers (Chap. 20); these factors also tell a part of the climate story. Other markers, such as speleothems (Chap. 14) or lake ostracods (Chap. 15) reveal changes in climate in more temperate areas.

Thus, a description of the global climate emerges from the local or regional climate reconstructions. Through coupling these reconstructions with dating, our knowledge of climate evolution progresses constantly. Nevertheless, this image is both fragmentary, because of the strong geographic and temporal disparity of our knowledge, and unclear, because of the uncertainties in the reconstructions that the paleoclimatologist tries to reduce. There is still a long way to go in terms of developing new indicators and improving those widely used in order to complete and refine this description.

The second volume of this book (Chaps. 22–30) focuses on the major processes and mechanisms explaining the evolution of past climate from geological to historical timescales, whereas last Chap. 31 examines future climate projections. First of all, we address, in the very long term, the interactions between tectonics and climate over the timescale of tens to hundreds of millions of years (Chap. 22). Then, we deal with the biogeochemical cycles that govern the concentrations of greenhouse gases in the atmosphere over the last million years (Chap. 23). And finally, we consider the interactions with ice caps (Chap. 24).

We will continue our journey simulating the climate evolution through time from the formation of the Earth (4.6 billion years ago) up to the future climates at scales from a few tens of thousands to a hundred of years. On this journey, it becomes obvious that the dominant processes, those that drive climate change, vary according to timescales: solar power, which increases by about 7% every billion years places its stamp on very long-term evolution, whereas at the scale of tens of millions of years, it is tectonics that sculpt the face of the Earth, from the high mountain ranges to the bathymetry of the ocean floor. Finally, 'the underlying rhythm of Milankovitch,' with a much faster tempo of a few tens of thousands of years can produce, if the circumstances permit, the glacial–interglacial cycles described in the preceding parts. On top of this interconnection of timescales, a broad range of processes and components of the climate system is superimposed. Through these chapters, we would like to highlight the need to model a complex system where different constituents interact at different timescales (Chap. 25). With the development of these models, the scope of investigation is vast. Indeed, ranging from recent Holocene climates (Chap. 30) to geological climates (Chaps. 26 and 27), how they evolve is underpinned by very different processes: from plate tectonics (Chap. 22) to orbital parameters (Chap. 28). The complexity of the system can also be seen in the abrupt reorganizations of the ocean–atmosphere system (Chap. 29). The capacity acquired in recent decades to replicate past climate changes using a hierarchy of models, and to compare these results with different types of data, has demonstrated the relevance of this approach coupling model simulations with data acquisition.

Nevertheless, the field of investigation of the Earth's past climates remains an important area of research with many questions being raised about the causes of climate reorganizations throughout the Earth's history. Even though several chapters clearly show recent breakthroughs in our understanding of past climate changes, and the sensitivity of our models to climate data has undeniably increased the extent to which we can rely on their outputs, we can legitimately question what they contribute to future climate. Chapter 10 addresses these issues. Will the ice caps, which have existed on Earth for only a short time relative to geological time, withstand human disturbance? And can this disturbance, apart from its own duration, have an impact on the rhythm of glacial–interglacial cycles?

At the end of these two volumes, you will have obtained the relevant perspective to project into the Earth's climates of the future. Indeed, by absorbing the most up-to-date knowledge of paleoclimatology in this book, you will be provided with the necessary objectivity to critically assess present and future climate changes. It will also give you the scientific bases to allow you to exercise your critical judgment on the environmental and climatic issues that will be fundamental in the years to come. Indeed, in the context of the Anthropocene, a period where man's influence has grown to become the major factor in climate change, the accumulated knowledge of the climate history of our planet gathered here is precious.

Gif-sur-Yvette, France

Gilles Ramstein
Amaëlle Landais
Nathaelle Bouttes
Pierre Sepulchre
Aline Govin

Contents

Volume 1

1. **The Climate System: Its Functioning and History** 1
   Sylvie Joussaume and Jean-Claude Duplessy

2. **The Changing Face of the Earth Throughout the Ages** 23
   Frédéric Fluteau and Pierre Sepulchre

3. **Introduction to Geochronology** 49
   Hervé Guillou

4. **Carbon-14** 51
   Martine Paterne, Élisabeth Michel, and Christine Hatté et Jean-Claude Dutay

5. **The $^{40}K/^{40}Ar$ and $^{40}Ar/^{39}Ar$ Methods** 73
   Hervé Guillou, Sébastien Nomade, and Vincent Scao

6. **Dating of Corals and Other Geological Samples via the Radioactive Disequilibrium of Uranium and Thorium Isotopes** 89
   Norbert Frank and Freya Hemsing

7. **Magnetostratigraphy: From a Million to a Thousand Years** 101
   Carlo Laj, James E. T. Channell, and Catherine Kissel

8. **Dendrochronology** 117
   Frédéric Guibal and Joël Guiot

9. **The Dating of Ice-Core Archives** 123
   Frédéric Parrenin

10. **Reconstructing the Physics and Circulation of the Atmosphere** 137
    Valérie Masson-Delmotte and Joël Guiot

11. **Air-Ice Interface: Polar Ice** 145
    Valérie Masson-Delmotte and Jean Jouzel

12. **Air-Vegetation Interface: Pollen** 151
    Joël Guiot

13. **Ground-Air Interface: The Loess Sequences, Markers of Atmospheric Circulation** 157
    Denis-Didier Rousseau and Christine Hatté

14 **Air-Ground Interface: Reconstruction of Paleoclimates Using Speleothems** 169
   Dominique Genty and Ana Moreno

15 **Air-Interface: $\delta^{18}O$ Records of Past Meteoric Water Using Benthic Ostracods from Deep Lakes** 179
   Ulrich von Grafenstein and Inga Labuhn

16 **Vegetation-Atmosphere Interface: Tree Rings** 197
   Joël Guiot and Valérie Daux

17 **Air-Vegetation Interface: An Example of the Use of Historical Data on Grape Harvests** 205
   Valérie Daux

18 **Air-Ground Interface: Sediment Tracers in Tropical Lakes** 209
   David Williamson

19 **Air-water Interface: Tropical Lake Diatoms and Isotope Hydrology Modeling** 213
   Florence Sylvestre, Françoise Gasse, Françoise Vimeux, and Benjamin Quesada

20 **Air-Ice Interface: Tropical Glaciers** 219
   Françoise Vimeux

21 **Climate and the Evolution of the Ocean: The Paleoceanographic Data** 225
   Thibaut Caley, Natalia Vázquez Riveiros, Laurent Labeyrie, Elsa Cortijo, and Jean-Claude Duplessy

Volume 2

22 **Climate Evolution on the Geological Timescale and the Role of Paleogeographic Changes** 255
   Frédéric Fluteau and Pierre Sepulchre

23 **Biogeochemical Cycles and Aerosols Over the Last Million Years** 271
   Nathaelle Bouttes, Laurent Bopp, Samuel Albani, Gilles Ramstein, Tristan Vadsaria, and Emilie Capron

24 **The Cryosphere and Sea Level** 301
   Catherine Ritz, Vincent Peyaud, Claire Waelbroeck, and Florence Colleoni

25 **Modeling and Paleoclimatology** 319
   Masa Kageyama and Didier Paillard

26 **The Precambrian Climate** 343
   Yves Goddéris, Gilles Ramstein, and Guillaume Le Hir

27 **The Phanerozoic Climate** 359
   Yves Goddéris, Yannick Donnadieu, and Alexandre Pohl

28 **Climate and Astronomical Cycles** 385
   Didier Paillard

29 Rapid Climate Variability: Description and Mechanisms 405
Masa Kageyama, Didier M. Roche, Nathalie Combourieu Nebout, and Jorge Alvarez-Solas

30 An Introduction to the Holocene and Anthropic Disturbance 423
Pascale Braconnot and Pascal Yiou

31 From the Climates of the Past to the Climates of the Future 443
Sylvie Charbit, Nathaelle Bouttes, Aurélien Quiquet, Laurent Bopp, Gilles Ramstein, Jean-Louis Dufresne, and Julien Cattiaux

About the Editors

Gilles Ramstein is a director of research at Laboratoire des Sciences du Climat et de l'Environnement (LSCE, France). His initial degree is in physics and since 1992 he has specialized in climate modeling.

He has been responsible for many French and European research projects on the Pleistocene, Cenozoic, and Precambrian eras. He has also been the advisor of many Ph.D. students who have explored and expanded the frontiers of paleoclimate modeling.

As a climate modeler, he studies very different climate contexts from 'Snowball Earth' episodes (717–635 Ma) to more recent, and occasionally future, climate situations.

The main research topics he focuses on are

• **Geological time from the Precambrian to the Cenozoic**:
– Investigation of relationships between tectonics, the carbon cycle, and the climate with an emphasis on the impact on the climate and the atmospheric CO_2 cycle of major tectonic events such as plate movements, shrinkage of epicontinental seas, mountain range uplift, and the opening/closing of seaways.

– Leading international collaborations on projects on monsoon evolutions and the dispersal of human ancestors during the Neogene periods.

• **From the Pleistocene to future climate**: In this framework, his major interests are interactions between orbital forcing factors, CO_2 and climate. More specifically, his focus is on the response of the cryosphere, an important component of the climate system during these periods, with an emphasis on the development of the Greenland ice sheet at the Pliocene/Pleistocene boundary and abrupt climate changes driven by ice sheet variations.

He has also published several books and co-edited the French version of 'Paleoclimatologie' (CNRS Edition) and contributed to an online masters program devoted to educating journalists on climate change (Understanding the interactions between climate, environment and society ACCES).

Amaëlle Landais is a research director at Laboratoire des Sciences du Climat et de l'Environnement (LSCE, France). Her initial degree is in physics and chemistry and, since her Ph.D. in 2001, she has specialized in the study of ice cores.

She has been responsible for several French and European research projects on ice cores working on data acquisition both in the laboratory and in the field, interacting extensively with modelers. She has been the supervisor of ten Ph.D. students and is deeply committed to supporting and training students in laboratory work.

Her main research interests are the reconstruction of climate variability over the Quaternary and the links between climate and biogeochemical cycles. To improve our understanding of these areas, she develops geochemical tracers in ice cores (mainly isotopes), performs process studies using laboratory and field experiments, and analyzes shallow and deep ice cores from polar regions (Greenland and Antarctica). Through numerous collaborations and improvement of ice core dating methods, she tries to establish connections with other paleoclimatic archives of the Quaternary.

Nathaelle Bouttes is a research scientist at the Laboratoire des sciences du climat et de l'environnement (LSCE/IPSL). Following the completion of her Ph.D. in 2010 on the glacial carbon cycle, she went to the University of Reading (UK) for a five-year postdoc on recent and future sea-level changes. She then spent a year at Bordeaux (France) with a Marie–Curie Fellowship on interglacials carbon cycle before joining the LSCE in 2016. Since then, she has specialized in understanding glacial–interglacial carbon cycle changes using numerical models and model–data comparison.

She is mostly using and developing coupled carbon-climate models to understand past changes of the carbon cycle, in particular the evolution of the atmospheric CO_2. She has been focusing on the period covered by ice core records, i.e., the last 800,000 years. She uses model–data comparison by directly simulating proxy data such as $\delta^{13}C$ to evaluate possible mechanisms for the orbital and millennial changes. She has been involved in several projects covering this topic as well as teaching and supervising Ph.D. students.

Pierre Sepulchre is a CNRS research scientist at the Laboratoire des sciences du climat et de l'environnement (LSCE/IPSL). He completed a Ph.D. on the Miocene climate of Africa in 2007, then went to UC Santa Cruz (USA) for a two-year postdoctoral position working on the links between the uplift of the Andes and atmospheric and oceanic dynamics. His lifelong research project at CNRS is to evaluate the links between tectonics, climate, and evolution at the geological timescales, focusing on the last 100 million years. Through the supervision of Ph.D. students and his collaboration with geologists and evolutionary biologists, he also worked at evaluating paleoaltimetry methods with the use of an isotope-enabled atmospheric general circulation model, as well as linking continental surface deformation, climate, and biodiversity in Africa and Indonesia. In recent years, he led the implementation and validation of a fast version of the IPSL Earth system model that allows running long climate integrations dedicated to paleoclimate studies.

Aline Govin is, since 2015, a research associate at the Laboratoire des Sciences du Climat et de l'Environnement (LSCE, Gif sur Yvette, France). She studied Earth Sciences at the Ecole Normale Supérieure of Paris (France) and obtained in 2008 a Ph.D. thesis in paleoclimatology jointly issued by the University of Versailles Saint Quentin en Yvelines (France) and the University of Bergen (Norway). Before joining the LSCE, she worked for five years as a postdoctoral fellow at the Center for Marine Environmental Sciences (MARUM, University of Bremen) in Germany.

Her research activities focus on the reconstruction of paleoclimatic and paleoceanographic changes by applying various types of geochemical and sedimentological tracers on marine sediment cores. She has mostly worked on the Earth's climatic changes of the last 150,000 years and is an expert of the last interglacial climate, which is an excellent case study to investigate the response of the Earth's climate to past warming conditions that could be encountered in the coming decades. Her research interests include the past variability of the deep North Atlantic circulation, the responses, and drivers of tropical monsoon systems (e.g., South American Monsoon), the development and calibration of paleo-tracers, the development of robust chronologies across archives, and the quantification of related uncertainties, as well as the comparison of paleo-reconstructions to climate model simulations of past climates.

She has authored around 30 scientific publications and has been involved in many French, German, and other international (e.g., Brazilian) projects.

The Climate System: Its Functioning and History

Sylvie Joussaume and Jean-Claude Duplessy

Climate plays an important role for mankind. It determines the conditions in which societies can develop as well as the resources available to them such as water and biological inputs (agriculture, forests, livestock). However, climate is a complex system. It is the result of interactions not only between the atmosphere, the oceans, landmasses and ice but also the biosphere: the living world. It varies depending on the timescale, and different mechanisms may come into play at different scales. The aim of this book is to show how a multi-disciplinary scientific community can now reconstruct, with increasing accuracy, the major features of past climates and discover how they are regulated by the geological evolution, geochemistry, physics and biology of our living planet, Earth.

Human living conditions are dependent on climate, but human beings, in turn, influence the climate system. They change the atmospheric concentration of greenhouse gases and aerosols, as well as the vegetation through deforestation and agriculture. For this reason, it is of major importance to society to understand how climate works and how man may be altering its course. This is a complex scientific problem because of the large number of feedbacks likely to occur and the study of past climates contributes to a better understanding of them by analyzing major climate changes provoked by natural causes.

Climate Change

Definition of Climate

Climate is defined by the statistics of the physical characteristics of the atmosphere. It differs from meteorology by focusing on statistics over several decades, by calculating the average state of the atmosphere and its variability from this average. In practice, climate is defined by the average conditions over a thirty-year period. Although this working definition makes sense while the weather is relatively stable, it becomes more difficult to apply during a period of rapid change. This was the case in the twentieth century during which two phases of rapid increase in the average temperature of our planet were detected by weather stations in the WMO network, one from 1910 to 1940 and the other from 1975 onwards (Fig. 1.1d). The period 1961–1990 is often taken as a reference.

Climate Changes in the Past

Climate is essentially variable, regardless of the time scales under consideration. Over past two millennia, historical chronicles and earliest instrumental measurements dating back to the seventeenth century have shown the existence of a very cold period in Europe from the sixteenth to the nineteenth century (the Little Ice Age), preceded by the Medieval Warm Period and another warm period during Roman times.

Geological data also show large upheavals in climate. Of course, this is over much longer periods than thirty years, but geologists strive to define a stratigraphic framework and precise geochronology to put these events into context within the history of our planet (Fig. 1.1). For example, about seven hundred and fifty million years ago, the Earth went through an intense glaciation phase; glaciers flowed down to sea level on every continent, even in low latitudes, to such an extent that our planet could be described as a snowball. Conversely, during the Mesozoic Era (25–65 million years), the conditions were hot, even at high latitudes. During the Cenozoic Era (from 65 million years), the glaciers grew slowly, first on the Antarctic continent and then on Greenland. For the past three million years or so, the Earth has experienced a succession of ice ages, marked by glaciers advancing over land at high and middle latitudes of

S. Joussaume (✉) · J.-C. Duplessy
Laboratoire des Sciences du Climat et de l'Environnement, LSCE/IPSL, CEA-CNRS-UVSQ, Université Paris-Saclay, 91191 Gif-sur-Yvette, France
e-mail: sylvie.joussaume@lsce.ipsl.fr

Fig. 1.1 a Variations in the climate of Earth over the past 4 billion years, estimated from geological data. **b** Changes in mean climate estimated from variations in the volume of glaciers and ice sheets present on the surface of continents for the last million years. **c** Variations in the average air temperature in the northern hemisphere from the year 1000 reconstructed from paleoclimate data, including historical data and the study of tree rings. **d** Variations in the average surface air temperature calculated from the global meteorological network data for the period 1860–2010. Note that the number of stations has varied during this period and was sparse throughout the late nineteenth century

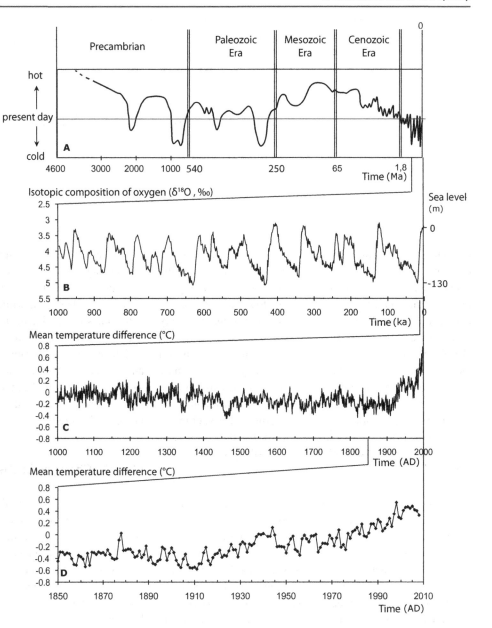

the northern hemisphere, separated by interglacial periods when the ice caps receded and remained confined to the Antarctic and Greenland.

We have been in an interglacial period, called the Holocene, for the past 11,000 years. The various aspects of the evolution of climate will be expanded upon later in this book.

The last million years is the best understood geological period, because climate can be reconstructed from detailed information provided by polar ice and marine and continental sediments. Over this period, a succession of oscillations between glacial and interglacial periods marks intervals which are the result of small changes in the Earth's orbit around the Sun. The glacial periods last almost ten times longer than the interglacial periods but they are interspersed with rapid warmings which follow the outbreaks of cold in the North Atlantic and neighboring landmasses. All these major climate shifts do not occur as a result of chance and it is the work of climatologists and paleoclimatologists to understand them by analyzing climate mechanisms and the causes of their variability.

Climate Mechanisms

The Sun is the major driver of climate. Received solar energy plays a key role in establishing climate conditions on the surface of our planet. But these depend critically on the composition of the atmosphere and energy exchanges between the surface of the planet and the atmosphere that surrounds it. The Earth radiation balance compares, for each point on the Earth's surface, the energy received from the Sun and that which is emitted back into space. Significant geographical differences drive wind and ocean currents which redistribute energy, influenced by the shape of the ocean basins and the relief of the land.

The Radiation Balance of the Earth

The Greenhouse Effect

A disk with a surface area of 1 m^2, located equidistant between the Earth and the Sun and intercepting solar radiation at a perpendicular angle, would receive an energy flow of 1368 W at the top of the Atmosphere (TOA). However, the Earth is a sphere whose surface area is four times greater than that of a disk with the same diameter. This is why, on average and over the course of a year, the solar flux intercepted by a unit area is four times lower. It corresponds to a power of 340 W/m^2 TOA with a known accuracy of roughly 1 W/m^2 (Fig. 1.2). Yet all this energy is not accessible to the Earth/atmosphere. A portion, about 36%, returns back to space after being reflected by the clouds, the suspended aerosols in the air, the Earth's surface and by the air molecules themselves. So, the real amount of energy absorbed amounts to 161 W/m^2. It is offset by an infrared flux emitted by the Earth and its atmosphere to space. In fact, the Earth behaves as a 'black body': it emits energy whose intensity is proportional to the fourth power of its absolute temperature (287 K), in accordance with Stefan's law. This radiation is almost entirely concentrated in the infrared range between 4 and 100 μm (microns), with a maximum intensity centered around 12 μm. Solar radiation also behaves like a 'black body' but at temperatures of around 6000 K, and covers a range of wavelengths from ultraviolet to near infrared, from 0.2 to 4 μm, and has a maximum intensity in the visible wavelengths of around 0.6 μm.

In the absence of any greenhouse effect, i.e. if the atmosphere were perfectly transparent to infrared radiation emitted by the Earth, the temperature in equilibrium with an average absorbed flow of 161 W/m^2, would be only −19 °C. In reality, water vapor, liquid water in clouds, carbon dioxide and other trace elements present in the air absorb a large portion of infrared radiation emitted by the surface, limiting the loss of energy towards space. Acting as 'black bodies', all these constituents re-emit infrared energy in all directions including towards the ground. This additional contribution means that the average surface temperature of the Earth is 14 °C, not −19 °C. This greenhouse effect is a natural phenomenon due in large part to the presence of water vapor, which contributes about 55% of the total greenhouse effect, to other greenhouse gases (carbon dioxide, methane, nitrous oxide) which account for 28%, with the remainder caused by clouds. Throughout the geological history of the Earth, the composition of the atmosphere has changed significantly and changes in the greenhouse effect have greatly contributed to past climate variations (see

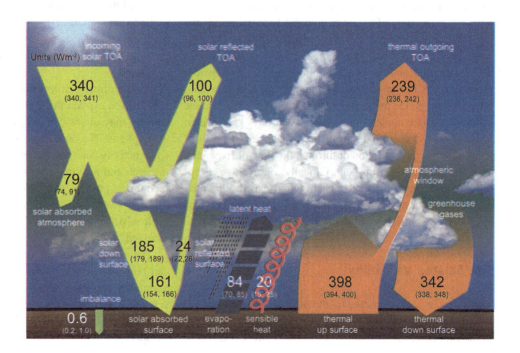

Fig. 1.2 Radiation balance of the Earth. The solar radiation incident at ground level is fully offset by infrared radiation emitted towards space (*Source* IPCC 2013)

Chap. 2, Volume 1 as well was Chaps. 22, 26 and 27, Volume 2). Since the beginning of the industrial era (about 1850), human activities have significantly increased the concentration of greenhouse gases already naturally present in the air and have introduced new ones, such as chlorofluorocarbons (CFCs) which are active agents of the greenhouse effect.

The Water Cycle

As solar radiation passes through the different layers of the atmosphere, part of it is absorbed by ozone in the stratosphere and by water vapor in the troposphere. About half of the incident energy reaches the surface of Earth, where it is partially compensated for by the loss of infrared radiation to the atmosphere. An energy surplus of 104 W/m^2 (Fig. 1.2) remains available at the surface. This energy warms up the surrounding air and causes evaporation of water from the surface of oceans and land, feeding into the water cycle on our planet. The water vapor is then transported by winds until it condenses as precipitation, releasing into the atmosphere the energy acquired at the surface during evaporation. Thus, the cycle of evaporation and precipitation of water takes energy from the surface of the oceans and land and redistributes it in the atmosphere. This transfer of latent heat cools the surface and warms up the atmosphere, thus lessening the differences in temperature between the upper and lower layers of the atmosphere, as well as between the equator and the poles. The water cycle thus plays a fundamental role in the redistribution of energy between the surface and the atmosphere.

Evaporation and condensation continuously renew the store of water vapor in the atmosphere. However, the amount of water vapor in the air at any given moment remains quite low. If it were completely condensed, the liquid layer thus formed would cover the Earth's surface in a layer 2.5 cm thick. Yet, on average, the water cycle involves the evaporation and the precipitation of water which would correspond to a layer of about 80 cm per year. The recycling time of water in the atmosphere is therefore very fast and the water vapor is completely renewed in ten days. The water, most of which evaporates from the oceans (86%), returns there either by precipitation or through the flow of rivers and streams after runoff from land. Globally, on average, evaporation and rainfall balance each other exactly, thereby maintaining a constant concentration of water vapor in air, as long as the average temperature of the air remains constant.

Sun-Related Variability

Variations in energy emitted by the Sun and the variations in the solar energy received by the Earth will affect the climate. In the first case, the solar activity cycles and the evolution of the Sun since the formation of the solar system modify the amount of energy it emits. In the second case, the slow variations of the movement of the Earth around the Sun influence the seasonal and geographical distribution of energy received in a given place on our planet.

Solar Cycles

In the mid-nineteenth century, the German astronomer, H. Schwabe, discovered spots on the Sun's surface that appear and disappear over an eleven-year cycle. When solar activity is more intense, marked by a greater number of spots, the Sun emits more energy. Since the 1980s, satellite measurements allow the estimation of variations in intensity of solar energy. These are around 0.1%, which corresponds to a very small perturbation (0.24 W/m^2) in the radiation balance of the Earth. Solar activity directly reflects changes in the Sun's magnetic field. The spots reappear in larger numbers when the magnetic field intensifies. Solar flares then become stronger; they eject a larger number of particles toward outer space and thus reinforce the solar wind. These electrically charged particles, mainly electrons and protons, reach the Earth's atmosphere where they cause magnetic storms—strong disturbances in the magnetic field—as well as magnificent auroras in the polar regions.

The influence of solar activity on climate has been debated for many years. In the second half of the seventeenth century, documented observations indicate an almost total disappearance of spots for a period of several decades, during the Little Ice Age (Fig. 1.3). At the end of the nineteenth century, the German astronomer H. Spörer and his English colleague W. Maunder linked these two phenomena, thus starting a controversy that persists today. The nature of the connection between the minimum solar activity (called the Maunder minimum) and a decrease in the intensity of solar radiation sufficient to induce a marked cooling that coincided with that time, still needs to be explained.

As the direct disruption in the solar radiation balance is too small to explain the phenomenon, it is believed that solar activity may affect climate through circulation in the upper atmosphere. Nevertheless, the link between variations in solar activity and the Earth's climate remains a subject of research and a source of controversy given the absence of a recognized physical mechanism. The relative role of external forcing (solar radiation) and internal/geological forcing (volcanism) in explaining the Little Ice Age still needs to be assessed.

The Sun exhibits variations over longer periods. These can be seen not only in the number of sunspots, but also in variations in solar diameter. This varies with a periodicity of 900 days, but this oscillation is influenced by solar activity. It is minimal when the activity is at its maximum. Like sunspots, solar diameter measurements started in the

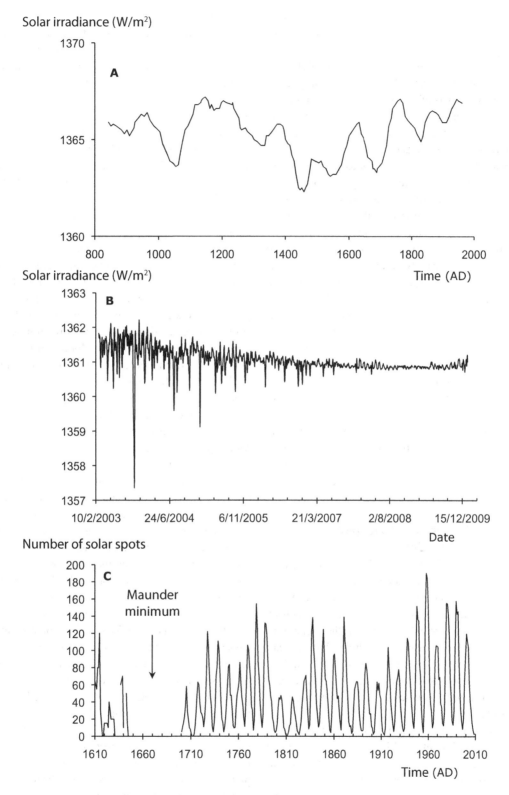

Fig. 1.3 a Variations in solar energy reconstructed from changes in the beryllium content of polar ice and from the modeling of stellar activity. It should be noted that the mean values calculated using stellar-activity modeling are higher, by about 5 W/m², compared with values measured using satellites. **b** Variations in energy emitted by the Sun, NASA satellite measurements. **c** Changes in the number of sunspots observed by astronomers since 1610

seventeenth century. They led to the discovery of a cycle of 80–90 years, called the Gleisberg cycle, which modulates the Schwabe cycle.

Solar wind plays only a minor role in the flow of charged particles received by the Earth. Most comes from galactic cosmic rays, which consists of electrons, protons, α particles (ionized helium nuclei) and heavier ions in very small quantities. It is isotropic and comes from everywhere in space. In periods of high solar activity, intense solar wind, through the magnetic field it creates, acts as a shield repelling the galactic cosmic radiation falling to Earth. This phenomenon inspired a geochemical method for determining variations in solar activity. Indeed, galactic cosmic rays, through spallation reaction on the atoms in the upper atmosphere, are responsible for the production of several cosmonucleides, the most well-known of which is Carbon-14. Less Carbon-14 is produced during intense solar activity. Measurements by geochemists on well-dated tree rings showed pseudo-periodic variations in the production of this cosmonucleide. This is attributed to fluctuations in solar activity, with periods of about 150–300 years (Suess cycles) and 2300 years (Hallstattzeit cycles). The existence of these cycles has been confirmed by the measurement of other cosmonucleides, such as beryllium-10, in polar ice. These are trapped in ice in Greenland, whose location in time can be determined simply by visually counting the annual layers or, for earlier periods, through more complex methods described in Chap. 9. The paleoclimatologists are now investigating if these periodicities can be reflected in geological records.

Long-Term Variations in the Movement of the Earth Around the Sun

The movement of the Earth around the Sun varies over time under the influence of the gravitational attraction of other planets (see Chap. 28, Volume 2). The orbit traveled by the Earth over a full year is almost exactly an ellipse with an eccentricity (the parameter which defines the degree of flattening of the ellipse with respect to a circle) that can vary over time. With periodicities close to 100,000 and to 400,000 years, the orbit goes from a circle with an eccentricity of zero to a slightly flattened ellipse with a maximum eccentricity of 6%.

The tilt in the axis of the Earth relative to the ecliptic plane is known as its obliquity and it influences the amount of sunshine received at different latitudes in different seasons. It is the reason for phenomena such as the polar night in winter and the midnight sun in summer at the highest latitudes. For this reason, the climate at high latitudes is especially sensitive to variations in the obliquity. With a periodicity of around 41,000 years, the obliquity angle oscillates between 22° and 25°, the current value being close to 23° 26′.

Because of the elliptical nature of the Earth's orbit, the distance between the Earth and the Sun varies at different times of the year. Currently, in the northern hemisphere, this distance is at its minimum in winter and at its maximum in summer, and the opposite is true for the southern hemisphere. In fact, the amount of solar radiation intercepted by the Earth decreases as the distance increases. This causes milder winters and cooler summers in the northern hemisphere, while the seasonal contrasts are accentuated in the southern hemisphere (although this impact is minor compared with the seasonal variations in high latitudes caused by obliquity).

Over the millennia, the position of the solstices and equinoxes slowly moves along the ellipse resulting in a variation in the solar energy received during each season. This movement of precession of the equinoxes is caused by a combination of two rotational movements. The first is the rotation of the Earth around an axis running through the poles which is perpendicular to the elliptic plane. A gradual shift in the orientation of the axis of rotation is caused by the attraction of the Sun and the Moon and traces out a circle over the North Pole in a cycle of approximately 26,000 years. The second is the elliptical orbit of the Earth around the sun which is superimposed on the first. The combination of these two movements results in a periodicity of the precession of the equinoxes of about 22,000 years. More specifically, the Earth's distance from the Sun fluctuates, not only due to the precession movement of the equinoxes, but also due to variations in the eccentricity of its orbit which varies according to a set of cyclical changes occurring over two proximate periods, one of 19,000 years and the other 23,000 years. Thus, approximately 10,000 years ago, the Earth reached its closest point to the Sun at the time of the summer solstice and not at the boreal winter solstice as it does today. At that time, the northern hemisphere received more solar energy in summer than it does today and obviously less in winter.

All of these modifications in the orbital parameters affect sunshine levels (still referred to as the insolation) at the different bands of latitude on Earth, and particularly the intensity of the seasonal cycle. Already, in 1924, the Serbian mathematician, Milutin Milankovitch, proposed that these slow variations of the movement of the Earth around the Sun could explain the glacial cycles. Indeed, as these slow variations induce a decrease in solar energy received in the summer at 60° N, snow, which has fallen in the winter, does not melt completely. Furthermore, it strongly reflects solar radiation, facilitating the snow to persist. Gradually, the snow accumulates and turns into an ice cap. This hypothesis has been debated for many years and was strongly opposed until the 1970s, when the cycles predicted by this theory were clearly confirmed by paleoclimate records in marine sediments and later in polar ice. We will see more precisely in Chap. 28 of

Volume 2 the state of our knowledge about the Milankovitch theory, or the 'astronomical theory' of paleoclimates.

The Sun's Evolution

Since the formation of the solar system, the Sun, like all stars of the same type, slowly consumes its hydrogen to produce helium, and the amount of heat it emits varies very slowly over long time scales. The standard stellar evolution models estimate that four billion years ago the luminosity of the Sun was 25–30% lower than it is today, and that it has increased more or less linearly over time. This model seems in accordance with observations made by astronomers of young stars. With the same Earth's atmosphere as today four billion years ago, the average temperature of the Earth would be below 0 °C, oceans would be frozen, and life would be impossible. Geological observations, however, indicate the presence of water in the liquid state and the first traces of life 3.5 billion years ago. This is the 'Pale Young Sun paradox' which is solved by assuming that the atmosphere had a very different chemical composition from that of today. Indeed, the elimination of carbon dioxide by the young Earth, and the low rates of weathering given the absence of continental crust, meant that the atmosphere during these ancient periods acquired an exceptionally high level of CO_2 and hence was responsible for a strong greenhouse effect, further enhanced by the presence of methane produced by bacteria. This will be discussed in detail in the chapter on the Precambrian (Chap. 26, Volume 2).

Reconstruction of the History of Atmospheric Composition

Although the Sun is the source of energy for the Earth, the energy made available depends essentially on the composition of the atmosphere: greenhouse gases and particles. Reconstructing the past history of the composition of the atmosphere is therefore an important element in understanding climate dynamics.

Again, the last hundreds of thousands of years constitute the best documented period because of the fossil air bubbles contained in the polar ice caps. The snow falling on the polar caps forms a porous firn, within which air circulates freely. Under the weight of accumulated snow, the pores gradually compact and the firn turns into ice that traps tiny air bubbles within it. This air keeps its original chemical composition, which allows the reconstitution of variations in the composition of the atmosphere over time, as long as we can find well-preserved ancient ice. The oldest ice is found in Antarctica, where a continuous recording of the greenhouse gas content (CO_2, CH_4) over the last 800,000 years has been established. These records show that the levels of carbon dioxide have not remained constant; they were high in warm periods, around 280 ppmv (280 cm^3 of CO_2 per m^3 of air) and only 200 ppmv during cold periods. Similarly, methane ranged from ∼700 ppbv (mm^3 per m^3 of air) in warm periods to less than 400 ppbv during cold ones, with a high temporal variability (Fig. 1.4).

Polar ice is the only direct recording of the composition of the atmosphere. As ice in the ice caps flows very slowly and is continuously renewed throughout geological time, it is impossible to reconstruct a record of carbon dioxide levels before a million years ago. For earlier periods, it is therefore necessary to use indirect empirical methods which have a much lower level of precision. These indirect "proxies" to reconstruct atmospheric CO_2 may be derived from stomata, boron isotope or alkenone (Chap. 27, Volume 2). For example, botanists observed that the stomata—pores through which leaves absorb carbon dioxide from the air—are smaller and fewer when carbon dioxide is high. This plant characteristic has been used as a means of establishing CO_2 levels for the past. However, the results were not clear-cut. For one, the fossil species being studied must be the same as the current species on which the empirical relationship between the levels of carbon dioxide and the number or diameter of the stomata is being established. Moreover, the relationship, which can only be determined in the current conditions, also depends on the availability of water to the plant, so it is not clear whether changes observed in fossil stomata are due to variations in CO_2 or in moisture.

As the CO_2 content of the air is governed by the partial pressure of this gas in the surface waters of the ocean, geochemists have tried to use carbon-13, a tracer of the oceanic phase of the carbon cycle, as a tool to reconstruct changes in atmospheric CO_2. One of the proposed markers is the $^{13}C/^{12}C$ ratio in foraminifera, microscopic animals in the form of plankton living in the surface waters of the oceans. These animals secrete a calcareous shell whose size is a few tenths of a millimeter and which are found in abundance in marine sediments. The $^{13}C/^{12}C$ ratio of planktonic foraminifera therefore depends on the isotopic composition of dissolved CO_2 in the surface waters, and indirectly on that of the atmosphere.

This isotopic approach can be compared against the independent records provided by the polar ice cores, so that the method can be evaluated over the last few hundred thousand years. The correlation is only proximate due to the complexity of the oceanic carbon cycle which depends in particular on the temperature of the sea water, on the primary production of the ocean, on the decomposition of organic matter and on the circulation of the bodies of water. Biologists came up with another method when they noticed that the fractionation of carbon isotopes during the absorption of carbon dioxide by seaweed depends on the dissolved carbon dioxide content and therefore the partial pressure of CO_2 in the seawater. This led them to the hypothesis that variations

Fig. 1.4 a Variations in atmospheric methane concentrations inferred from variations in methane concentration in the air bubbles trapped in the ice cores drilled at Dome C (EPICA). **b** Variations in atmospheric carbon dioxide concentrations inferred from variations in carbon dioxide concentration in the air bubbles trapped in the ice cores drilled at Dome C (EPICA). **c** Air temperature variations in Antarctica inferred from changes in the isotopic composition of the ice cores drilled at Dome C (EPICA). **d** Changes in mean climate of the Earth estimated from variations in the isotopic composition of oxygen in benthic foraminifera, a proxy for variations in the volume of glaciers and ice caps on land surfaces over the last million years

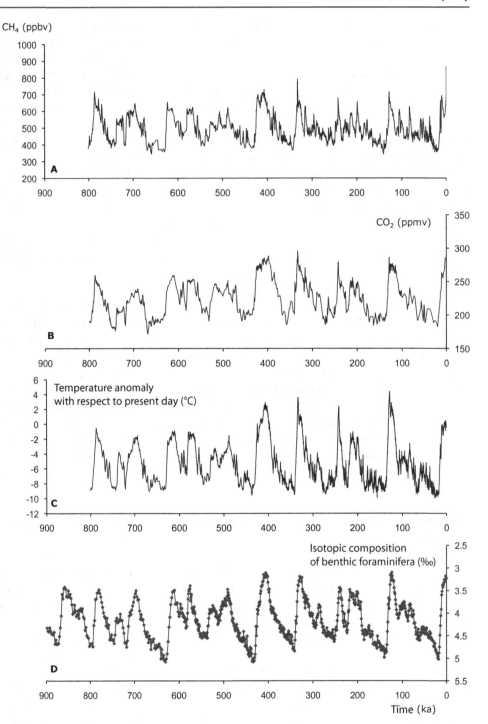

in the $^{13}C/^{12}C$ ratio in specific compounds formed during photosynthesis, such as in alkenones found in marine cores, would reflect changes in the CO_2 composition of the surface waters of the ocean and of the air. However, the relationship obtained depends on the ratio of surface to volume of cells performing photosynthesis, which introduces a new uncertainty in the reconstructions. Finally, it is clear that even isotopic methods, which use accurately measurable geochemical parameters, produce only rough estimates of the carbon dioxide composition of the air and its variations.

At long time scales ($>10^6$ years), the changes in the levels of CO_2 in the atmosphere are determined by the relative extent of degassing by volcanoes and mid-ocean ridges on the one hand, and the consumption of CO_2 by chemical erosion of silicates on the other. This means that the key role is played by plate tectonics. Thus, a gradual reduction of

CO_2 in the air may be due to either a lower degassing rate or an increase in erosion of the surface of the continents. The latter depends on a complex set of parameters, themselves related to climate, such as air temperature, precipitation, continental runoff and vegetation. Geochemists therefore try to reconstruct the changing partial pressure of atmospheric CO_2 using models; CO_2 emissions are estimated using geological data on the speed of movement of the plates; consumption of the gas is taken into account in simplified models by coupling the carbon cycle to climate and by considering the geographical context resulting from plate tectonics. For example, the breaking-up of the arid supercontinent Rodinia, into a multitude of small humid continental masses, 800–700 million years ago, led to the creation of basaltic regions, easily erodible chemically. This resulted in a significant drop in carbon dioxide levels in the air which may explain the great glaciations of the period.

Airborne dust also plays an important role in the radiation balance of the atmosphere, mainly by intercepting solar radiation and thereby reducing the amount of energy reaching ground level. Dust levels have varied considerably in the past, as is evidenced in polar ice. Falling snow brings down atmospheric dust with it which then remains trapped in the ice. The more the air is charged with dust, the more of it the snow absorbs. In this way, strong atmospheric dust levels during the glacial periods of the Quaternary have been demonstrated. The dust came from continental erosion which was then transported by winds. They gave rise to huge accumulations of very fine particles. These created the loess present in China, and in smaller quantities, in Western Europe (Chap. 13, Volume 1).

The Atmosphere

The Main Features of Atmospheric Circulation

The net balance between the radiation received from the Sun and that emitted into space does not have a uniform distribution. The net energy flux varies, depending on the latitude, geographic regions and season. Solar radiation decreases significantly between the equator and the poles, but there is little difference in emitted infrared radiation. The result is a surplus of energy in the tropics and a deficit in the north and south latitudes above 40°. Heated at the equator, cooled at the poles, the atmosphere and the ocean are activated and carry the excess energy from tropical regions to the deficient higher latitudes. According to currently available measurements, the two fluids of the planet contribute with relatively similar amplitude to this transport (Fig. 1.5).

A strong circulation in the atmosphere traveling from the equator to the poles is established in order to ensure the transport of energy necessary for the thermal balance of the planet. The warmer air, and therefore lighter, rises above the equator, before diverging and heading at high altitudes towards the poles. Above the polar regions, on the contrary, cold, dense air descends toward the surface, and travels toward the equator, which forms a large loop between the equator and poles. This mechanism, described in 1735 by the English scientist George Hadley, would happen if the Earth was turning very slowly. In reality, this large convection cell remains confined between the equator and the subtropical regions, where it forms the so-called 'Hadley' circulation. Associated with the Hadley circulation, low-pressure belts

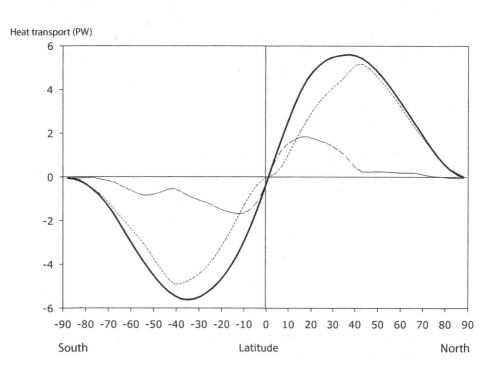

Fig. 1.5 Average transport of energy by the atmosphere (thin dotted) and ocean (dashed), and total transport (solid line). Positive transport towards the north and negative towards the south is recorded

predominate in the equatorial regions, while high pressures belts predominate in subtropical regions. A convergence of winds towards the equator, the trade winds, is thus observed in the tropics at sea level. The trade winds are dry at the start of their journey, since they are powered by the descending branch of the Hadley cell. Like the Harmattan over Africa, they maintain desert conditions on tropical continents. Over the ocean, there is high evaporation of surface waters heated by solar radiation and the trade winds pick up this vapor and carry it towards the low latitudes.

At the equator, with its ascending branch of warm moist air, low-altitude convergence is manifested by strong convective activity and heavy rainfall. These allow the development of a lush, tropical rainforest on land, while over the great ocean basins, convective activity is focused along a narrow longitudinal strip of one hundred kilometers wide: the intertropical convergence zone. It is in this area that storms and rain are concentrated: the 'doldrums', dreaded by sailing vessels in the past, and now by solo sailors.

Beyond 30° latitude, the flow of air, deflected eastward by the Coriolis force, reaches such speeds that it becomes unstable and breaks into eddies and meanders. Large meanders in this western circulation appear as vast oscillations, usually between three and six of them, which encircle the Earth. Depressions and anticyclones succeed each other in the middle latitudes between 30 and 60° north and south, creating very variable weather conditions. This is the cause of the 'temperate' climate prevailing in Western Europe. By mixing the hot subtropical air and cold polar air, these vortices take over the transfer of the excess energy from the tropics to the poles from the Hadley circulation. However, this circulation is affected by the contrast between land and oceans, and by the presence of mountains, both favoring the anchoring of global planetary waves whose intensity and position change over time. These waves impact on the geographical distribution of climate and cause, for example, a warmer climate on the west coasts than on the east coasts of the continents of the northern hemisphere. The contrast between the climate of Canada and that of France is a striking example.

There have been very significant changes in the intensity and location of the winds in the past, particularly during glacial-interglacial oscillations. They are evidenced by the presence, more or less marked, of pollen or desert dust transported to the ocean, sometimes very far from the coast, where they contribute to marine sedimentation. In the marine environment, when the wind blows parallel to the coast, it causes upwelling of deep cold water. Variations in their intensity, reflecting that of the wind, result in variations in the temperature of surface water that paleo-oceanographers have managed to reconstruct (see Chap. 21).

For recent periods, historical records provide information, sometimes subtle, on the variability of the winds and storms. For example, during the Little Ice Age, variations in the position of the winds were detected in the Pacific Ocean by analyzing the travelling time of galleons transporting goods between Manila (Philippines) and Acapulco (Mexico). The General Archive of the Indies held in Seville relates that the journey could take between less than three months and more than four months. The routes were always the same: departing from Manila, the galleons went east, allowing themselves to be carried by the stable westerly winds. For the return, they headed west, catching the northeast trade winds, and the duration of the journey was determined, in the end, by the location of the opposing winds (from the southwest) that they encountered as they approached Manila. Historians were thus able to show the existence of a period of about forty years in the middle of the seventeenth century when the headwinds were very common due to a northward shift of the large depressions. Changes in the strength and direction of winds are therefore an important manifestation of past climate changes.

Water Vapor, Clouds and Rainfall

Water in the atmosphere, in the form of vapor in the air, or as a liquid or ice in clouds, plays an important role in climate dynamics. Firstly, the amount of water vapor contained in the air is a function of increasing temperature as defined by the Clausius-Clapeyron relationship that links the saturation level of water vapor in air to temperature. When air temperature increases, its water vapor content increases. Furthermore, water vapor is the most important greenhouse gas in the atmosphere. Any increase in its concentration in the air in turn induces further warming of the atmosphere, engaging a positive feedback mechanism which amplifies the original disturbance.

When the water vapor content exceeds the threshold of saturation, water vapor condenses, causing cloud formation. Clouds have a particularly complex role, because of two opposite effects. On the one hand, they reflect part of the solar radiation, which has the effect of cooling the surface of the Earth. On the other hand, they have a greenhouse effect that causes it to heat up. These two effects are not completely balanced. On average, with current climate conditions, the reflectivity effect is more important that the greenhouse effect, and so, overall, clouds cool the Earth. But in the case of climate change, will clouds play a moderating or amplifying role? Low-level clouds and clouds in the upper layers of the troposphere are very different. Low-level clouds are usually thicker, and reflect solar radiation, but being near the surface, they have little impact on the greenhouse effect. Conversely, high altitude clouds like cirrus, are much thinner and are very cold. They let solar radiation pass through, but they contribute strongly to an increase in the greenhouse effect. It is not clear if global warming will be accompanied by more low-level clouds with a moderating effect, or high-level clouds with an amplifying effect. The answer is complicated further by the fact that it depends on changes in the general circulation in the atmosphere.

Through coalescence, micro-droplets of water grow and turn into rain, snow or ice, depending on the temperature. How precipitations are distributed reflects the main features of the general circulation of the atmosphere: ascendance and heavy rains at the equator, subsidence associated with a lack of rainfall in the sub-tropics, rainfall in the mid-latitudes linked to the passage of depressions. On average, excess in the evaporation rate over the oceans is offset by a surplus of rain over land. This transfer of water from oceans to land is particularly apparent during the seasonal phenomenon of the 'monsoon', well known in South East Asia but also in Africa. In summer, when the land warms up, low-pressure heat expands and causes a convergence of ocean winds towards the mainland. Charged with humidity, the winds rise and discharge a large amount of rain over land.

Reconstructing Changes in Precipitation

There are many signs to be found on the surface of our planet of how precipitation has changed over a given area. The accumulation of ice in the polar ice caps is directly dependent on the supply of snow. Glaciologists have shown that snowfall was half as abundant during glacial periods as it is today. At lower latitudes, fluctuations in rainfall are recorded in lake sediments. Deposits left above the current water level are evidence of phases of intense rain, particularly in the beginning of the interglacial periods. The levels of the lakes drop when rains abate. For example, an aerial view of Lake Chad shows tracks of the various shorelines that record the regression of the lake over past millennia. About 6000–8000 years ago, it occupied an area of 340,000 km^2 (equivalent to more than 2/3 of the area of France). By 2000, it was reduced to 1500 km^2, or less than 1% of its maximum size.

Lakes are not the only records of rainfall fluctuations on land. In limestone terrain, variations in the growth of concretions in caves are another indicator of fluctuations in the supply of groundwater by rain. Dating using geochronological methods (see Chap. 14) can detect slowdowns or arrested growth during dry periods, followed by recovery when the groundwater supply resumes in wet periods.

Monsoons are a prime example of intense rainfall affecting large areas, both in Africa and Asia. Rain falls during the summer months when the overheated land masses are the source of low pressures towards which the humidity-charged oceanic air masses converge. The intensity of the monsoons has fluctuated considerably during the Quaternary. This has resulted in enormous variations in the volume of water flowing in the major rivers which are fed by the rains, such as the Niger and the Nile in Africa, or the great rivers that drain the Himalayas and discharge into the Bay of Bengal. Variations in freshwater inputs to the ocean have been so significant that they have resulted in large fluctuations in salinity in areas close to the mouths of these rivers. Paleoceanographers are able to reconstruct these changes through isotopic and micropaleontological analysis of marine sediments deposited near the river mouth (Fig. 1.6, Chap. 21, Volume 1).

Land surfaces contain many other traces of major changes in the hydrological cycle in the past. During periods of great aridity, dry winds have facilitated the creation of dunes which became established when the rains returned. The vegetation growing in different regions is as much determined by air temperature as it is by precipitation and water availability. This is what causes the variation in the thickness and density of the annual rings of trees. Pollens found in lake sediments, peat bogs and in marine sediments close to the coast are used by geologists to reconstruct the major vegetation types that developed throughout the various geological periods allowing them to infer the temperature and humidity conditions that then prevailed (Chap. 12, Volume 1).

Modes of Variability of the Atmosphere

Atmospheric circulation is very variable. Over short timescales, variability is dominated by the duration of depressions, usually a few days. Over longer time scales, the circulation shows variability patterns over periods of up to several years. In Europe, the variability is dominated by fluctuations in the system caused by low-pressure from Iceland and high pressure from the Azores. This dipole oscillates between a 'positive' phase which is marked by a strengthening of the low and high pressures, stronger westerly winds bringing rain, and high temperatures in northern Europe, and the 'negative' phase where pressures and westerly winds subside, moving the rainy zone to the south of Europe (Fig. 1.7). The 'North Atlantic Oscillation' (often designated by the acronym NAO) occurs at all time scales and explains about a third of the variability in weather conditions in Western Europe, especially in winter. Positive and negative phases tend to predominate for ten years or more, which makes the NAO particularly interesting in the study of the climate of Europe. This mode seems to be caused by the atmosphere alone, and yet it has an influence over ocean circulation. The mechanisms that allow atmospheric circulation to present an oscillation over such a long period are still not fully understood.

Did the North Atlantic Oscillation exist in the past? Measurements of atmospheric pressure, in particular ones

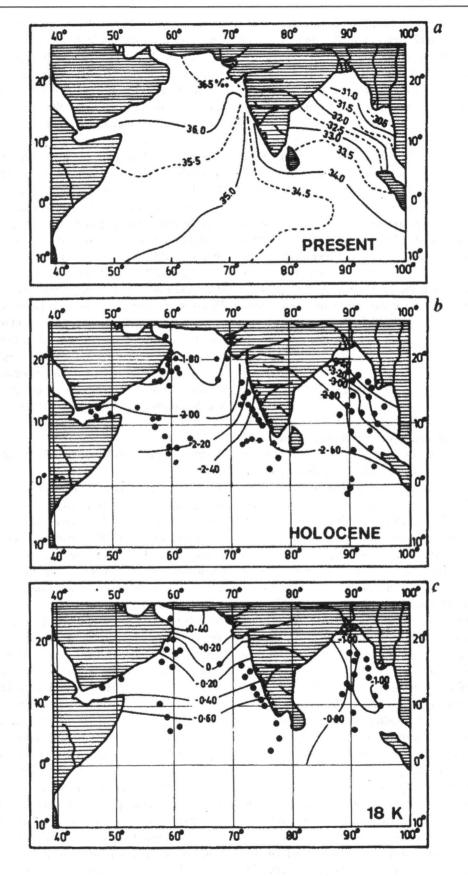

◀ **Fig. 1.6** Impact of variations in the intensity of the monsoon between the last glacial maximum and current times on the hydrology of the North Indian Ocean. **a** The current salinity of the Bay of Bengal where the rivers draining the Himalayas flow is much lower than that of the Arabian Sea where evaporation is dominant. **b** The isotopic composition of oxygen in planktonic foraminifera in recent marine sediments is a good record of these changes because there is little variation in the temperature of the northern Indian Ocean. **c** The isotopic composition of oxygen in planktonic foraminifera in marine sediments deposited during the last glacial maximum dated by ^{14}C at around 18,000 years BP shows a considerable decrease in the flow of rivers into the Bay of Bengal and thus in the intensity of monsoon rains

carried out in Reykjavik and Gibraltar, allow scientists to reconstruct past trends as far back as 1850. Comparison between winter air temperatures measured in weather stations shows that during the positive phase of the NAO, temperatures are higher than average in Western Europe and the southeast of Northern America, but below average in Greenland, the Labrador Sea, in northwestern Africa and the Eastern Mediterranean. The opposite is true in the negative phase of NAO.

These teleconnections between different sites are used by paleoclimatologists to reconstruct changes in the NAO in the pre-instrumental period. Tree rings provide a good record of climate conditions, especially of periods of extreme heat or drought (Chap. 16, Volume 1). Recordings obtained from trees of North America and Europe show alternating phases, some positive, some negative, since the beginning of the eighteenth century, with short periodic elements (8 and 2.1 years) but also multi-decadal ones (70 and 24 years).

Snowfall and ice accumulation in the western part of Greenland are abnormally low during the positive NAO

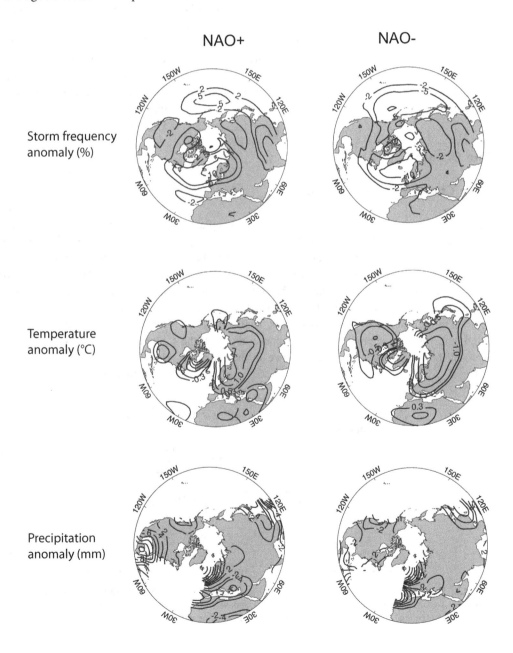

Fig. 1.7 The positive and negative phases of the NAO and their impacts on the distribution of temperature, precipitation and winds

phase, and this has allowed the reconstruction of the NAO and its intensity over the last 350 years with, again, the detection of multi-year and multi decadal intervals. Teleconnections associated with the NAO are also recorded in marine sediments. Over the last ten thousand years, there has been a tendency towards cooling of the surface waters of the eastern North Atlantic Ocean while there has been a contrasting warming of the subtropical western Atlantic and the eastern Mediterranean. This trend is seen as a sign of a continued weakening of the NAO during the Holocene.

However, on these time scales, it becomes difficult to distinguish between a change in the NAO expressed over several millennia and a long-term climate trend, driven by the slow fluctuations in orbital parameters. Indeed, cyclic variations in the precession (with a cycle of 21,000 years) were responsible for an increase in the winter incident solar radiation in the tropics 10,000 years ago, followed by its progressive decrease accompanied by a drop in the difference in atmospheric pressure between the tropics and the high northern latitudes during the Holocene. This is an example of the changes in insolation changes predicted by astronomical theory.

The Oceans

Main Characteristics of the Oceans

The oceans cover two thirds of the surface of the planet. With an average depth of 3900 m, they have a very high thermal inertia, much greater than that of the atmosphere. A layer 3 m deep of ocean surface waters has the same heat capacity as the 10 km troposphere. This feature explains why coastal regions have a much less contrasted climate than regions in the interior of large landmasses. It also plays an important role in determining the response time of the atmosphere-ocean system to a disturbance in the radiation balance.

The atmosphere and the oceans exchange momentum through the friction exerted by winds at the air-sea interface. They are thus responsible for the great marine currents, well-known to ocean-going sailors. The atmosphere and the oceans also exchange energy and water. Energy exchanges, through solar radiation, the infrared flow, turbulent eddies at the surface, and sensitive and latent heat, impact on the temperature of surface waters of the ocean. Water exchanges, through evaporation and precipitation, have an impact on salinity: evaporation increases the salinity of seawater while rain, conversely, decreases salinity. These interactions, which create variations in the temperature and salinity of seawater, ultimately determine its density. Density is indeed inversely proportional to temperature and directly proportional to salinity. Density differences are then the cause of the movement of large, deep-water masses in the world's oceans.

These physical interactions, shown schematically in Fig. 1.8, are supplemented by exchanges of matter, such as, for example, of carbon dioxide or sulfur compounds, which interact with the biogeochemical cycles of the different elements. In this way, physics, chemistry and biology are very closely linked in the ocean.

Oceanic Circulation

In the tropics, winds cause large ocean anticyclonic circulation, called 'vortices' or 'gyres', turning in a clockwise direction in the northern hemisphere and counterclockwise in the southern hemisphere. However, there is a marked asymmetry between the eastern and western sides of the ocean basins. For example, to the east of the North Atlantic, the Canary Current spans a much wider area and has a much lower intensity than the current on the western side, the Gulf Stream, which is very intense and is only a little more than 100 km wide. This strengthening of currents along the western edges of ocean basins is not specific to the Atlantic

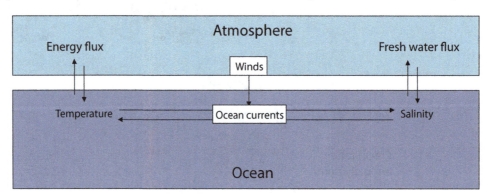

Fig. 1.8 Interactions between the atmosphere and the oceans: winds drive the surface currents that transport temperature and salt in the oceans. The exchange of energy and water between the atmosphere and oceans condition the temperature and surface ocean salinity that change the density of the water, thereby causing ocean currents

Ocean. It is also observed in the warm current of Kuro-Shivo in the North Pacific and in the Agulhas Current which runs along the South African coast in the southern hemisphere. This phenomenon is a complex result of the increasing strength of the Coriolis force with distance from the equator.

Differences in density also lead to large movements of the oceanic water masses. The densest waters are in the polar regions where sea ice forms. Already dense because of their low temperature, the waters receive a further load from two salt inputs: one coming from the flow of currents from the subtropics, where evaporation is intense and the other from salt released during sea ice formation. These dense waters tend to drop due to gravity below the warmer and less salty waters, then to spread out at the bottom of the ocean where the temperature, around 0–2 °C, varies little from the poles to the equator. These dense water masses are the starting point of the great global circulation loop of the ocean, called thermohaline, since denser waters tend to sink below less dense waters and the density of water masses is dependent only on temperature and salinity. This mechanism plays an important role in ocean circulation, as it contributes over 75% to the formation of all the masses of deep waters of the world ocean. Paradoxically, this process of downwelling deep water occurs only in a very small fraction of the surface of the oceans: in the Labrador Sea, the Norwegian and Greenland Seas and in some regions of North Atlantic as well as at the edge of the Antarctic continent, particularly in the Weddell Sea. In overall, the combination of the thermohaline circulation and the circulation caused by winds makes up the ocean meridional overturning circulation.

Downwelling waters in the North Atlantic descend to a depth of 2000–3000 m up to latitude of around 60° south, where water masses undergo a slow movement of ascent to the surface. Carried by the Antarctic Circumpolar Current that runs from west to east around the southern polar continent, the deep waters from the North Atlantic then spread out in the South Pacific and the Indian Ocean. The return part of this great circulation loop occurs through warm currents near the surface. They pass between the Indonesian islands, cross the Indian Ocean, circumnavigate Africa by the Agulhas Current, and then up towards the North Atlantic with the Gulf Stream and the North Atlantic Drift. But, while the return by the warm currents takes a few decades or even up to a hundred years, it takes several hundred to a thousand years from the time the cold waters sink in the North Atlantic to their arrival in the center of the Pacific, showing the slowness of this gigantic mixing achieved by the deep circulation.

Reconstructing Ocean Circulation in the Past

Paleoceanography reconstructs past ocean circulation by analyzing sediments that have settled in more or less regular layers on the ocean floor, the uppermost layers corresponding to the most recent deposits. This discipline has flourished thanks to several scientific and technical developments. One of these developments is in the methods of coring and drilling which allow cores to be brought to the laboratories which have relatively undisturbed sediments reliably recording the conditions in the ocean at the time the sediment was deposited. Furthermore, in-depth analysis of fauna (foraminifera) and flora (diatoms, coccoliths) fossils that lived in the illuminated area or in the water column permit the reconstruction of the temperatures of surface waters (see Chap. 21). Finally, new geochemical methods based on the analysis of stable and radioactive isotopes of elements present in marine sediments have provided dating methods, stratigraphic markers and tracers of large marine currents (see Chaps. 4, 6 and 7, Volume 1).

It is now possible to draw up temperature maps of surface waters of the ocean at critical times of the history of climate on Earth, such as during the Last Glacial Maximum with a ^{14}C age close to 18,000 years (in other words, a calendar age of about 20,000 years, Chap. 2) or during interglacial periods of the Quaternary. Paleoceanography also allows the reconstruction of the movements of fronts separating surface water masses with very different characteristics: a descent of polar waters in lower latitudes, driving out temperate waters, is accompanied by a strong cooling also felt by adjacent coastal areas. Conversely, their retreat is accompanied immediately by a significant warming.

Marine sediments also contain markers for the conditions that prevailed at depths, at the water-sediment interface. The most important of these are benthic foraminifera, microscopic animals with a calcareous shell whose isotopic composition is a particular reflection of the temperature and dissolved carbon dioxide content of the deep waters of the ocean. Measurements by oceanographers show that at a certain depth in the ocean, the physical and chemical characteristics of the waters are almost constant over a distance of several dozen kilometers. By taking sediment cores at different depths in an ocean basin, it is possible to reconstruct the features of large deep water masses, to deduce the main features of their circulation within a given period and to track their variations over time. Paleoceanography therefore allows the reconstruction of the main features of changes in the ocean in three dimensions.

El Niño, Interplay Between the Atmosphere and the Oceans

Interactions between the atmosphere and the oceans can lead to climate variability modes. El Niño is a perfect example. Every two to ten years, an abnormal situation occurs in the Pacific, manifested by the appearance of unusually warm waters along the coast of Peru and major disruptions in the tropical rainfall pattern.

In 'normal' periods, the water temperature is around 28–29 °C in the western tropical Pacific, while it does not exceed 20–25 °C in the east. This strong asymmetry in temperature between the east and west maintains the atmospheric circulation which, in turn, maintains the temperature gradient. The warmer waters provide the heat and humidity necessary for the development of strong convective activity over the western Pacific, which develops the air ascendance associated with the Hadley circulation, while the air descends over the cold waters of the eastern Pacific. This asymmetry between the eastern and western Pacific is associated with a circulation called Walker. In turn, the trade winds at the surface, which blow from east to west, maintain the east-west temperature gradient.

In the eastern Pacific, they cause a surface current deflected by the rotation of the Earth to the right in the northern hemisphere and to the left in the southern hemisphere, which drives out the surface water on both sides of the equator, and causes an upwelling of cold water to compensate. The trade winds also propel surface waters towards the west, where the accumulation of water inhibits the upwelling process; heated by the Sun, these waters reach the highest ocean temperatures, thereby favoring intense convective activity. This situation is called *La Niña* when the differences between east-west are particularly strong.

During an *El Niño* event, the circulation of both the ocean and the atmosphere change simultaneously following a series of mutual actions and reactions in which it is impossible to distinguish which of the ocean or the atmosphere triggers the phenomenon. In particular, the water of the central Pacific is warmed to 28–29 °C, which has the effect of moving the high convective activity towards the east. A decrease in the strength of the trade winds in the western Pacific follows, and possibly even a reversal of their direction. With weaker trade winds, the surface current weakens and the warm waters of the western Pacific flow back towards the east, a backlash that warms the central Pacific and interrupts the upwelling of cold waters at the coasts of South America. But *El Niño* starts a series of waves in the ocean that eventually pans out and restores the so-called 'normal' situation.

Because of their well-recorded and varied consequences, the existence of variations in the intensity and frequency of *El Niño* events in the past are known. In the Andes, the arrival of the warm waters on the Pacific coast brought heavy, sometimes catastrophic, rainfall that caused floods or even huge landslides that geologists are able to date. Prehistoric sites also bear traces of these events, and archaeologists have established that certain Andean civilizations developed during periods when El Niño events were rare or weak and regressed with the return of torrential rains. However, the most precise way to reconstruct the sequence of these events is by analyzing the geochemistry of the corals abundant in the waters of the equatorial Pacific Ocean and located at the heart of the phenomenon. For example, Tarawa Atoll located near the International Date Line (180° meridian) usually has a dry, almost desertic climate. When an *El Niño* event occurs, the warm waters reach it, atmospheric convection becomes intense locally and heavy rains fall on the entire atoll and the coral reef that surrounds it. Corals, animals with a calcareous skeleton with recognizable annual bands, record these rainy passages. Analysis of them has allowed the number of *El Niños* in the last century to be counted and to show that their frequency has changed over the last hundred years. Conversely, near Australia, which is usually in a region of warmer water, *El Niño* events are characterized by decreased rains and cooling, which were reliably recorded by the coral reefs of New Guinea or Fiji. These recordings show that *El Niño* events have also existed during periods of glacial climate, and confirm that their frequency and intensity have varied in the past, the twentieth century being a period during which they were particularly strong (Fig. 1.9).

The Terrestrial and Marine Biosphere

The biosphere, defined as all living organisms, also intervenes in the operation of the climate system. Some theories even propose that, throughout geological time, the biosphere has contributed to the regulation of climate in order to create conditions compatible with life.

The Geographical Distribution of the Biosphere

On land, the biosphere is mainly made up of vegetation which is distributed according to the critical climate characteristics which are sunshine, temperature and precipitation. Rainforests can only develop if temperature and humidity conditions are favorable for the twelve months of the year. They are replaced by a dry forest or savannah if the soil water content decreases over several months. The savannah itself becomes increasingly sparse as aridity increases eventually becoming a desert. North of the tropics, seasonal

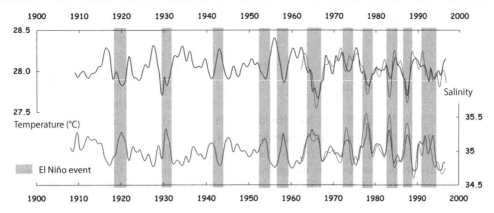

Fig. 1.9 Changes in the temperature and salinity of surface waters in the archipelago of Fiji (Western Pacific) reconstructed from changes in the isotopic composition of coral living in coastal areas. The period 1960–1995 for which instrumental measurements are available was used as a calibration period (Courtesy of Dr. Anne Juillet LSCE)

changes in temperature and humidity dominate the climate. Vegetation takes the form of deciduous forests in humid temperate regions, of drought-resistant flora in the Mediterranean area or prairie grass in dry areas with strong seasonal contrasts. Even further north is the area of the boreal forest (taiga), consisting of birch and conifers, and finally the tundra, where trees cannot grow. The mapping of current vegetation is therefore closely linked to major climate zones.

In the oceans, the marine biosphere depends on the temperature and salinity, but also on the amount of light and nutrients available to enable the production of phytoplankton which forms the basis of the ocean food chain. However, the areas where the production of phytoplankton is abundant are very limited. Apart from some coastal fringes of tropical regions, the Southern Ocean and the North Atlantic are the only areas capable of producing enough nutrients to continuously feed a wide range of living matter. Most of the ocean areas consist of essentially sterile water, hence the bright blue color of tropical waters.

The Role of the Biosphere

On land, vegetation alters the exchanges of energy, water and momentum. Vegetation permits greater solar energy absorption than bare soil. Indeed, the albedo (reflection power) of vegetation cover is 10–15% compared to 35% for bare soil. Trees also enhance the evaporation from the surface, through transpiration by the foliage and through pumping water from the soil by the root system. Finally, a tree creates an obstacle at the surface and increases turbulence close to the ground, hindering the wind more efficiently than bare soil. Based on numerical simulations, these physical effects of vegetation seem to have reinforced the intensification of monsoon rains during the mid-Holocene, 6000 years ago.

During the 1980s, the discovery of past variations in carbon dioxide and methane in air trapped in Antarctic ice has directed the spotlight onto the role of the biosphere, hitherto neglected. Indeed, these changes that show disruption of biogeochemical cycles cannot be explained solely by the physical exchanges between the atmosphere and oceans. The biosphere must be taken into account. The land-based biosphere absorbs carbon dioxide by photosynthesis, but emits it by respiration during the life of the plant and also later, during its decomposition in the soil through the action of bacteria. On average, as long as the climate remains constant, the absorption and emission of carbon dioxide by the land-based biosphere balances out and biological activity recycles atmospheric carbon.

In the oceans, phytoplankton also absorbs carbon dioxide through photosynthesis. This carbon is then reused to form the tissues of other living organisms in the food chain as well as organic waste of all kinds. In this way, phytoplankton is responsible for a rapid recycling of carbon in the surface waters of the oceans: the absorption carried out during photosynthesis is offset by the constant emission of carbon dioxide caused by the respiration of algae, zooplankton and fish, as well as by the oxidation of waste. A fraction of this carbon, about 10%, is subtracted from this recycling. Fecal pellets, dead tissue and other waste sink as a result of their weight and take some of the carbon absorbed at the surface of the oceans with them to the ocean floor. Most of this 'marine snow' dissolves or decomposes through the action of bacteria before reaching the bottom of the ocean, releasing carbon organic matter which is added to the dissolved carbon in the deep ocean waters. A tiny fraction of this 'marine snow', around 1% of the carbon taken from the surface, is deposited in the ocean abysses, forming sediment in which carbon is trapped for millions of years.

Biosphere of the Past and Paleoclimates

For a long time, geologists have known that the land-based and marine biospheres varied greatly in the past in response to the slow process of evolution (on the scale of millions of years, Chap. 27, Volume 2) but also to changes in the global climate and in the general circulation of the atmosphere and oceans, particularly at the pace dictated by the astronomical paleoclimate theory (Chaps. 28, 30, Volume 2).

During their reproduction cycle, most plants produce tiny grains with a shape and decoration characteristic of their species. These are the pollen (the male fertile element of the flower) and spores (vegetative structure for propagation or reproducing) which have a very hard outer shell. They are transported by wind or streams and stored well if they land in a low-oxygen environment. They have been found in lake sediments and bogs, where they provide a reasonably accurate picture of the vegetation once covering the vicinity of the lake or bog. In this way, it has been discovered that, twenty thousand years ago, the vegetation of France was that of a polar steppe, at the peak of the last glaciation.

Land sediments contain other fossils which also provide information on the local climate: diatoms, mollusks, ostracods (tiny crustaceans with a calcareous shell) living in freshwater, larger remains of plants (stem and leaf), charcoal remains from large natural fires, fossil remains of larger animals, such as bones of mammals found in archaeological sites.

The link between the climate and plant fossils is so close that statistical methods have been developed to quantify this link, based on current observations. The resulting relationships, called 'transfer functions' are used to estimate past climate conditions if the same plant associations are found in ancient sediments as those known today (see Chap. 12, Volume 1).

Leaving aside the abundant fossils present in coastal areas and in coral reefs, the marine biosphere leaves many traces of its diversity in marine sediments: fish otoliths and teeth, aragonitic shells of pteropods (marine snails), calcareous shells of planktonic and benthic foraminifera, coccoliths (calcareous sheets secreted by microscopic algae, called coccolithophorids and living in warm or temperate waters), siliceous skeletons of marine diatoms (algae living in cold waters rich in silica) and radiolarians (microscopic animals living in deep water). The relationship between the temperature of the sea water and abundance of various species of foraminifera fossils and diatoms is close, which helps to establish transfer functions to estimate marine paleotemperatures with an accuracy close to 1–2 °C (Chap. 21, Volume 1).

It is important to be aware that transfer functions are only valid insofar as the species discovered in ancient sediments, both land and marine, are the same species as those that are present today. Even if only a rough estimate of the conditions that prevailed in the geological past is required, it is important to remember that species evolve and are likely to adapt to very different environments over millions of years. Thus, in the Jurassic era, 150–200 million years ago, well-developed coral reefs in warm waters harbored a variety of mollusks such as *Pholadomya*, *Tridacna* and *Astarte*. Their distant descendants can be found today in very different environments: *Pholadomya* buried in the mud in warm coastal waters, in a reef environment for *Tridacna* and in polar waters for *Astarte*. This shows how the reconstruction of paleoenvironments requires a cautious approach and the comparison of the various clues found in fossils.

The Cryosphere

Water in the form of ice or snow is the cryosphere. Glaciers and ice caps cover approximately 11% of the surface of the Earth. Spread out over the oceans, the water they contain would increase the sea level by about 77 m. Sea ice, which is formed by freezing seawater, covers approximately 7% of the oceans, but is only a few meters thick. As it floats, its melting does not cause the sea level to rise. On land, snow cover varies greatly with seasons.

The Role of the Cryosphere

The main property of the cryosphere is its albedo. It can reach 80–90% for fresh snow and barely drops below 50% when surface ice melts or when the snow covers trees. This property introduces the second major positive feedback loop of the climate system. Where snow or ice melts out, the ground absorbs a larger fraction of the incident solar energy. Containing more energy, it heats up, thus facilitating the melting of the remaining snow and ice. The process thus becomes amplified. In the Milankovitch theory of ice ages (see Chap. 28, Volume 2), the decrease in incoming solar radiation, due to the slow variations of the orbit of the Earth around the Sun, triggers the formation of ice caps thanks to this positive feedback. Indeed, following the decrease of sunshine, snow accumulated in the winter does not melt completely, in turn less energy is absorbed and the snow starts accumulating until ice caps are formed. As we shall see later (Chap. 25, Volume 2) the mechanism is actually more complex, and involves feedback from the ocean and boreal biosphere in response to changes in insolation, even if this positive feedback does constitute the factor triggering the start of an ice age.

However, the role of the cryosphere is not limited to its impact on the surface albedo. On the oceans, sea ice cuts off the ocean from the atmosphere, and blocks the air-sea exchanges of water, salt and other chemicals. When it is forming, sea ice eliminates salt which makes the surrounding sea water denser causing it to sink to depths, thus feeding the ocean thermohaline circulation. Conversely, when it melts in summer, the salinity of surface water decreases abruptly. On land, glaciers are not static. They flow under their own weight at a speed of up to several hundred meters per year. When they reach the coast, the ice breaks into icebergs that are propelled by winds and currents, and melt when they arrive in warmer water, causing a drop in salinity of surface waters of the ocean. Such changes are likely to bring about important feedbacks, such as stopping deep-water formation and the thermohaline circulation (see Chap. 29, Volume 2).

A climate change favoring the flow of glaciers could promote the destabilization of ice sheets, such as the one covering West Antarctica. Indeed, it rests on a base that would normally be covered by the sea, and the ice sheet is in contact with the ocean on all sides. Such a mass of ice is unstable. Because of this instability, gigantic floating ice platforms, several hundred meters thick, surround the entire part of the ice sheet in contact with the ocean.

The sheets covering the Ross and Ronne Seas, each with a surface area similar to that of France, calve huge icebergs, tens of kilometers in length, into the Weddell Sea. Without any rocky terrain, they are in direct contact with the ice cap and their speed often reaches several meters per day. What helps to safeguard the West Antarctic ice sheet is the presence of several islands around it. These anchor points slow down the ice flow. Ice therefore progresses slowly until that last barrier which, once crossed, leaves the field open to the calving of icebergs resulting from the fragmentation of the ice platform. The future of the West Antarctic ice cap in response to ongoing climate change is a real concern for the centuries to come.

The Cryosphere in the Past and Paleoclimates

The variability in the cryosphere is considerable, no matter what time scale is considered. Over the span of a season, snow cover and sea ice show the biggest variations. Measurements taken since the mid-twentieth century show that the area of sea ice each year goes from 15 million km^2 in early spring to just 6 million km^2 at the end of the summer. However, since 1980, a clear decrease in this range has been observed and it has reduced to nearly 3 million km^2 in 2012 summer. A similar trend in the retreat of mountain glaciers is observed during the twentieth century, whether in the Alps, or in the mountains of Africa or South America.

On a geological scale, the variability of the cryosphere is even greater. It is seen in the geological traces left by glaciers recording their passage and their broadest expansion (moraines, boulders streaked by friction marks as they were transported by the glacier over the surface bedrock on which it rested, vast continental shields like Canada eroded away). Geological observations have thus led us to believe that some 750 million years ago, all continents were covered with glaciers, and that the oceans were probably covered with perennial sea ice at the same time.

About 450 million years ago, during the Ordovician, a gigantic ice cap covered the Sahara where even today one can see striated rocks, glacial valleys, remains of moraines and channels that collected the water from the melting ice in summer and brought it to North Africa. Flying over western Mauritania, one can recognize the sandy bed of a great river that came into being beside an ice cap and traced out many meanders before flowing further north, to the seas bordering the glaciated African continent. These rivers were covered by icebergs that melted slowly and released stones they carried. They can be found today in exposed terrains, in Morocco, Galicia and even in the Armorican massif, south of Caen.

About two hundred million years ago, a long, globally warm era commenced (Jurassic and Cretaceous), during which time glaciers appear to have been rare, if they had not completely disappeared. The wide variety of flora and fauna reflects a variety of environmental conditions, from temperate in Japan, Siberia and Australia to very hot in America, Africa and Eurasia. There are no known tracks of large glaciers, even on the Antarctic continent, although, of course, as the continent is now covered with a thick ice cap it is only accessible to geological observations at its periphery.

The glaciations of the Quaternary are the culmination of a long process of cooling of the Earth that began more than thirty million years ago, firstly with the development of an ice sheet on Antarctica, and then on Greenland. For the last million years, glacial-interglacial oscillations have dominated the climate of our planet. While today it is the Southern hemisphere which is the most glaciated with Antarctica containing about 28 million km^3 of ice for only 1 million km^3 of ice in Greenland, it was the Northern hemisphere that was the most glaciated at the height of the last glaciation 20,000 years ago, with 50 million km^3 of ice over Canada (Laurentide Ice Sheet, 4 km thick) and northern Europe.

Throughout the last ice age, the Laurentide Ice Sheet extended beyond the American continent to reach the Atlantic continental shelf and the Labrador Sea. This cap could become unstable, suddenly releasing huge numbers of icebergs which invaded the entire North Atlantic where they melted. The sudden appearance of these armadas of icebergs, designated as 'Heinrich events' caused a significant decrease in the salinity of the water at high latitudes, preventing downwelling of dense water in winter, thus stopping the thermohaline circulation and the transport of warm water by the Gulf Stream and North Atlantic Drift. The result was a freezing of Europe and a disruption of the climate over almost the entire planet. The behavior of the ocean-atmosphere-cryosphere interactions during glacial climate periods is complex. In addition to the Heinrich events that occurred during the last glaciation at intervals of eight to ten thousand years, the paleoclimate record inferred from the isotopic analysis of Greenland ice indicates a sudden warming (>10 °C) between Heinrich events, over periods of less than a few centuries that end with a slower cooling leading to a return to glacial conditions. These are the Dansgaard-Oeschger events which occur every two to three thousand years and which are not explained by any one unanimously accepted theory (instability of the European ice sheets, internal oscillation in ocean circulation, amplification of a weak solar forcing). The climate of the ice ages appears to have been much more variable than the climate we have known for the past 10,000 years, but the conditions for stability of the climate system remain a research topic that is far from fully understood (Chap. 29, Volume 2).

The Lithosphere: Over Large Timescales

The surface of the Earth, which makes up the lithosphere, also intervenes over long time scales of the order of a million years or more. Plate tectonics determine the position of the continents, the relief, the shape of the ocean basins, as well as the level of carbon dioxide emitted by volcanoes (Chap. 2, Volume 1 and Chap. 22, Volume 2). For example, the glaciation of the Sahara during the Ordovician is linked with the movement of the African plate, which had a polar location at that time. Pioneer studies suggested that the Antarctic ice cap was triggered 34 million years ago, when the American and Australian continents had sufficiently drifted from the Antarctic to allow the establishment of the circumpolar current that isolated Antarctica from the mid-latitudes. More recently, it has been shown that the long trend decrease of atmospheric CO_2 was pivotal in the triggering of Antarctica ice sheet. The uplift of the Himalayas, through its impact on the weathering and erosion of surface rocks, has likely contributed to increased consumption of atmospheric carbon dioxide and to the global cooling observed since the Eocene. The uplift of Tibetan Plateau as well as the Paratethys shrinkage amplified the Asian monsoon. The major tectonic phenomena therefore have a major impact on global climate on a geological scale (Chap. 2 Volume 1 and Chaps. 22, 26 and 27, Volume 2).

The lithosphere also interacts with the atmosphere through volcanism. During violent volcanic eruptions, large amounts of dust and gas are ejected into the atmosphere, up to several tens of kilometers in altitude. The emitted sulfur dioxide combines with the water vapor present in the stratosphere to form micro droplets of sulfuric acid, which is very effective in reflecting solar rays and causes a significant cooling of the surface of the Earth, of around 0.5–1 °C on average. In addition, these very small droplets can remain in the stratosphere for several years before falling into the troposphere where they are eliminated by rain. For example, the most intense volcanic eruption of the last two centuries, that of the Indonesian volcano Tambora in 1815, projected about 150 km^3 of debris and gases into the atmosphere and was followed by two exceptionally cold years. The year 1816 was even deemed to Canada and New England to have been 'the year with no summer'. Recently, after the eruption of the Philippine volcano Pinatubo in June 1991, one of the largest of this century, the weather stations recorded a significant lowering of temperature for several months.

Although the influence of volcanic eruptions on climate is limited mostly to a cooling for a relatively short time, rarely exceeding a few years, geologists think that unusually intense eruptions, called trapps, could have contributed to large-scale disruptions in climate and mass extinctions through their emissions. For example, it is estimated that the fissure eruptions that continued for several hundred thousand years to form the gigantic Deccan plateau, could have been of sufficient magnitude severally increased the carbon dioxide content of the air and to warm it by 3–4 °C until the weathering of rocks consumed the excess. Further in the past, 250 million years ago, huge fissure eruptions in Siberia are often cited as one of the factors responsible for global warming that marked the end of the Permian (Chap. 27, Volume 2).

The Climate System

Atmosphere, oceans, cryosphere, biosphere, and the lithosphere, all contribute to climate changes through a complex set of actions and reactions, both physicochemical and biological (Fig. 1.10). The atmosphere and the ocean, the two fluids that transport the excess energy received in the tropics to the poles, are central in climate dynamics, and this is true for all periods of time ranging from a few years to millennia. Strongly coupled with the mechanical effect of wind on the surface of the sea and through the exchange of heat and water, together they can cause fluctuations in natural climate phenomena, the scale of which can be glimpsed in *El Niño* for example. Their interactions can be strongly modified by changes in the land surface (vegetation, snow, ice). By their impact on the albedo and the water cycle, these changes modulate the exchange of heat and water between the two hemispheres, and between the oceans and land. Finally, past

Fig. 1.10 The climate system and the exchanges of energy, water and carbon that affect it

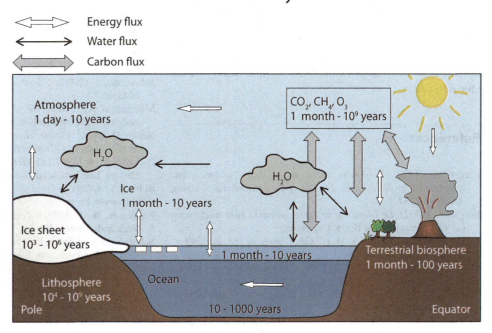

climates provide many examples of fundamental reorganizations of the climate system.

For periods extending beyond a few thousand years, changes in insolation due to slow changes in the Earth's orbit around the Sun, the formation of ice sheets, the sinking of bedrock and changes in the composition of air also become drivers of climate change. Drivers which are however themselves dependent on other components of the climate system: the atmosphere, oceans and biosphere, through the underlying energy, water and carbon cycles. The formation of ice sheets is in effect controlled by the temperature of the air and by the accumulation or melting of snow. The concentration of carbon dioxide depends on the gas exchanges between the atmosphere and oceans, themselves governed by winds and temperature; it also varies with the activity of marine phytoplankton, itself strongly controlled by ocean circulation. As a result, these interconnections can transform the slow variations of the movement of the Earth around the Sun into spectacular glaciations. Over millions of years, plate tectonics become dominant, changing not only the geography of the Earth's surface, but also the composition of the atmosphere.

For a long time, scientists believed that climate changes of very large amplitude were phenomena that occurred slowly relative to the scale of human life, governed mainly by phenomena occurring over large time scales, such as the evolution of the Sun measurable over hundreds of millions of years, plate tectonics, the effects of which are felt gradually over millions of years, or the astronomical forcing with cycles of a few dozen millennia. The recent discovery of large iceberg armadas (Heinrich events), which can tip the climate into a glacial state in only a few decades shows that this is not the case. As soon as the calving of icebergs stops, the climate warms suddenly, a new rapid change that Neanderthals and Cro-Magnon witnessed. And between two Heinrich events, Dansgaard-Oeschger events are another manifestation of quick, abrupt climate change. All these observations show that the climate system as a whole is unstable due to multiple feedbacks that can result from the slightest disturbance.

We are not immune today to a brutal, unexpected change in climate, since human activity has reached a level that significantly disrupts the radiation balance of the atmosphere. The role of paleoclimatology is to document climate variability at all time scales and to help to highlight the mechanisms that come into play so as to understand the resulting changes. As such, paleoclimate information is analyzed in the Intergovernmental Panel on Climate Change assessment reports to outlines our understanding of climate change and mechanisms as well as to evaluate climate models. For example, the 5th Assessment Report reported how the current atmospheric concentration of carbon dioxide is unprecedented in the past 800,000 years, and how the rate of sea level rise since the mid-nineteenth century has been larger than the mean rate for the previous two millennia. It also informed on high sea levels during the last interglacial climate and proved the capability of models to reproduce past warm and cold climates.

Find out more

References to studies mentioned in this chapter presenting the climate system will be detailed and reported in the various chapters of this book. We indicate below some general literature which readers may want to consult to deepen their knowledge of the behavior and history of the climate system.

References

Berger, A. (1978). Long-term variations of daily insolation and quaternary climatic changes. *Journal of the Atmospheric Sciences, 35*, 2362–2367.

Berger, A. (1992). *Le Climat de la Terre, un passé pour quel avenir* (479 p). Bruxelles: De Boeck Université.

Broecker, W. S., & Peng, T. H. (1982). *Tracers in the sea* (690 p). Palisades, N.Y.: Eldigio Press.

Crowley, T. J., & North, G. R. (1991). *Paleoclimatology* (349 p). Oxford: Oxford University Press.

Duplessy, J. C., & Morel, P. (1990). *Gros Temps sur la planète* (296 p). Paris: Éditions Odile Jacob.

Holland, W. R., Joussaume, S., & David, F. (1999). Modeling the Earth's climate and its variability. In *École des Houches* (565 p). Amsterdam: Elsevier.

Hurrell, J. W. (2003). *The North Atlantic oscillation: Climatic significance and environmental impact* (279 p). Washington, D.C.: American Geophysical Union.

IPCC (Ed.). *Climate change (2013): The physical science basis; Contribution of Working Group I to the fifth assessment report of the Intergovernmental Panel on Climate Change.* Cambridge, United Kingdom and New-York, USA: Cambridge University Press.

Johnson, G. C., Mecking, S., Sloyan, B. M., & Wijffels, S. E. (2007). Recent bottom water warming in the Pacific Ocean. *Journal of Climate, 20*, 5365–5375.

Joussaume, S. (2000). *Climat, d'hier à demain* (143 p). Paris: CNRS éditions.

Mélières, M. A., & Maréchal, C. (2015). *Climate change: Past, present and future* (416 p). Hoboken: Wiley.

Neelin, D. (2011). *Climate change and climate modeling* (282 p). New York: Cambridge University Press.

Philander, G. S. (1989). *El Niño, La Niña, and the southern oscillation* (293 p). London: Academic Press.

Rohling, E. J. (2017). *The oceans: A deep history*. Princeton: Princeton University Press.

Ruddiman, W. F. (1997). *Tectonic uplift and climate change*. New York and London: Plenum Press.

Schneider, S., & Londer, R. (1984). *The co-evolution of climate and life* (563 p). San Francisco: Sierra Club Books.

Steffen, W. Sanderson, R.A., Tyson, P. D., Jäger, J., Matson, P. A., Moore III, B., Oldfield, F., Richardson, K., Schellnhuber, H. J., Turner, B. L., Wasson, R. J. (2005). *Global change and the earth system, IGBP Series* (336 p). Berlin: Springer.

Tomczak, M., & Godfrey, J. S. (2005). *Regional oceanography: An introduction*. PDF version available at: http://www.cmima.csic.es/mirror/mattom/regoc/pdfversion.html.

Wang, B. (2006). *The Asian monsoon*. Springer-Praxis books in *Environmental Sciences* (779 p).

The Changing Face of the Earth Throughout the Ages

Frédéric Fluteau and Pierre Sepulchre

The face of the Earth has changed dramatically over the last 4.5 billion years. The growth and emergence of the continental crust transformed a largely ocean-covered planet in its early days into a planet with land masses. Under the action of mantle dynamics, the first continental crusts merged with island arcs to constitute the first continents at the end of the Archean and the Paleoproterozoic (see Fig. 2.1 for a chart of geological time). These continents gathered to form the first documented supercontinent (Bleeker 2003) around 1.5 Ga, before breaking up and then coalescing again. Each phase has profoundly changed the Earth's surface. Vast mountain ranges got uplifted and ocean basins formed, then disappeared, through erosion or later geological events for the former, by posterious closure linked to tectonics for the latter. The climatic upheavals that have marked the Earth's history are strongly linked to these paleogeographic events through direct or indirect coupling between the different solid, liquid and gaseous envelopes that are depicted in volume 2. Here we provide an overview of the major geological stages of the early Earth before detailing the paleogeographic changes of the modern Earth in terms of continental distribution and paleotopography.

Paleogeographic Reconstructions

Paleogeographic reconstruction requires to quantify the changes in location of the continents, as well as their coastlines and topography, through the geological ages. Here we describe the techniques and tools used to retrieve such information, as well as the uncertainties inherent in each method.

Continental Drift

In 1915, the German meteorologist, Alfred Wegener, laid the foundations for continental drift in a book entitled *Die Entstehung der Kontinente und Ozeane* (The Origin of Continents and Oceans). This theory is based on several arguments, in particular on the complementarity of the continents bordering the Atlantic Ocean and the continuity of the terrain and their deformations. But Wegener was not the first to make these observations. A Dutch cartographer, Abraham Ortelius, observed this complementarity of the Old and New Worlds in his atlas *Theatrum Orbis Terrarum* published in 1570 and questioned the cause of this rupture and this expansion towards the west in his book *Thesaurus Geographicus* published in 1596 (Romm 1994). Two hundreds and fifty years later, the French geographer, Antonio Snider-Pellegrini, published "The Creation and its Unveiled Mysteries" (1858) in which he drew the position of the continents after the closure of the Atlantic Ocean. However, the theories of Wegener were not limited to the complementarity of the continents bordering the Atlantic Ocean and geological continuity, he also postulated on the basis of the paleontological continuity of flora (*Glossopteris*) and fauna (*Mesosaurus, Lystrosaurus, Cynognathus*), and paleoclimatic continuity. Glacial sediments from the Carboniferous era were discovered at the end of the nineteenth century in Africa, India and Australia. Wegener showed that by closing the Atlantic and Indian oceans, all these outcrops would form a coherent cluster close to what was the South Pole at the time. All these arguments supported the idea of a supercontinent, the Pangea. This revolutionary theory of the continental drift was strongly rejected by the Earth sciences community who preferred to believe in their fixity, and who criticized Wegener for the absence of mechanisms

F. Fluteau (✉)
Université de Paris, Institut de physique du globe de Paris, CNRS, 75005 Paris, France
e-mail: fluteau@ipgp.fr

P. Sepulchre
Laboratoire des Sciences du Climat et de l'Environnement, LSCE/IPSL, CEA-CNRS-UVSQ, Université Paris-Saclay, 91191 Gif-sur-Yvette, France

© Springer Nature Switzerland AG 2021
G. Ramstein et al. (eds.), *Paleoclimatology*, Frontiers in Earth Sciences,
https://doi.org/10.1007/978-3-030-24982-3_2

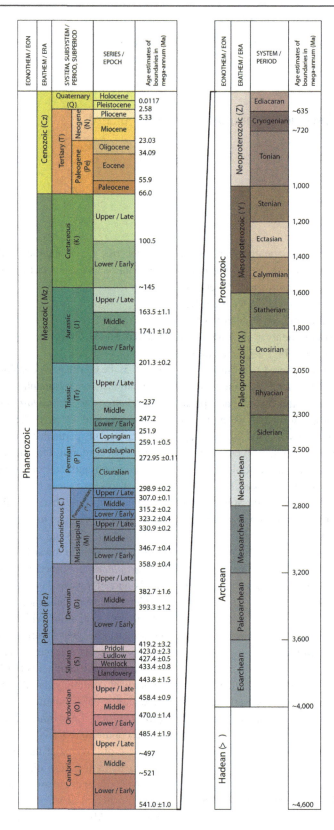

Fig. 2.1 Divisions of the geological time approved by the U.S. Geological Survey Geologic Names Committee (2018)

demonstrating his theory. The paleontologists of the early twentieth century clung to the concept of fixed geography, and invented the theory of "continental bridges," land ties that would have linked the continents, to explain the migration of fauna and flora. Nevertheless, some geologists believed in continental drift and supported it, such as Alexander Du Toit and Emile Argand. A decisive argument in favor of Wegener's theory was established by Arthur Holmes. He was the first to establish the basics of thermal convection of the Earth's mantle (Holmes 1929). But here again, this argument did not receive the expected positive response from the Earth science community, and it was not until the 1960s that Holmes was seen as a pioneer of plate tectonics.

From the 1940s, exploration of the ocean floor would revolutionize Earth sciences. The mapping of the morphology of the ocean floor, and the discovery of the Atlantic Ridge under the initiative of Bruce Heezen and Marie Tharp (Heezen 1962; Heezen and Tharp 1965), led Harry Hess[1] to propose a theory of the expansion of the ocean floor (Hess 1962). Half a century after Wegener, Earth sciences were experiencing a revolution, following the discovery of magnetic anomalies (Chap. 7). An American oceanographic campaign mapped the magnetic field around the Juan de Fuca Ridge, off the North American coast, in the eastern Pacific Ocean. By subtracting the ambient magnetic field from the magnetic data, alternately positive and negative anomalies were found (Raff and Mason 1961). Based on these observations, both Morley, and Vine and Matthews (Vine and Matthews 1963), proposed the theory of the renewal and expansion of the ocean floor in 1963.

These marine magnetic anomalies result from the acquisition of a thermoremanent magnetization by the iron and titanium oxide particles in the oceanic crust subjected to the Earth's magnetic field, after the oceanic crust (basaltic lava), emitted at the ocean ridges at a temperature of about 1100 °C, begins to cool. When basalt reaches the Curie point (the magnetic particle-dependent temperature, about 570 °C for the ferromagnetic magnetite crystals contained in the basalt), the direction of the field is fossilized by the magnetic carriers of the oceanic crust. Above the Curie point, the material is paramagnetic, and each magnetic carrier behaves like a small compass that follows the direction of the magnetic field without storing it. As soon as the temperature of the rock passes below the Curie temperature, the magnetic carriers aligned along the lines of force of the Earth's magnetic field are permanently frozen in this direction. The Earth's magnetic field can also be recorded in sedimentary rocks if they contain magnetic particles (Chap. 7).

Sea floor spreading is the cornerstone of plate tectonics, but to complete this theory, temporal constraints needed to be integrated, enabling the understanding of the pace at which ocean ridges opened. During this same period, the geophysicists Cox and Doell, and the geochemist Dalrymple established the first timetable of magnetic reversals for the last 4 million years (Chap. 7), using a new technique of isotopic dating with the potassium and argon elements (Cox et al. 1964) (Chap. 5). By comparing the magnetization polarity and the age of basalt samples taken at sea, it is clear that the age of the oceanic crust increases with distance from the line of the ridge. From the 1960s onwards, the DSDP (Deep-Sea Drilling Project) and ODP (Ocean Drilling Program) scientific missions were launched with the aim of drilling for sediment deposits on this oceanic crust. These sedimentary cores allow the sequences of magnetic polarities fixed by period to be established, thanks to the fossil content, and thus, to also date the marine magnetic anomalies. The speed of opening of the ocean ridges can then be determined. In 1968, Heirtzler and his group quantified the speed of opening of the South Atlantic Ocean by analyzing a marine magnetic sequence dating back to the Pliocene (3.35 Ma). In the following years, the kinematic parameters of all the oceans were determined one after another, and the evolution of the different ocean basins could then be traced. These parameters reflect the movements of one lithospheric plate relative to another one that is arbitrarily fixed. These movements being defined on a quasi-spherical surface, they can be expressed by an angle of rotation about an axis passing through the center of the Earth and defined by the longitude and latitude of its pole.

Moving on from the oceans to the land, in the early 1950s, some scientists studied the natural remanent magnetization of rocks. At all points of the globe, the magnetic field is defined by a vector collinear to the field lines. The magnetic field of the Earth originates from convective movements within the outer core, which consists of liquid iron (about 90 wt%), nickel (about 4 wt%), along with some lighter elements, such as silicon, sulfur and oxygen. The movements within the conductive core induce electric currents which, in turn, generate a magnetic field. The Earth's magnetic field functions like a self-excited dynamo. To compensate for the energy losses associated with the electrical resistivity within the Earth's core, energy input obtained from the conversion of ohmic dissipation into heat, from the gravitational energy and from the release of latent heat during the crystallization of the inner core ensures the thermal equilibrium of the milieu and permits the functioning of the geodynamo, as well as its continuity over geological time.

---

[1]When B. Heezen presented their findings to Princeton in 1957, Harry Hess stood up and said: "Young man, you have shaken the foundations of geology!" (Yount 2009).

It is therefore important to check the time at which the geodynamo was set up. To do this, it is necessary to find very old rocks that would have registered and retained a primary magnetization. Samples taken from the Matachewan Dikes dated at 2.5 Ga or basalts from the Fortescue Basin of the Pilbara Craton (Australia) dated at 2.7 Ga yielded a primary remanent magnetization which suggests the presence of a dipolar magnetic field at the end of the Archean (Tarduno et al. 2014). Knowing this, it is technically possible to measure the direction of magnetization from this time onwards. But what about rocks older than this? Archean formations are highly likely to have undergone a complex geological history, and, in particular, one or more episodes of metamorphism erasing the primary magnetic signal in favor of a more recent secondary magnetization. To overcome this problem, a technique based on the measurement of the magnetization carried by an isolated mineral (single feldspar, quartz phenocrysts) was developed to show the presence of a magnetic field as far back as 3.5 Ga. Finally, zircons dated between 3.3 and 4.2 Ga discovered in the Jack Hills metaconglomerate showed a magnetic signal carried by magnetite and considered to be primary. The paleointensity of the magnetic field (not its direction) was measured and it varies between 12 and 100% of the value of the current field at the equator possibly suggesting the presence of a terrestrial magnetic field as far back as the Hadean era (Tarduno et al. 2015).

The geomagnetic field **H** at any point on the surface of the globe can be defined by two angles, the declination D (angle between the magnetic north and the geographic north (counted positive east of true north)) and the inclination I (the angle between the horizontal plane and the direction of the field **H** (counted positive if downward)). At the first order, the present-day magnetic field can be represented by the field produced by a magnetic dipole inclined to the Earth's axis of rotation by 11.5° and slightly off-centered by about 500 km from the center of the Earth. Differences between the current magnetic field and this theoretical field exist locally, as evidenced by the contribution of non-dipolar terms (quadrupole, octupole). In the even stronger hypothesis of a perfectly dipolar, axial and centered magnetic field, a simple mathematical relation connects the magnetic inclination measured at a point of the Earth's surface to the latitude of this point. In other words, knowing the inclination of the magnetization vector fossilized by the magnetic minerals at a site at different times in the past allows us to calculate the successive paleolatitudes of this site by assuming that the geomagnetic pole has always coincided with the geographical pole, itself supposed fixed. Conversely, from a set of samples distributed over a continental land mass, we can calculate the position of the magnetic pole associated with the continent or geological mass under consideration, always assuming the axial and centered dipole magnetic field: this is therefore a "virtual" geomagnetic pole (VGP), since the reference is a continental land mass whose past position is not known.

The successive positions of the VGP over time track the path of the apparent drift of the magnetic poles, with the continent being studied fixed at its current position. The first apparent polar wander path (APWP) over a period of 600 Ma was established in 1954 by Creer and his research group, based on magnetic measurements of samples taken from geological formations in Great Britain (Creer et al. 1954). The virtual magnetic poles constituting the APWP approximate, supposing a geocentric axial dipole, the successive paleopositions of the rotation axis with respect to the continent from which the paleomagnetic pole was determined (Fig. 2.2). In the 1950s and 1960s, the existence of a central axial dipole was challenged, calling into question the importance of the APWP. In 1964, however, Irving was able to verify by means of climate indicators the hypothesis of a central axial dipole. Indeed, the climate of the Earth is, first and foremost, a function of insolation. The result is that the distribution of climate indicators (e.g. evaporites for subtropical zones, coral reefs for tropical areas, glacial sediments at high latitudes), tends to be symmetrical on either side of the equator.

To check if the distribution of the paleoclimate indicators discovered within geological formations changed over time, Irving replaced the continents (and the sites of the paleoclimatic indicators) in their paleopositions, using the virtual magnetic poles, and applied the hypothesis of a geocentric

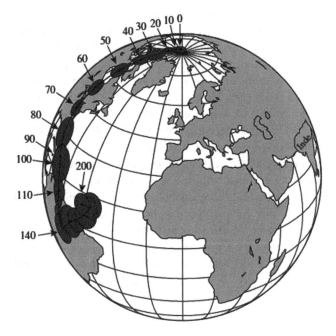

Fig. 2.2 The apparent polar wander of the magnetic poles of India (according to data from Besse and Courtillot 1991). The age of the magnetic pole (in Ma) is indicated. The ellipses represent the uncertainty (95%) on the magnetic pole position at a given age

axial dipole. Thus reconstructed, the latitudinal distribution of the paleoclimate indicators is statistically identical to their present distribution, thus supporting the hypothesis of the axial centered dipole. Other studies carried out in the 1960s made it possible to validate this hypothesis, first for the Plio-Pleistocene and then gradually as far back as the Precambrian. In 1968, Le Pichon proposed a model of movement of rigid land masses relative to each other, thereby reconstructing seafloor spreading for all the Cenozoic (Le Pichon 1968).

The trajectories of the APWP make it possible to place the continents in their original position with respect to the axis of rotation of the Earth, in other words, in terms of paleolatitude and orientation. However, paleomagnetism is not enough, since it does not specify the paleolongitude of the continents because of the spherical symmetry of the magnetic field (assuming an axial and centered dipole magnetic field). Kinematic parameters of the oceans are used to pin down the position of a continent relative to another. However, this method can only be used when the ocean separating two continents is bordered by passive margins and especially when these kinematic parameters are known (available for the last 170 million years only). If one of these oceanic margins is active, the positioning of one plate relative to another is nevertheless possible by traveling across one or more other intermediate continents separated by passive margins. Nevertheless, there are cases where the deformation of the continental land masses by tectonics (in a collision zone, for example), the presence of active margins, or the age of reconstruction prevent the use of this method combining paleomagnetism and ocean kinematics. In these cases, paleogeographic reconstructions depend on paleomagnetism and the APWP.

These trajectories of the apparent polar wander are not all of equivalent quality. To overcome this disadvantage, Besse and Courtillot (1991, 2002) and Torsvik et al. (2012) proposed calculating artificial trajectories of APWP where the poles of all the plates of a certain age are integrated into a single referential (i.e. transferred from one lithospheric plate to another using the kinematic parameters of the oceans). This method is applicable to the last 320 million years (despite the absence of kinematic data for the oceans beyond 170 Ma since all the continents were then combined into a supercontinent, the Pangea). Before 320 million years, for all continents and regardless of the period for those land masses surrounded by active margins, reconstructions rely solely on the trajectories of APWP, although geological and/or paleontology arguments may provide constraints on the relative position of the masses. Of course, there is an overall increase in uncertainties with age.

The now well-known movements of large lithospheric masses are part of the classical theory of plate tectonics. However, the whole Earth can also tilt relative to the axis of rotation in response to the heterogeneities of masses in the mantle, modifying the tensor inertia of our planet (the maximum axis of inertia is aligned with the axis of rotation of the Earth). In the paleomagnetic reference system, the APWP is thus caused by the movement of the plate due to plate tectonics and to the overall movement of the continents.

In 1972, Morgan proposed using hot spots associated with convective plumes from the D″ transition zone as a definitive frame of reference. Deep mantle convective plumes are assumed to be fixed because of the sufficiently slow movements in the lower mantle. The movements of some lithospheric plates can thus be positioned within the "hot spots" reference frame. Unlike magnetic poles, paleolatitude and paleolongitude are constrained (by a frame of reference associated with the mantle), but this method can only be used for the last 130 million years. Before this time, the traces of hot spots on the ocean floor cannot be traced, erased by the subduction zones, while the few remaining traces are not constraining enough.

The differences in latitude and in rotation of the continents between the paleomagnetic and hotspot reference frames make it possible to isolate the movement of the global drift of the crust-mantle couple with respect to the axis of rotation or the true polar wander from the perspective of a continent. The amplitude of the global drift generally does not exceed 1° per million years over the last 130 million years (Besse and Courtillot 2002). Nevertheless, there are some rapid events (occurring over a few million years) during which a global drift of about 10° of all the continental masses has been observed, for example in the Paleocene (Moreau et al. 2007). Even with a lack of data in the hot spot reference, it is possible to isolate the true polar wander by determining common drifts in APWP of different continents. Events of great amplitude have also been suggested, for example a 90° movement between the Lower and Middle Cambrian (Kirschvink et al. 1997) or during the Ediacaran (Robert et al. 2017). These events result in a continental drift rate of the order of 10° per Ma, therefore much higher than the maximum continental drift rate caused by plate tectonics of about 2° per Ma (Seton et al. 2012) but perfectly compatible with the theoretical maximum speed of the true polar wander of about 10° per Ma, taking into account a viscoelastic Earth (Greff-Lefftz and Besse 2014; Robert et al. 2017).

The Paleomagnetic Tool, Tests and Uncertainties

To understand the difficulty of obtaining robust paleogeographic reconstructions, it is important to look at the paleomagnetic tool used (Chap. 7). The direction of the magnetic field is measured in a laboratory using a magnetometer, an instrument capable of detecting natural remanent magnetizations of less than 10^{-12} A/m, on cylindrical samples of approximately 10 cm^3 taken from sedimentary or volcanic formations and as perfectly oriented in space as possible. This orientation in space of the sample collected in the field is essential in order to determine the direction of the magnetization vector (represented by inclination and declination) and then to calculate the position (longitude and latitude) of the virtual magnetic pole associated with it. To obtain a reliable virtual magnetic pole, several criteria must be met. First of all, one needs to have a sufficient number of samples. The number of samples collected is usually eight per site, and the number of sampling sites within a geological formation is at least six sites. As in the processing of the signal, increasing the number of sites and samples per site improves the signal-to-noise ratio (thus reducing the sources of errors). The uncertainty on the virtual magnetic pole is of the order of a few degrees (1° = 111 km) in the best case, which still represents a few hundred kilometers. The displacement of continent of less than several hundred kilometers is therefore difficult to detect, which, taking into account the average speed of plate drift ($\sim$ 4 cm/year), represents a time span of about ten million years. The dating of the sampled geological formations must also meet a standard of robustness in order to produce a reliable virtual magnetic pole. The primary character of the magnetization of the ferromagnetic carriers must be verified. The magnetization signal must be synchronous or recorded in the millions of years (<5 Ma) after diagenesis (in the case of sedimentary formations). This is to ensure that the magnetization age is essentially the same as the "stratigraphic" age of the formation. When these two ages differ, a subsequent re-magnetization has occurred (in response, for example, to the burial at several kilometers deep, which has the effect of imprinting a new magnetic signal in the sample masking partially or wholly the primary magnetization).

Tests have been developed to evaluate the quality of the measured primary natural remnant magnetization (NRM). The first is the reversal test. On average, the magnetic field experiences several reversals per million years (there are, nevertheless, periods of several tens of millions of years during which no reversal is observed, Chap. 7). Over a time period of less than a few million years (<5 Ma), the calculated virtual magnetic pole must be statistically identical, regardless of the polarity of the magnetic field. This test increases the confidence in the primary character of the NRM and ensures that the secular variation of the magnetic field resulting from the movements of molten iron in the outer core has indeed been averaged. The secular variation of the magnetic field gives rise to a rapid drift of the magnetic pole which can move the virtual pole by more than 10° from its mean position. Thus, the direction provided by a single lava flow cooled in less than one year does not correspond to the average direction of the dipole. A second test consists of checking the reliability of the magnetic pole determined from sampling sites for which the dip in the geological layers differs. Indeed, the geological formations can be subjected to deformations. To calculate the virtual magnetic pole, the paleo-horizontality of the sampled sedimentary formation that prevailed at the moment of the acquisition of the magnetic signal must be restored. Taking into account the tectonic (tilt) correction for each sampling site must have the effect of clustering the set of magnetic directions obtained for each site. If this structural correction does not have the desired effect on the data, this means that the magnetization was acquired during or after deformation.

However, these tests are not sufficient to certify the robustness of the data that can be affected by various biases. Regional tectonics in areas of active collisions such as in the Alps or by rifting in Afar could lead to horizontal block rotation that is well recorded by paleomagnetism. These regional tectonics can mask the large drifts of plates or larger masses. A second bias is related to the geometry of the magnetic field. The calculation of a virtual magnetic pole is based on the assumption of a geocentric axial dipole. In the case of samples collected from sedimentary formations, the magnetic measurements are carried out on cylinders of approximately 10 cm^3. Given the low accumulation rate of sediments and the effects of diagenesis (such as compaction), a cylinder of 10 cm^3 may represent a time period of several thousand years, and the measured magnetic direction is in some sense an "average" direction of this time period. For a sampling site, the magnetic direction is the average of the magnetic directions of the samples taken from a unit of a sedimentary formation. It is therefore possible that the time interval associated with the mean magnetic direction represents several thousand years, or even more. The variations of the magnetic field are thus smoothed, and the magnetic direction measured is indeed that produced by a geocentric axial dipole field in the study area under consideration.

In the case of volcanic series, the thermoremanent magnetization (TRM) is acquired during the cooling of the lava flow. It thus represents a (quasi-) instantaneous photography of the Earth's magnetic field. To overcome the effect of the secular variation, the average direction of a large number of sites must be measured, so as to tend towards the direction which would be obtained with an axial geocentric dipole. However, the presence of a persistent quadrupole term in the paleomagnetic data has been identified as a possible source of errors. The presence of quadrupole terms of about a few percent (the level generally observed) implies an error of a few degrees in latitude at the equator. This does not therefore significantly affect the virtual magnetic pole which is calculated assuming the axial geocentric dipole hypothesis. There is also a possibility that the magnetic field could include an octupole component affecting Asia during the Tertiary and Pangea during the Permian. If we consider an octupole term equal to 10% (resp. 20%) of the dipolar component, the maximum error committed is $\sim 7°$ (resp. 13°) in the mid-latitudes. Finally, a last source of error is related to the preservation of the magnetic signal in the sedimentary series. Indeed, during diagenesis, the magnetic carriers can undergo a decrease in the inclination of the magnetic carriers. The inclination measured is therefore less than the actual inclination of the magnetic field that prevailed at the time this detrital remanent magnetization was acquired. This inclination flattening can be a source of error in paleoreconstructions (Cogné et al. 2013).

The Topography of the Earth

Mountain ranges and high plateaus play an important role in atmospheric circulation. The uplift of the Himalayas and the Tibetan plateau is undoubtedly the most characteristic example of the relationship between topography and the evolution of atmospheric circulation and climate. The hypsometric curve of the Earth reveals that today, reliefs higher than 2 km represent about 10% of the land area. Still, this curve reflects the orogenic context of the Earth that has prevailed since the beginning of the Cenozoic only, and one expects that this curve evolved through time.

The great areas of high-altitude terrain ($\sim$>2 km) typically developed during the formation phases of a supercontinent (for example, in the Carboniferous, during the formation of the supercontinent Pangea), while periods of break-up of the supercontinents (the Mesozoic for example) are characterized by more modest reliefs (although low reliefs of less than 2 km in general, can form in case of continental rifting). In the oceans, seafloor reliefs are mostly dominated by ocean ridges and some high oceanic plateaus rising above the abyssal plains.

The altitude of the continents, plains, collision mountain ranges (Himalayas, Alps) and the zones of the East African Dome result first of all from the (quasi-) isostatic equilibrium of the lithospheric column marked by density heterogeneities on the underlying asthenosphere. The bathymetry of the ocean floor is controlled by the cooling and by the progressive thickening of the oceanic lithosphere formed at the dorsal ridges in isostatic equilibrium on the asthenosphere.

However this scheme does not apply everywhere, suggesting other mechanisms. Some reliefs are not at isostatic equilibrium, as is the case, for example, of volcanic islands that develop on an oceanic crust (like Hawaii) or continental crusts previously subjected to the weight of an ice cap (like the Scandinavian region). The charge brings about a flexure whose wavelength is related to the elastic rheology of the crust. After the rapid melting of an ice cap, the deformation gradually fades over time with a rate depending on the viscosity of the underlying mantle.

Topography can also be controlled by dynamic processes[2] related to movement of matter and heat transfer within the viscous mantle. Upward or downward movements of mantle matter or dipping of a lithospheric plate into the mantle create mass anomalies that can induce long-wave crustal deformations. This is known as dynamic topography (Husson 2006) as opposed to isostatic topography. The uplift of the southern African plateau and the Colorado plateau during the Cenozoic is explained by movements of mass anomalies in the mantle. Dynamic topography also helps to explain the history of the Western Interior Seaway that connected the Arctic Ocean to the Gulf of Mexico across the North American continent during the Cretaceous, but also the flooding followed by the exondation of part of Australia at the same time, or the flooding of the Sunda shelf in Indonesia during the Pleistocene (Sarr et al. 2019).

Restoring the past topography of the Earth is undoubtedly the most difficult part of paleogeographic reconstructions. It involves determining the spatial expanse of the terrain, its age and altitude. The reliefs bring about deformations and/or structural and petrological markers which, when fixed in time, are used by geologists to constrain the tectonic event. These markers are not always easy to detect because more recent events often mask earlier events. A phase of continental accretion caused the India-Asia collision in Southeast Asia during the Triassic period, but this orogeny remains uncertain, partly because this collision erased part of the previous geological history. Determining the paleoaltitude of mountains is therefore crucial in order to model pre-Quaternary paleoclimates. Several methods (flora, sediment, oxygen isotopes, cosmogenic isotopes) have been developed, but only the most commonly used will be depicted here. It is very important to differentiate between absolute methods, which makes it possible to estimate the paleoaltitude of a relief at a given time and methods estimating the vertical velocity of the rocks in these mountains. The relative methods reflect a balance between the vertical movements linked to a geodynamic event (a collision for example) and those due to erosion which denudes the surface, favoring the rise of deep-set rocks by isostatic readjustment.

To measure an altitude or to estimate a paleoaltitude, a reference level is essential. The surface of the oceans provides this reference, but this fluctuates over geological time. However, the amplitudes in these variations are less than the uncertainty obtained on paleoelevations, regardless of the methods used. Nonetheless, eustatic variations are crucial when examining the regions between the continental shelf (> −200 m) and the vast lowland plains (<200 m).

---

[2]Although dynamic, collision mountain ranges and intracontinental rifting are essentially isostatic equilibrium processes.

The Evidence from Flora

The assemblage of plant fossils has been used as a marker for paleoelevation for almost half a century. This method is based on the relationship between the vegetation type and the average temperature at a given location. It is broken down into several variants. The first method relies on finding an assemblage of modern plant taxa equivalent to that of the fossil site (or at least sharing the highest possible number of taxa with it), but it assumes that there was no adaptation by these taxa to climatic variations in the past (Su et al. 2019). This method is particularly well adapted to the last 10 million years. A second method is based on the physiognomy of the leaves of plants (size, shape, thickness, type of leaf margin), synonymous with the adaptation of the plant to a given climatic context. A relationship between the physiognomy of current plant leaves and the mean annual temperature has been established. This method assumes, however, that the response of the leaf physiognomy to the climate has been constant over time. This tool is calibrated on dicots and does not take into account all biogeographic provinces (e.g., Australia) or all taxa (such as conifers). Moreover, it is only applicable as far back as the Upper Cretaceous, since flowering plants only appeared during the Cretaceous. However, the main advantage of this approach is that it avoids any systematic recognition of taxa because only the morphological characteristics count (which implies optimal fossilization conditions). To determine altitude, it is necessary to know the average temperature at sea level of a site of the same age. The difference in temperature between these two sites divided by the vertical gradient of the temperature indicates the paleo-altitude. However, the temperature lapse rate varies from 4 to 10 °C/km. This depends on the latitude, the humidity of the air mass, the continentality of the site and the topography itself. The choice of this parameter is therefore decisive. Uncertainty remains high (Peppe et al. 2010), around 700–1000 m for paleo-altitudes estimated at 3–4 km.

To overcome the problem of the vertical gradient, a method based on the preservation of moist static energy h in the atmosphere has been developed (Forest et al. 1995). The parameter h is the sum of a thermodynamic parameter, the enthalpy H of humidity and the potential energy gZ and has the advantage of a distribution that is relatively zonal, especially in the mid-latitudes of the northern hemisphere. This method requires an atmospheric circulation where horizontal movements predominate over vertical movements, to ensure the conservation of moist static energy. The humid enthalpy H was calibrated on current leaf indices (similar to the mean temperature in the previous method). To determine the paleo-altitude of a site, its humid enthalpy H and the humid enthalpy H_0 of a contemporary reference site located at sea level must be known. The uncertainty on paleoaltitude is only slightly lower than for the preceding method.

The plant-based methods are also subject to uncertainties because of the impact of vegetation on the climate that are ignored. The vegetation cover affects the radiative balance of the Earth through its albedo and the water balance through evapotranspiration (e.g. Otto-Bliesner and Upchurch 1997), thereby could introduce a bias and cause paleo-altitudes to be mis-estimated.

The Evidence from Erosion Sediment

By restoring the mass of deposited sediments in a basin to the original relief, it is possible to calculate paleo-altitudes. This presumes that the geometry of the sediment source is known and that this has not changed over time, that there are no sediment losses due to subduction or to incorporation during more recent orogenic events and finally it assumes that the relief has always been in isostatic equilibrium. These considerations considerably limit the use of this method to a few endorheic basins that do not have subduction zones (intracontinental), such as the Tien Shan range in Asia during the Miocene. Its application to older orogeny is complex.

Stable Isotope Paleoaltimetry

Water oxygen is made of different stable isotopes (^{16}O, ^{17}O, ^{18}O, e.g. Chaps. 11, 14, 15, 16, 21). The ratio of heavy to light isotopes, when compared to a global reference value (namely the Vienna standard mean ocean water, VSMOW), is noted $\delta^{18}O$. It can be measured in various surface waters (ice, ocean, lakes, rivers), carbonates from pedogenic or lacustrine sediments, or biogenic archives. At the global scale, is has been observed that $\delta^{18}O$ in rainfall decreases from low to high latitudes, and from coastal to inland areas. On continents, $\delta^{18}O$ measured directly in precipitation or in rivers along different elevation transects also has been shown to scale with altitude, with $\delta^{18}O$ decreasing as elevation increases. These observations have been explained by the Rayleigh-type distillation process that occurs theoretically in a cooling air parcel ascending along a mountain range: As air rises and cools, water vapor condenses then precipitates. During these steps, heavy isotopes are more favorably removed from the air parcel, progressively depleting water vapour along the way. Ultimately, rainfall is more and more depleted in heavy isotopes with elevation, i.e. $\delta^{18}O$ decreases (Fig. 2.3). This theoretical framework, together with regional measurements of $\delta^{18}O$ have led to determine

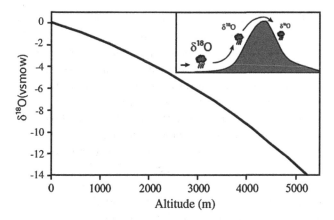

Fig. 2.3 Evolution of the isotopic ratio of oxygen ($\delta^{18}O$) as a function of altitude

"isotopic lapse rates", i.e. expected change in $\delta^{18}O$ as a function of elevation (Mulch 2016). These lapse rates depend on many parameters, including initial relative humidity and temperature of the air parcel.

The relation between $\delta^{18}O$ and elevation has been used to estimate paleoelevations since the late 90s (e.g. Chamberlain 1999; Quade et al. 2007; Rowley and Garzione 2007; Mulch 2016). Oxygen isotopic ratios can be measured in lacustrine or soil carbonates, and used to infer the paleo-rainfall $\delta^{18}O$ using a relationship involving the temperature of carbonate formation (Kim and O'Neil 1997) and assuming thermodynamical equilibrium between the water and calcite. Reconstructing rainfall $\delta^{18}O$ of dated records thus allowed to infer past elevations and numerous studies have been carried out to reconstruct uplift history of mountain ranges worldwide, including the Andes, the north American cordillera and the Tibetan Plateau.

Still, several processes can cause bias in the results. The isotopic fractionation depends not only on temperature but also on physical processes that alter the simple model of air parcel ascent and depletion. Large-scale climate conditions (e.g. greenhouse climate) and dynamics can modify moisture advection and can also lead to mixing of air masses with different origins and isotopic signatures. Moreover, changes in atmospheric dynamics related to the peculiar conditions of the warm climates of the Paleogene have been shown to alter the isotopic lapse rates. Such potential biases have been highlighted for the reconstruction of the altitudes of the Tibetan plateau (e.g. Botsyun et al. 2016, 2019; Li and Garzione 2017) and the Andes (Poulsen et al. 2010), with either extensive datasets of river $\delta^{18}O$ measurements or the use of isotope-enabled general circulation models. Lastly, the isotopic anomaly recorded in pedogenic carbonates is not from rainwater, but from runoff waters. In the case of the Himalayas, for example, measurements of $\delta^{18}O$ carried out on various modern rivers can differ significantly from meteoric waters. The $\delta^{18}O$ of the pedogenic carbonates is more a reflection of the averaged $\delta^{18}O$ of the drainage basin than of the meteoric waters. For lake carbonates, $\delta^{18}O$ is affected by evaporation on the surface of the lake. Depending on the data location, these biases can add up to a significant uncertainty, close to that observed with the plant-based methods (see Mulch 2016 for a review).

As for the estimates of ocean paleo-temperatures (Chap. 21), the clumped isotope (Δ_{47}) measurement (Eiler 2007; Bonifacie et al. 2017) appears to be the next promising tool in paleoaltimetry. The propensity to form molecules containing bonds between heavy isotopes of carbon (^{13}C) and of oxygen (^{18}O) in a carbonate (mainly in the form of $^{13}C^{18}O^{16}O$) is more probable, the lower the temperature is at the formation of the pedogenic carbonate (Chap. 21). There is therefore a relationship between this quantity and the formation temperature, which makes it possible to work back to paleo-altitudes by establishing the vertical temperature lapse rate (which still may have varied in the past) (Quade et al. 2007). Moreover, the measurements independent of the formation temperature through Δ_{47} and of the $\delta^{18}O$ isotopic ratio of these carbonates, make it possible to trace back to the $\delta^{18}O$ of the water and thus to isolate the effect of altitude from parasitic influences, such as climate, season, or latitude. Burial does not appear to affect the measurement of Δ_{47} in general, an advantage in the case of orogeny. This method makes it possible to obtain paleo-altitudes with a lower uncertainty than other methods. This isotopic tool was used to constrain the evolution of paleo-altitudes and the speed of uplift of the Altiplano, a high plateau in the Andes in South America, during the Miocene (Garzione et al. 2014).

Eustatic Variations and Ocean Gateways

Ocean circulation plays an important role in the climate system by transporting heat from low to high latitudes, particularly through surface currents. However, this transport depends on the distribution of the continents and the shape of the ocean basins. At present, deep waters are formed mainly in the North Atlantic Ocean, specifically in the Norwegian Sea and the Labrador Sea. Therefore, the configuration of deep ocean circulation must have been different in the past. At the end of the Cretaceous, the North Atlantic Ocean was not yet open and exchanges between the central Atlantic Ocean and the Arctic Ocean were very limited. Deep water formation could therefore be transferred to the Pacific Ocean, based on numerical models simulating the large-scale circulation of the ocean and atmosphere (e.g. Otto-Bliesner et al. 2002).

Changes in geographical configuration through continental drift, or more radically, through the opening or closing of ocean basins, are therefore likely to cause profound changes in ocean circulation. Either shallow or deep, ocean gateways play a crucial role in water masses circulation and associated heat and salt fluxes (Ferreira et al. 2018). During the Cenozoic, the position of the continents has not drastically changed but some ocean gateways have been opened or closed thereby impacting ocean circulation. The Tasman seaway opened during the Late Eocene, the Drake passage during the Late Eocene/Oligocene, the Fram Strait during the Oligocene. Conversely the East Tethys seaway closed during the Early/Middle Miocene, the Central American Seaway during the Late Miocene, the Indonesian throughflow during the Late Miocene/Pliocene. The timeline of these opening/closures are therefore crucial for paleoclimate reconstruction purposes, and can be constrained through the use of paleoceanographic markers and sea level records.

Sea level changes profoundly change the face of the Earth. Periods of high sea level, such as during the Cenomanian ($\sim$95 Ma) studied by Eduard Suess at the end of the nineteenth century, caused the flooding of large continental areas. Vast shelf seas, generally shallow (<200 m), covering up to 30% of the land area, were thus observed during the Phanerozoic period. In some cases the combination of a high sea level and a flexure of the lithosphere can result in the formation of shelf seas several hundred meters deep. In other cases, flooding or drying is controlled by dynamic topography.

The dynamics of these epicontinental seas are still not well understood because there is no analog of such large shallow water basins today. These epicontinental seas would have favored climates with little thermal contrast due to the high calorific capacity of the water and their thermal inertia. Conversely, periods with low sea levels would have favored more contrasted climates. Analysis of the sedimentary facies makes it possible to locate the coastlines. These have changed with eustatic variations (change of sea level), but also with continental uplift and subsidence, the flexure of the lithosphere in response to the build up of ice sheet or its melting or variations in the sedimentary fluxes. Changes in coastlines inferred from the sedimentary facies do not therefore constitute a direct marker of sea level.

In 1977, Peter Vail and his team, geologists with the American oil company Exxon, produced a curve of sea level changes. Vail showed that the geometry and position of the sedimentary units deposited on the continental shelves and in the basins had varied according to sea level, the subsidence of the area of deposition, the sedimentary flow and the carbonate production. The geometry and position of the sedimentary units are the main influencers of variations in sea level, as subsidence, deposition and carbonate production are considered to be less variable. The determination of sedimentary facies from core samples taken by drilling and the location of seismic reflectors on the continental margins and adjacent basins make it possible to reconstruct the arrangement of sedimentary deposits and to infer eustatic variations. Several curves have since been produced (Haq et al. 1987; Haq and Schutter 2008). The curves obtained (Fig. 2.4) show nested eustatic cycles. The largest eustatic variations, about 200 m, with a timescale of tens of million years, are related to life cycle of supercontinents (Wilson cycle), plate reorganization, dynamic topography and crustal production variations. Secondly, significant variations (several tens of meters) but with a lower timescale (from $\sim$2 to $\sim$10 Ma) are likely due to regional tectonic forcing. Thirdly, the sea level fluctuations on the timescale of 0.5 to $\sim$2 Ma are not well understood, it could be related to the climate changes or ice volume variations. Even more rapid variations on the timescale of tens to hundred years are clearly related to glacioeustatism whereas postglacial rebound acts on timescale of thousands to hundreds of thousands of years.

However, this interpretation of the high frequency variability found on an eustatic graph at the scale of geological time has been called into question. Moreover, the covering of coasts observed in sedimentary systems may not necessarily translate into eustatic variations.

A different method from that of Haq et al. (1987) was proposed by Miller et al. (2005). It determines eustatic variations using an inverse model by calculating the effects of the sediment load, compaction and the oceanic load necessary to simulate the deformation of a basin (subsidence or uplift) located on a passive margin. The variations in sea level obtained by this method do not exceed 100 m, which is half that proposed by Haq et al. (1987) for the Phanerozoic. Moreover, the eustatic variations obtained by these two methods may be out of phase with each other. Unlike the curve by Haq et al. (1987), which is based on numerous records, Miller's sea level variation curve is based on a small number of sites on the eastern margin of North America. The deformation undergone by this basin caused by internal dynamic processes originating in the mantle is not corrected. This basin may have recorded a "dynamic" topography with long wavelength radiation (as opposed to the topography linked to isostasy of a lithospheric column) partially skewing the eustatic signal. Debates still exist regarding the exact evolution of sea-level throughout the Phanerozoic, as a more recent study, based on a full geodynamical model, provided a large range of sea-level fluctuations for the last 500 Ma (Vérard et al. 2015) (Fig. 2.4).

To understand these discrepancies, one must analyze the mechanisms of sea level change. Variations in the position of coastlines can come from several different sources: variation in the volume of water (the content) or variation in the

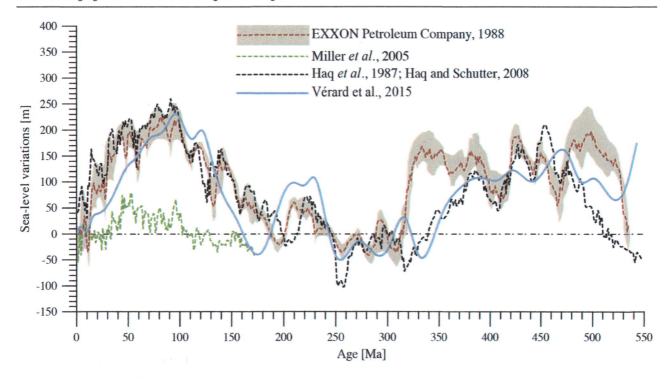

Fig. 2.4 Variations in sea level throughout the last 550 million-years, according to various reconstructions. Red curve is from the EXXON company, retrieved in Vérard et al. (2015); black curve is from Haq et al. (1987) and Haq and Schutter (2008); green curve is from Miller et al. (2005), blue curve is from Vérard et al. (2015). The figure is modified from Vérard et al. (2015)

capacity of ocean basins (the container). Changing the volume of water in the ocean involves a process of contraction or thermal expansion, capture or release of this water by another reservoir (e.g., ice caps), subducted down with hydrated minerals in the mantle or released through degassing at oceanic ridges. Although the volume of water present in the atmosphere in vapor and liquid form ($\sim 13{,}000$ km^3) is fundamental to the functioning of the climate system, it is negligible compared to the volume of water contained in the ocean basins (~ 1347 million km^3). Variations in water volumes in ocean basins on a time scale of between 1 and 100 ka is due to the formation or melting of ice caps. The quantity of water held in the form of ice can lead to eustatic variations of more than 100 m. At the geological time scale, the presence of ice at the poles is an episodic phenomenon, because it implies the presence of a continent in a near-polar position and climate conditions allowing the formation of permanent snow cover. For example, glaciation covered the southern part of the Gondwana continent (which corresponds to the southern part of South America, the southern part of Africa, the southern part of the Arabian Peninsula, parts of India and Australia and the whole of Antarctica, these areas having been united at one time) for about 70 million years (335–265 Ma), marked by alternating phases of growth and melting of the ice cap. The Gondwana ice sheet at its climax may have sequestered a volume of water of approximately 200 m (compared with 70 m for the Antarctic cap of today). Finally, water storage in lakes, rivers (~ 0.26 million km^3) and in underground reservoirs (~ 9 million km^3) makes a very marginal contribution.

The most frequently cited mechanism to explain sea level changes over the scale of geological time is oceanic crustal production rates and extruded oceanic plateaus. In the 1980s, the oceanic production rate was estimated for the last 180 Ma. For the lower Cretaceous it was estimated to be twice as high as today, while a drop of about 50% over the last 50 million years was found. However these estimates have been called into question following new estimates of accretion rates and normalized fluxes which suggested variations in the fluxes with an amplitude of 30% or less around the present value (Cogné and Humler 2006). Such estimates could challenge the role of the ridges in sea level variations. The Wilson cycles, i.e. the construction followed by the fragmentation of a supercontinent, have been identified as a possible mechanism to explain the first-order eustatic variations. Indeed, fragmentation periods produce young oceanic crust. The volume occupied by the ridges at the expense of the old crust lost by subduction reduces the total volume of the oceans and raises their level, and conversely during the period of supercontinent construction, the total volume of the ridges decreases and the level of the oceans drops. In other words, the average age of the oceans reflects the first-order eustatic variations. It is thus a major challenge to develop plate motion models than can provide

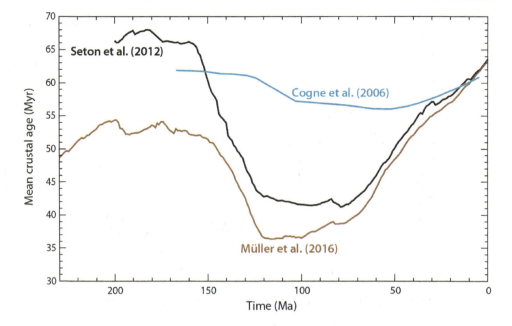

Fig. 2.5 Mean crustal age through time according to different models. Retrieved and modified from Müller et al. (2016)

estimates of the seafloor age evolution through time. Müller et al. (2016) have provided a synthesis of the different model results since the pioneering works of Cogné et al. (2006), that shows how including new assumptions can make the chronology of seafloor evolution change. Figure 2.5 depicts some of them.

An Overview of the Changing Face of Earth Through the Ages

To reconstruct the paleogeography of the past, the locations of the large plates as defined by Morgan in 1968 and a number of smaller continental masses need to be known. This is a difficult task for the periods prior to 1.1 Ga (1100 Ma) due to the smaller number of paleomagnetic data available. The position of the paleoshorelines depends on the available lithological information. Using available data, we sketch the portrait of the Earth since the Archean. The main features of paleogeographic evolution are shown in the atlas (Fig. 2.6).

Currently, the oldest observable geological units in a rocky outcrop are more than 3 billion year-old, but they are rare and do not allow to estimate the area of land present at that time. However, the dating of a large number of zircons, an ubiquitous mineral in many rocks (igneous, metamorphic and sedimentary), has allowed to retrace the main periods of continental crust production. Zircons older than 4 Ga discovered in Archean geological formations testify to the existence of a continental crust, even ephemeral, a few hundred million years after the formation of the Earth. Around 3.5 Ga, the first stable continental land masses appeared. Analysis of the oldest cratons reveals that these were formed by the amalgamation of modest-sized scraps of continental crust, island arcs, accretion prisms and oceanic volcanic plateaus. The first major peak of continental crust production dates back to 2.7 Ga, initiated by an avalanche in the mantle leading to the formation of a large number of mantle plumes. This 2.7 Ga event resulted in an increase in the number of cratons, that is to say, a permanent continental crust, and thus a sharp increase in the continental surface (Hawkesworth et al. 2017). However, the existence of a supercontinent at this period remains unlikely (Bleeker 2003).

Reconstructing the paleogeography of the Archean is very uncertain because paleomagnetic data are not sufficient to constrain the position of all cratons in space and time. However their geological histories can provide additional information. The outlined scenarios are based on the similarity between the geological series preserved on each craton (similarity in the lithological sequences, synchronism of metamorphic and magma events affecting the cratons, continuity of magmatic intrusions). However, this combination does not always produce a unique scenario. For the period between 2.7 and 1.8 Ga, Bleeker (2003) counts no fewer than 35 pieces of this paleogeographical puzzle. Internal heating twice as strong at this time is conducive to an organization of the plates more fragmented than today, and makes a vast and unique supercontinent unlikely. However, it is not impossible that groupings of small cratons occurred during the Late Archean. Indeed, the presence of numerous dykes dated between 2.4 and 2.1 Ga could testify for the dislocation of these ephemeral land masses.

The following period was marked by the formation of the first supercontinent, Columbia (also referred to as "Nuna"), through the assembly of small cratons causing numerous

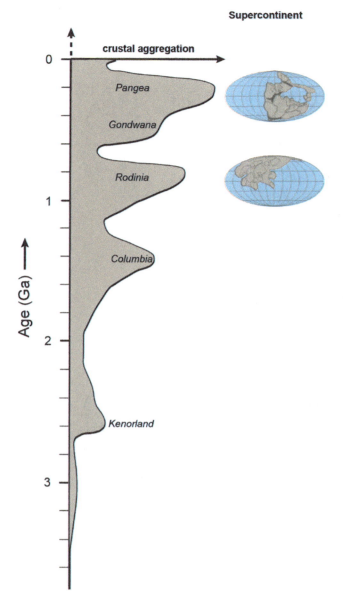

Fig. 2.6 Crustal aggregation states (supercratons and supercontinents) through time [adapted from Bleeker (2003)]. Position of the major continents and land masses in the past [according to Li et al. (2008) for the Neoproterozoic, McElhinny et al. (2003) for the Paleozoic, Atlas Tethys and Peritethys, as well as reconstructions by the author for various studies on the Mesozoic, Fluteau et al. (2001, 2006)]. AFR = Africa; NAM = North America; SAM = South America; AMZ = Amazonia; ANT = Antarctic; ARA = Arabia; AUS = Australia; BAL = Baltic; NCH = Northern China; SCH = Southern China; EAN = Eastern Antarctic; IND = Indochina; INDIA = India; IRA = Iran; KAL = Kalahari; KAZ = Kazakhstan; LHA = Lhasa; LAU = Laurussia; MAD = Madagascar; NUB = Nubia; QAN = Qantiang; RDP = Rio de Plata; SFCG = Sao Francisco-Congo; SIB = Siberia, WAF = West Africa, WAN = West Antarctica. The thick lines represent the assumed location of the ridges (divergent lithospheric plates); the thick black lines with triangles represent the assumed location of the subduction zones (convergent lithospheric plates). The shaded areas indicate the active orogenic zones for the period being studied. The names of the main orogenies are indicated in italics. For the Cretaceous, the main shelf seas are represented in white (in a simplified way)

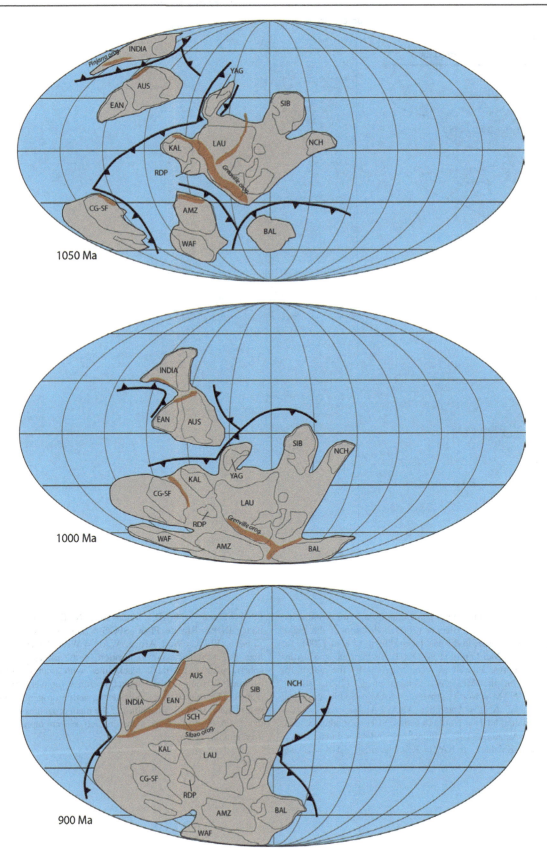

Fig. 2.6 (continued)

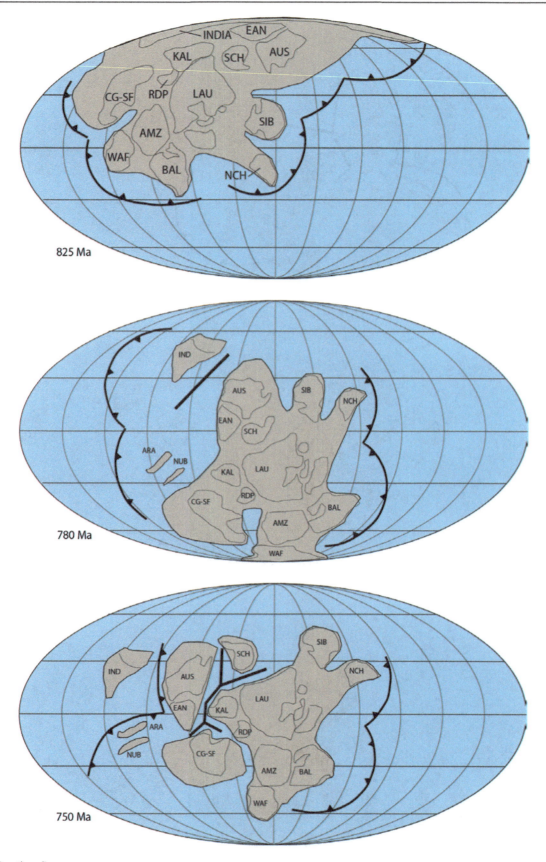

Fig. 2.6 (continued)

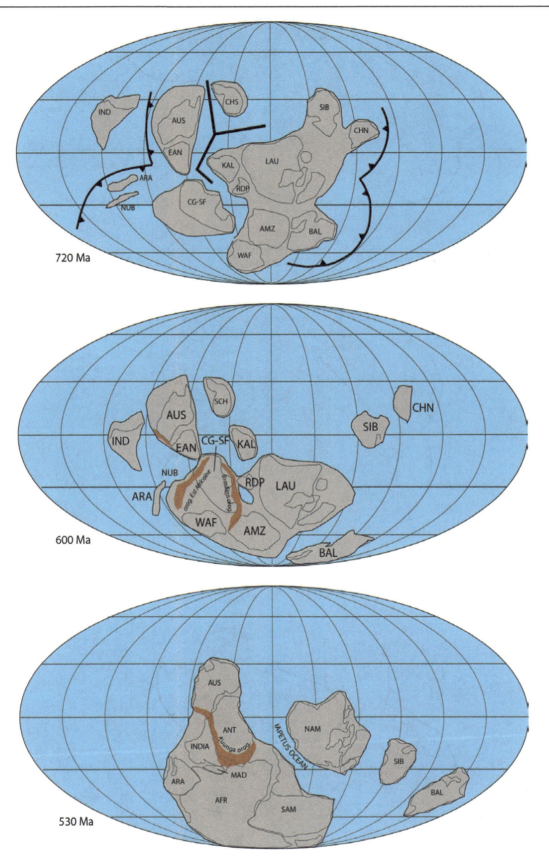

Fig. 2.6 (continued)

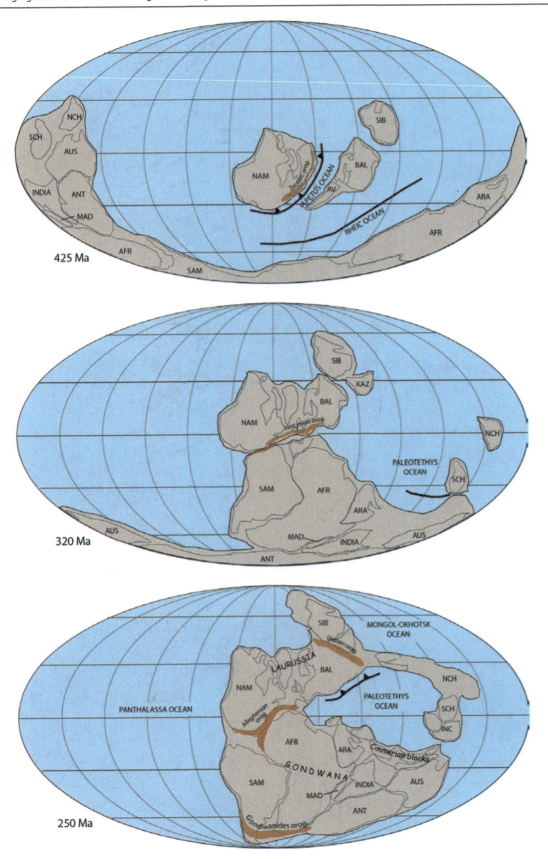

Fig. 2.6 (continued)

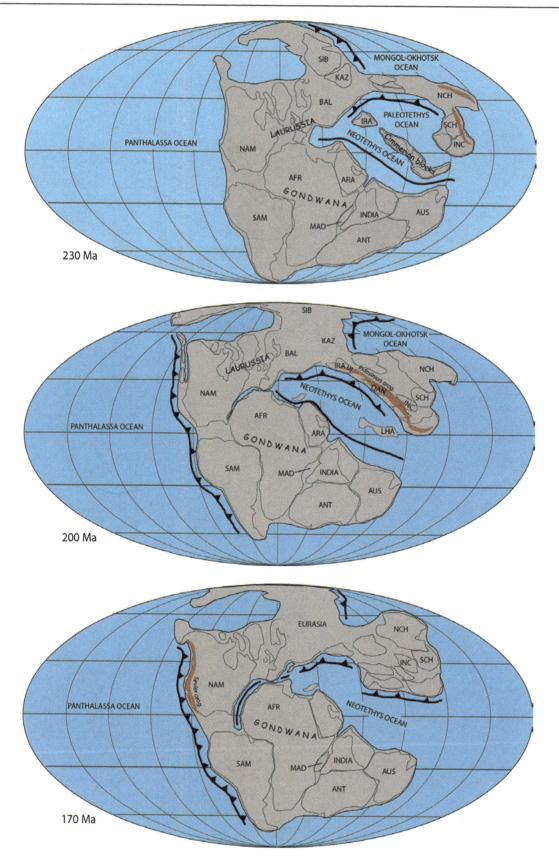

Fig. 2.6 (continued)

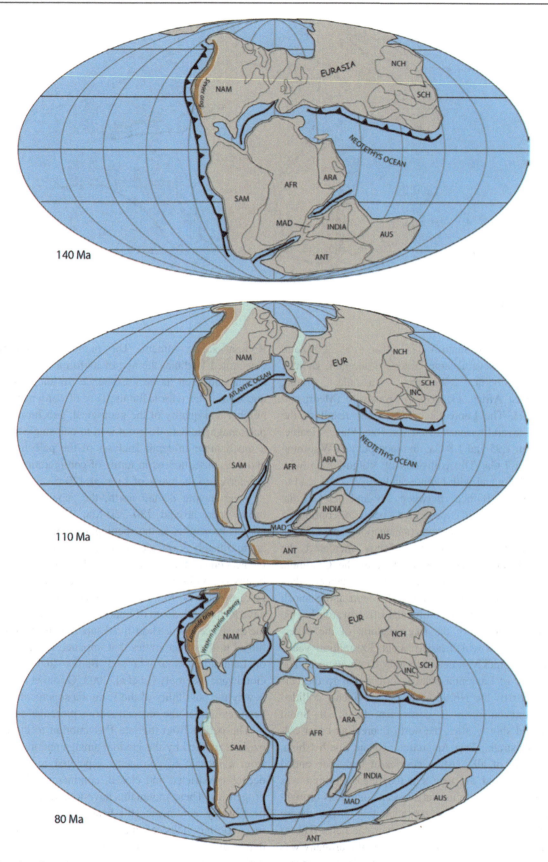

Fig. 2.6 (continued)

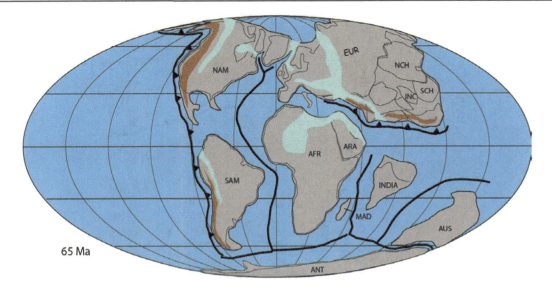

Fig. 2.6 (continued)

orogenies (Rogers and Santosh 2002; Zhao et al. 2002) and the structuring of several large cratons (Laurentia, Baltica, Greenland, Siberia, Western Australia, India, Amazonia-West Africa, Congo-Sao Francisco) (Meert and Santosh 2017). The Laurentia craton is the result of the assembly of several terranes (Superior, Rae, Slave, Hearne, Nain) between 1.95 and 1.8 Ga, and incorporated Wyoming province by 1.7 Ga. The collisions of Volgo-Uralia with Sarmatia by 2 Ga and with Fennoscandia by 1.7 Ga lead to the formation of Baltica. Between 1.8 and 1.3 Ga, the Columbia supercontinent experienced a period of continuous accretion along the active margins (subduction zones). These magmatic accretionary belts significantly increased the emerged land surface. A reconstruction of the Columbia supercontinent is made possible through paleomagnetic data of some cratons at 1.5–1.4 Ga interval (Meert and Santosh 2017). The development of continental rifts towards 1.5 Ga marked the beginning of the fragmentation of the supercontinent Columbia which lasted about 300 million years. However, very quickly, a new assemblage commenced. The southern coast of Laurentia (which corresponds to the current east coast of the North American continent) collided with the Amazon mass (a constituent block of South America), and shortly after, the north Laurentian coast collided with Australia, East Antarctica and the South China. The unification of all existing cratons at that time ended around 950 Ma and the new supercontinent Rodinia was formed. These various collisions between continents brought about the existence of several large mountain ranges, whose geological signature is found in metamorphic belts. This is the case in North America, where the Grenville orogeny was dated to 1 Ga, and in South China with the Sibao orogeny. However, there is still no consensus on the history of the Rodinia supercontinent. The number of continents that constituted Rodinia, the age of its formation or of its dislocation, and even its very existence for some, remain open questions. The reason for the lack of consensus is that the number and quality of the geological, geochronological and paleomagnetic data does not lead to one single solution. For a much more in-depth analysis of the paleomagnetic data and the consequences in terms of configurations, the reader is referred to the work of Li et al. (2008).

After having drifted north, the supercontinent Rodinia broke apart around 780–750 Ma. The dispersion of the continents resulted in the opening up of ocean basins oriented approximately north-south. Due to the spherical symmetry of the geocentric dipole magnetic field, the width of these ocean basins is poorly constrained. At the end of the Neoproterozoic, around 600 Ma, a new supercontinent, Pannotia, could have formed. Made up of Laurentia and Gondwana domains, the existence of this short-lived supercontinent could only be linked to the uncertainties that impaired the age of break-up of Laurentia, Amazonia and Baltica and the timing of Gondwana assembly (Li et al. 2008; Oriolo et al. 2017). This is why the possibility of the Pannotia supercontinent has not been retained in the reconstructions presented in this book.

The period from the late Precambrian to the early Paleozoic is marked by the gradual amalgamation of Gondwana from a mosaic of continents separated by oceans. These oceans constricted and closed, continents collided, causing orogenesis. These mountains have long disappeared, but the present-day continents still bear relics of them, such as metamorphic belts of high pressure, magmatism and/or deformations. These events, that have been dated by isotopic methods, have been extensively discussed in Cawood and Buchan (2007).

The Nubie craton (NUB) and several small blocs were amalgamated between 750 and 600 Ma. This event was the first phase of the East African orogeny dated to between 750 and 600 Ma. A little further east, the convergence of the Congo-Sao Francisco (CG-SF) and Amazon (AMZ) cratons led to the closure of the Adamastor ocean, and their collision around 650 Ma caused the Brazilian orogeny. This assembly of cratons formed part of the future Gondwana continent. At the end of Precambrian, around 550 Ma, Laurentia and Amazonia separated, marking the opening of the Iapetus ocean, while the cratons of Australia, Antarctica, West Antarctica and India assembled to form another part of the future Gondwana continent. This event is reflected in the Kuunga and Pinjarra orogens. The closure of the ocean separating the Kalahari and Congo-Sao Francisco cratons leads to their collision shortly after 520 Ma, instigating the Damara/Zambezi orogeny. This event occurred approximately synchronously with the final amalgamation of the Gondwana continent around 530 Ma. This final stage in the construction of the Gondwana continent brought about a new orogenic phase in Eastern and Southern Africa (which was superimposed on the older one dated at 600 Ma). In the final step in the construction of Gondwana, the passive margin that bordered the western part of this continent became an active margin (the oceanic crust was subducted under the Gondwana continent) over nearly 18,000 km, as evidenced by the ages of magma activity between 550 and 500 Ma. This active margin is linked with the Ross-Delamerian orogeny in Australia and Antarctica, the Saldanian orogeny in Southern Africa and Pampean orogeny in South America. The final assembly of Gondwana and the establishment of an active margin mark the end of the orogens along the sutures of the Gondwana mosaic. It is clear that the height of the reliefs remains very hypothetical, but the intensity of the collisions and geological evidence suggest that the reliefs were high.

The Gondwana continent would continue to exist for nearly 400 million years until it ended during the Cretaceous. At the beginning of the Ordovician (~ 480 Ma), the Avallon plate, originally located close to the South Pole (its remnants can be found in northeastern North America and Western Europe), separated from Gondwana, causing the birth of a new ocean, the Rheic Ocean (Nance et al. 2012). The Avallon plate migrated towards the north and collided with Baltica at the end of the Ordovician around 440 Ma. At the end of the Silurian (~ 420 Ma), the collision between Laurentia and Avalonia + Baltica completed the disappearance by subduction of the Iapetus Ocean and leads to the Taconic orogeny in Laurentia. A new continent, Laurussia, was formed. This collision brought about the Caledonian/Acadian orogeny, which affected Scandinavia, Greenland, Western Europe and the northeast part of North America. The Rheic Ocean was gradually subducted underneath Laurussia, inexorably bringing Laurussia and Gondwana closer together. The closure of this ocean continued until the Devonian (~ 360 Ma), after which period a generalized collision occurred along the suture between Gondwana and Laurussia (Matte 1986). On the Laurussia side, this resulted in the Alleghanian orogeny in North America and the Hercynian (or Variscan) orogeny in Europe. The Alleghanian orogeny affected the eastern margin of the North American continent (Canada and the United States) and extended south through the state of Texas in the United States to Mexico with the Ouachita orogeny. The Hercynian orogeny can also be observed on the Gondwana side through the Mauritanide orogeny in West Africa (Villeneuve 2008). The dating of deformation and metamorphism associated with this collision ranges from 340 to 270 Ma. The Hercynian orogeny affected much of western and central Europe. This event is dated to between 340 and 290 Ma. At the end of the Carboniferous, these reliefs could have been as high as those of the Himalayas at the present time (Matte 1986). The amalgamation of Gondwana with Laurussia would go on to form the Pangea supercontinent, which would exist until 170 Ma. This supercontinent reached its maximum size with the accretion of the Siberian-Kazakhstan plate in response to the closure of the Uralian ocean at the northeastern margin of Pangea during the Permian, giving rise to the uplift of the Urals. The active margin on the south of Gondwana experienced a new orogenic cycle that occurred between the late Carboniferous (~ 310 Ma) and the Upper Triassic (~ 220 Ma). The Gondwanide orogeny affected Australia, southern Africa and southern regions of South America, already deformed by the Ross-Delamerian, Saldanian and Pampean orogenies during the Cambrian.

The Lower Permian marked the beginning of the separation of an assemblage of plates, called the Cimmerian plate, from the northeastern margin of Gondwana, leading to the opening of a new ocean, Neotethys (also known as Meso Tethys) (Metcalfe 2002). The Cimmerian plate (southern China, Indochina, Lhasa, Qiangtang and others) drifted northward, closing the Paleo Tethys Ocean by subduction under the northeastern margin of Pangea (the eastern part of Laurussia and Kazakhstan) and under Tarim and North China, while to the south the Neotethys Ocean continued to open. This string of plates could have, for a while, isolated the Paleo Tethys Ocean from the Panthalassa Ocean.

The configuration of the Pangea is full of uncertainties (this debate is not shown in the maps of Fig. 2.6). The Pangea reconstruction is well constrained by geological and geophysical data for the Late Triassic-Early Jurassic at the beginning of continental breakup. This is not the case for the pre-Late Triassic. Bullard et al. (1965) used to rotate the two APWPs in a common frame permit to restore the paleopositions of two landmasses during these periods. Doing that, it results in a large continental overlap between Gondwana and Laurussia in excess of 1000 km. To solve this discrepancy, a large disconnecting dextral fault operated during the Permian and Triassic periods was proposed (Irving 1977), resulting in a relative movement of Laurussia to the north and of Gondwana to the south. Based on paleomagnetic data, the total dislocation length of this detachment could have been several thousand kilometers, but could be much less if observations in the field of the cumulated dislocations of the known disconnecting faults in operation at that time are considered. Moreover, the non-dipolarity of the magnetic field (strong influence from an octupolar field) during this period but more likely the lack of fidelity in the magnetic field recorded in the red sandstones caused by inclination shoaling may be "distorting" the paleomagnetic data (Domeier et al. 2012).

At the beginning of the Jurassic, around 200 Ma, the Pangea supercontinent broke into two vast continents, Laurasia (Laurussia + Siberia + other smaller plates) to the north and Gondwana to the south. This separation marked the beginning of the opening of the central Atlantic Ocean. The opening directed the drift of North America northwestwards. This initiated the subduction of the Farallon and Kula plates (northeast region of the Pacific Ocean) under the western margin of North America, resulting in the accretion of small heterogeneous blocks and island arcs to the North-American continent during the Jurassic. As this margin deformed, this was the Sevier orogeny which lasted until the Lower Cretaceous.

Further east, the Paleo-Tethys Ocean became totally subducted around 200 Ma and the Cimmerian plates collided with the North China and Tarim plates. This continental mosaic was separated from the southern margin of Siberia by the Mongol-Okhotsk Ocean. Triangular in shape, this ocean closed like a scissors during the Jurassic and at the end of this period ($\sim$150 Ma), the small continent formed by the mosaic of small plates collided with the southern margin of Siberia. The Eurasian continent was formed. The multiple collisions between the small plates more than likely caused deformation, but probably did not create any significant reliefs. The opening of the central Atlantic Ocean continued. The Panthalassa, Neotethys and Central Atlantic oceans all become connected in the subtropical zone of the northern hemisphere.

The Middle Jurassic ($\sim$170 Ma) marked the beginning of the break-up of Gondwana. Madagascar, India, Australia and Antarctica separated from the Africa-South America duo. During the Lower Cretaceous ($\sim$130 Ma), South America and Africa started to split and the South Atlantic Ocean opened between Patagonia and Southern Africa. This event could mark the beginning of the deformation of the western central Andes margin in South America (Torsvik et al. 2009). The complete separation of the two continents and the connection with the central Atlantic Ocean only occurred at the end of the Lower Cretaceous period, about 30 million years later. At the beginning of the Lower Cretaceous ($\sim$110 Ma), India began its drift northwards. The Neo-Tethys Ocean was subducted under the southern margin of Eurasia, while in the south of India, the Indian ocean opened (McKenzie and Sclater 1971). Jagoutz et al. (2015) concluded that the exceptional rate of convergence exceeding 140 mm yr^{-1} is due to the existence of a double northward dipping subduction zones between the Indian and Eurasian plates during the Cretaceous. Around 90 Ma, Madagascar and India separated, the Carlsberg ridge is formed and marked the beginning of the opening of the northwestern part of the Indian Ocean.

The Cretaceous is characterized by a high sea level which led to the formation of vast shelf seas starting at the Albian ($\sim$100 Ma) to the Maastrichtian ($\sim$65 Ma). The flooding of the continents reached its maximum at the beginning of the Upper Cretaceous, around 95 Ma, when a large part of Europe was inundated. At the height of the marine incursion, a shallow sea developed over North Africa across the current Sahara Desert, temporarily linking the Neo-Tethyan Ocean to the South Atlantic Ocean. In North America, during the Upper Cretaceous, a sea passage, the Western Interior Seaway, was established between the Arctic Ocean and the Gulf of Mexico, while a new orogenic phase affected the margin of this continent (Laramide orogeny). A sea passage formed in western Siberia linking the Arctic Ocean and the Neo-Tethyan Ocean, and only disappeared during the Eocene.

The beginning of the Cenozoic is marked in the northern hemisphere by the opening of the third and last part of the Atlantic Ocean, the northern part. North America and Eurasia separated. However the Arctic Ocean remains almost isolated from the rest of the oceans during the Early Cenozoic favouring the deposits of black shales due to poorly oxygenated water (Jakobsson et al. 2007) until the deepening of the Fram Strait during the Late Eocene, $\sim$36 Myr (Poirier and Hillaire-Marcel 2011). In the southern hemisphere, Australia and Antarctica are definitively separated during the Eocene. The Antarctic migrated to its polar position. The circum-Antarctic basin was formed. In the western part of North America, subduction geometry evolved, deformation progressed eastward, the Rocky Mountains lifted up, while some more coastal reliefs were lowered due to a change in the pattern of constraints in this region. In South America, the uplift of the Andean Cordillera appears to have accelerated towards the end of the Cenozoic. India collided with Asia at the beginning of the Cenozoic ($\sim$50 Ma). India's drift to the north continued after the collision at a rate of 5–6 cm/year. Part of the crustal thickening is accommodated by the play of large right-lateral deformations reactivating old sutures between the plaques constituting the Asian mosaic, and by the uplift of the Himalayas and other chains farther north. These great disconnecting systems laterally eject plaques such as South China and Indochina (Tapponnier et al. 2001). This deformation of the southern margin of Asia is in fact generalized throughout the southern margin of Eurasia, in response to the Africa-Europe overlap resulting in the Alpine orogeny. This deformation contributed to the establishment of sea basins, sometimes deep, in Eurasia. These basins were connected to each other and extended from Western Europe to the foot of the Tibetan zone (during the process of uplifting) to form a vast epicontinental sea, the Paratethys, connected to the Atlantic Ocean and to the Tethys. The intense deformation of the southern margin of Eurasia during the Miocene gradually isolated the Paratethys from the rest of the oceans. Supplied only by runoff waters and subjected to the deformation of the substratum, this sea was gradually reabsorbed. Today, the Caspian Sea and the Black Sea are the only descendants of the Paratethys.

The Arabia-Eurasia collision during the Early Miocene lead to the closure of the Neo-Tethys sea and the uplift of the Zagros collisional wedge (Pirouz et al. 2017). In the Middle Miocene, the northern part of the Arabian Peninsula became exposed, marking the birth of the Mediterranean Sea. This would dry up almost completely at the end of the Miocene (the Messinian crisis) for a few hundred thousand years, due to an upheaval in the Betic zone (Gibraltar region) combined with a slight decrease in sea level. During the late Miocene, the northward drift of Australia initiated the closure of the Indonesian seaway, while the closure of the Central American Seaway linked to a tectonic uprising permanently isolated the Atlantic and Pacific waters.

Box: Focus on the uplift of the Tibetan Plateau.

> Since 50 Ma, the average convergence speed between India and Asia has remained around 5 cm per year. The Himalayan range occurred due to the deformation of the northern edge of the Indian sub-continent when it was subducted under the Asian continent. To the south of the Tsangpo suture is the Himalayan domain: on Indian crust with remnants of the Tethyan sedimentary cover and accretion prism associated with subduction of the Tethyan Ocean before collision. North of the suture, the Tibetan plateau has sedimentary layers from the Asian paleomargin. These are two very different realms with different Cenozoic geological histories, although related to the same event. During the millions of years after the collision, a proto-Himalayan chain must have developed on the remains of the Andean chain associated with the subduction of the Tethys, but the extent of this is not known. The Tibetan plateau was still a low-lying area.

> Two contrasting opinions have been proposed to explain the uplift of the Tibetan plateau: a "soft Tibet" model and a "staggered model". The "soft Tibet" model considers the uplift to be the result of an isostatic rebound caused by the "sinking" of the crustal root of a thickened Tibetan proto-plateau. The plateau reached its stable altitude of about 5000 m, and began to creep, as evidenced by the normal faults that line the southern plateau. The other theory (Tapponnier et al. 2001) considers, on the contrary, that the uplift took place in successive stages. Part of the India-Asia collision is absorbed by the lateral extrusion of landmasses (South China, Indochina), which caused old sutures inherited from the accretion of small landmasses during the Paleozoic and Mesozoic to protrude, profoundly modifying the paleogeography of South-East Asia (resulting in the closure of the Indonesian passage). These large strike-slip faults, separating the landmasses, connect with small perpendicular faults parallel to each other. The associated reliefs isolate small endorheic sedimentary basins which gradually fill up with the output from the dismantling of the chain "like a bathtub". The uplift of the Tibetan plateau would have occurred in successive stages since 50 Ma. The southern part of the Tibetan plateau rose during the Eocene, the central part during the Oligocene-Miocene and the northeastern part is currently uplifting.

Conclusion

The Earth moved from being entirely composed of an oceanic crust to the emergence of the first shreds of continental crust during the Archean, then to its episodic growth, mainly during the Precambrian. The paleogeography of this period is uncertain, given the small number of paleomagnetic data available, but it seems that the first consolidation of landmasses into a supercontinent (followed by a break-up phase) dates back to that time. It is only from the end of the Precambrian that we have global paleogeographic reconstructions, but there are no unique and definitive solutions for much of the Paleozoic, as uncertainties remain on the arrangement of some continents and the dimensions of the oceans. Paleogeographic reconstructions become more reliable at the end of the Paleozoic, and data from ocean kinematics makes it possible to constrain the relative positions of some continents relative to others. The distribution of the continents, either as they grouped together or dispersed, represents an important climate forcing through direct and undirect effects, while a second trigger of long-term climate change is mountain uplift. In that respect, the last 60 million-years have been crucial, since most of present-day mountain ranges have uplifted during this time interval. The Himalayan orogeny and the uplift of the Tibetan plateau are certainly the two most studied geological events over the past three decades in terms of impact on climate, but other mountain ranges such as the Andes, the Rockies and the east African dome very likely have played important roles on the Cenozoic climate change. Some clues on this point are provided in Volume 2, Chap. 22. Still, the paleoelevations of the reliefs remain shrouded in uncertainty, and efforts both on the modeling and the paleoelevation indicator sides will need to be pursued to improve our understanding of paleogeography evolution of Earth through time, and ultimately better quantify its impact on long-term climate changes.

References

Besse, J., & Courtillot, V. (1991). Revised and synthetic apparent polar wander path of the African, Eurasian, North American and India plates, and true polar wander since 200 Ma. *Journal of Geophysical Research: Solid Earth, 95*, 4029–4050.

Besse, J., & Courtillot, V. (2002). Apparent and true polar wander and the geometry of the geomagnetic field over the last 200 Myr. *Journal of Geophysical Research: Solid Earth, 107*, EPM 6-1–EPM 6-31.

Bleeker, W. (2003). The Late Archean record: A puzzle in ca. 35 pieces. *Lithos, 71*, 99–134.

Bonifacie, M., Calmels, D., Eiler, J. M., Horita, J., Chaduteau, C., Vasconcelos, C., et al. (2017). Calibration of the dolomite clumped isotope thermometer from 25 to 350 °C, and implications for a universal calibration for all (Ca, Mg, Fe) CO_3 carbonates. *Geochimica et Cosmochimica Acta, 200*(1), 255–279.

Botsyun, S., Sepulchre, P., Donnadieu, Y., Risi, C., Licht, A., & Caves Rugenstein, J. K. (2019). Revised paleoaltimetry data show low Tibetan Plateau elevation during the Eocene. *Science, 363*. http://doi.org/10.1126/science.aaq1436.

Botsyun, S., Sepulchre, P., Risi, C., & Donnadieu, Y. (2016). Impacts of Tibetan Plateau uplift on atmospheric dynamics and associated precipitation $\delta^{18}O$. *Climate of the Past, 12*, 1401–1420.

Bullard, E. C., Everett, J. E., & Smith, A. G. (1965). The fit of the continents around the Atlantic. *Philosophical Transactions of the Royal Society London, 258A*, 41–51.

Cawood, P. A., & Buchan, C. (2007). Linking accretionary orogenesis with supercontinent assembly. *Earth-Science Reviews, 3–4*, 217–256.

Chamberlain, C. P., Poage, M. A., Craw, D., & Reynolds, R. C. (1999). Topographic development of the Southern Alps recorded by isotopic composition of authigenic clay minerals, South Island, New Zealand. *Chemical Geology, 155*, 279–94.

Cogné, J.-P., Besse, J., Chen, Y., & Hankard, F. (2013). A new Late Cretaceous to present APWP for Asia and its implications for paleomagnetic shallow inclinations in Central Asia and Cenozoic Eurasian plate deformation. *Geophysical Journal International, 192*, 1000–1024.

Cogné, J. P., & Humler, E. (2006). Trends and rhythms in global seafloor generation rate. *Geochemistry, Geophysics, Geosystems, 7*, Q03011. https://doi.org/10.1029/2005GC001148.

Cogné, J. P., Humler, E., & Courtillot, V. (2006). Mean age of oceanic crust drives eustatic sealevel change since Pangea breakup. *Earth and Planetary Science Letters, 245*, 115–122.

Cox, A., Doell, R. R., & Dalrymple, G. B. (1964). Reversals of the earth's magnetic field. *Science, 144*, 1537–1543.

Creer, K. M., Irving, E., & Runcorn, S. K. (1954). The direction of the geomagnetic field in remote epochs in Great Britain. *Journal of Geomagnetism and Geoelectricity, 6*, 163–168.

Domeier, M., van der Voo, R., & Torsvik. T. H. (2012). Paleomagnetism and Pangea: The road to reconciliation. *Tectonophysics, 514–517*, 14–43.

Eiler, J. M. (2007). "Clumped-isotope" geochemistry—The study of naturally-occurring, multiply-substituted isotopologues. *Earth and Planetary Science Letters, 262*, 309–327.

Ferreira, D., Cessi, P., Coxall, H. K., de Boer, A., Dijkstra, H. A., Drijfhout, S. S., et al. (2018). Atlantic-Pacific asymmetry in deep water formation. *Annual Review of Earth and Planetary Sciences, 46*, 327–352.

Fluteau, F., Besse, J., Broutin, J., & Ramstein, G. (2001). The Late Permian climate. What can be inferred from climate modelling concerning Pangea scenarios and Hercynian range altitude? *Palaeogeography, Palaeoclimatology, Palaeoecology, 167*, 39–71.

Fluteau, F., Ramstein, G., Besse, J., Guiraud, R., & Masse, J. P. (2006). The impacts of the paleogeography and sea level changes on the mid cretaceous climate. *Palaeogeography, Palaeoclimatology, Palaeoecology, 247*, 357–381.

Forest, C. E., Molnar, P., & Emanuel, K. A. (1995). Palaeoaltimetry from energy conservation principles. *Nature, 374*, 347–350.

Garzione, C. N., Auerbach, D. J., Jin-Sook Smith, J., Rosario, J. J., Passey, B. H., Jordan, T. E., et al. (2014). Clumped isotope evidence for diachronous surface cooling of the Altiplano and pulsed surface uplift of the Central Andes. *Earth and Planetary Science Letters, 393*, 173–181.

Greff-Lefftz, M., & Besse, J. (2014). Sensitivity experiments on true polar wander. *Geochemistry, Geophysics, Geosystems, 15*, 4599–4616.

Haq, B. U., Hardenbol, J. A. N., & Vail, P. R. (1987). Chronology of fluctuating sea levels since the Triassic (250 million years ago to present). *Science, 235*, 1156–1166.

Haq, B. U., & Schutter, S. R. (2008). A chronology of paleozoic sea-level changes. *Science, 322*(5898), 64–68. https://doi.org/10.1126/science.1161648.

Hawkesworth, C. J., Cawood, P. A., Dhuime, B., & Kemp, T. I. S. (2017). Earth's continental lithosphere through time. *Annual Review of Earth and Planetary Sciences, 45*, 169–198.

Heezen, B. C. (1962). The deep-sea floor. In S. I. Runcorn (Ed.), *Continental drift* (pp. 235–288). New York: Academic Press.

Heezen, B. C., & Tharp, M. (1965). Tectonic fabric of the Atlantic and Indian Oceans and continental drift. In P. M. S. Blackett, E. Bullard, & S. K. Runcorn (Eds.), *A Symposium on Continental Drift, Philosophical Transactions of the Royal Society A, 1033*, 90. https://doi.org/10.1098/rsta.1965.0024.

Heirtzler, J. R., Dickson, G. O., Herron, E. M., Pitman, W. C., & Pichon, X. L. (1968). Marine magnetic anomalies, geomagnetic field reversals, and motions of the ocean floor and continents. *Journal Geophysical Research, 73*, 2119–2136.

Hess, H. H. (1962). History of ocean basins. In A. E. J. Engle, H. L. James, & B. L. Leonard (Eds.), *Petrologic studies: A volume in honor of A. F. Buddington* (pp. 599–620). New York: Geological Society of America.

Holmes, A. (1929). A review of the continental drift hypothesis. *Mining Magazine* 1–15.

Husson, L. (2006). Dynamic topography above retreating subduction zones. *Geology, 34*(9), 741–744.

Irving, E. (1964). *Paleomagnetism and its application to geological and geophysical problems.* Hoboken: Wiley.

Irving, E. (1977). Drift of the major continental blocks since the Devonian. *Nature, 270,* 304–309.

Jagoutz, O., Royden, L., Holt, A. F., & Becker, T. W. (2015). Anomalously fast convergence of India and Eurasia caused by double subduction. *Nature Geoscience, 8,* 475–479.

Jakobsson, M., Backman, J., Rudels, B., Nycander, J., Frank, M., Mayer, L., et al. (2007). The early Miocene onset of a ventilated circulation regime in the Arctic Ocean. *Nature, 447,* 986–990.

Kim, S. T., & O'Neil, J. R. (1997). Equilibrium and nonequilibrium oxygen isotope effects in synthetic carbonates. *Geochimica et Cosmochimica Acta, 61,* 3461–3475.

Kirschvink, J. L., Ripperdan, R. L., & Evans, D. A. (1997). Evidence for a large-scale reorganization of early cambrian continental masses by inertial interchange true polar wander. *Science, 277*(5325), 541–545.

Le Pichon, X. (1968). Sea-floor spreading and continental drift. *Journal Geophysical Research, 73,* 3661–3697.

Li, L., & Garzione, C. (2017). Spatial distribution and controlling factors of stable isotopes in meteoric waters on the Tibetan Plateau: Implications for paleoelevation reconstructions. *Earth and Planetary Science Letters, 460,* 302–314.

Li, Z. X., Bogdanova, S. V., Collins, A. S., Davidson, A., De Waele, B., Ernst, R. E., et al. (2008). Assembly, configuration, and break-up history of Rodinia: A synthesis. *Precambrian Research, 160,* 179–210.

Matte, P. (1986). Tectonics and plate tectonics model for the Variscan Belt of Europe. *Tectonophysics, 126,* 329–374.

McElhinny, M. W., Powell, C. M., & Pisarevsky, S. A. (2003). Paleozoic Terranes of Eastern Australia and the drift history of Gondwana. *Tectonophysics, 362,* 41–65.

McKenzie, D., & Sclater, J. G. (1971). The evolution of the Indian Ocean since the Late Cretaceous. *Geophysical Journal International, 24,* 437–528.

Meert, J. G., & Santosh, M. (2017). The Columbia supercontinent revisited. *Gondwana Research, 50,* 67–83.

Metcalfe, I. (2002). Permian tectonic framework and palaeogeography of SE Asia. *Journal of Asian Earth Sciences, 20,* 551–566.

Miller, K. G., Kominz, M. A., Browning, J. V., Wright, J. D., Mountain, G. S., Katz, M. E., et al. (2005). The Phanerozoic record of global sea-level change. *Science, 310,* 1293–1298.

Moreau, M.-G., Besse, J., Fluteau, F., & Greff-Lefftz, M. (2007). A new global Paleocene-Eocene apparent polar wandering path loop by "stacking" magnetostratigraphies: Correlations with high latitude climatic data. *Earth and Planetary Science Letters, 260,* 152–165.

Morgan, W. J. (1968). Rises, trenches, great faults and crustal blocks. *Journal Geophysical Research, 73*(308), 1959–1982.

Morgan, W. J. (1972). Deep mantle convection plumes and plate motions. *AAPG Bulletin, 56,* 203–213.

Mulch, A. (2016). Stable isotope paleoaltimetry and the evolution of landscapes and life. *Earth and Planetary Science Letters, 433,* 180–191. https://doi.org/10.1016/j.epsl.2015.10.034.

Müller, R. D., Seton, M., Zahirovic, S., Williams, S. E., Matthews, K. J., Wright, N. M., et al. (2016). Ocean basin evolution and global-scale plate reorganization events since Pangea breakup. *Annual Review of Earth and Planetary Sciences, 44*(1), 107–138.

Nance, R. D., Gutierrez-Alonso, G., Duncan Keppie, J., Linnemann, U., Brendan Murphy, J., Quesada, C., et al. (2012). A brief history of the Rheic Ocean. *Geosciences Frontiers, 3*(2), 125–135.

Oriolo, S., Oyhantçabal, P., Wemmer, K., & Siegesmund, S. (2017). Contemporaneous assembly of Western Gondwana and final Rodinia break-up: Implications for the supercontinent cycle. *Geoscience Frontiers, 8,* 1431–1445.

Otto-Bliesner, B. L., & Upchurch, G. R. Jr. (1997). Vegetation-induced warming of high-latitude regions during the Late Cretaceous period. *Nature, 385,* 804–807.

Otto-Bliesner, B. L., Brady, E. C., & Shields, C. (2002). Late Cretaceous Ocean: Coupled simulations with the National Center for Atmospheric Research climate system model. *Journal of Geophysical Research: Atmospheres, 107*(D2), ACL 11-1–ACL 11-14.

Peppe, D. J., Royer, D. L., Wilf, P., & Kowalski, E. A. (2010). Quantification of large uncertainties in fossil leaf paleoaltimetry. *Tectonics, 29*(3). https://doi.org/10.1029/2009TC002549.

Pirouz, M., Avouac, J. P., Hassanzadeh, J., Kirschvink, J. L., & Bahroudi, A. (2017). Early Neogene foreland of the Zagros, implications for the initial closure of the Neo-Tethys and kinematics of crustal shortening. *Earth and Planetary Science Letters, 477,* 168–182.

Poirier, A., & Hillaire-Marcel, C. (2011). Improved Os-isotope stratigraphy of the Arctic Ocean. *Geophysical Research Letters, 38,* L14607.

Poulsen, C. J., Ehlers, T. A., & Insel, N. (2010). Onset of convective rainfall during gradual Late Miocene rise of the Central Andes. *Science, 328,* 490–493.

Quade, J., Garzione, C., & Eiler, J. (2007). Paleoelevation reconstruction using pedogenic carbonates. *Reviews in Mineralogy and Geochemistry, 66,* 53–87.

Raff, A. D., & Mason, R. G. (1961). Magnetic survey off the west coast of North America, 40° n. latitude to 52° n. latitude. *GSA Bulletin, 72,* 1267–1270.

Robert, B., Besse, J., Blein, O., Greff-Lefftz, M., Baudin, T., Lopes, F., et al. (2017). Constraints on the Ediacaran inertial interchange true polar wander hypothesis: A new paleomagnetic study in Morocco (West African Craton). *Precambrian Research, 295,* 90–116.

Rogers, J. J. W., & Santosh, M. (2002). Configuration of Columbia, a Mesoproterozoic supercontinent. *Gondwana Research, 5,* 5–22.

Romm, J. (1994). A new forerunner for continental drift. *Nature 367,* 407.

Rowley, D. B., & Garzione, C. N. (2007). Stable isotope-based paleoaltimetry. *Annual Review of Earth and Planetary Sciences, 35,* 463–508.

Sarr, A. C., Husson, L., Sepulchre, P., Pastier, A. M., Pedoja, K., Elliot, M., et al. (2019). Subsiding Sundaland. *Geology, 47*(2), 119–122. https://doi.org/10.1130/G45629.1.

Seton, M., Müller, R. D., Zahirovic, S., Gaina, C., Torsvik, T., Shephard, G., et al. (2012). Global continental and ocean basin reconstructions since 200 Ma. *Earth Science Reviews, 113,* 212–270.

Su, T., Farnsworth, A., Spicer, R. A., Huang, J., Wu, F.-X., Liu, J., et al. (2019). No high Tibetan Plateau until the Neogene. *Science Advances, 5*(3), eaav2189. https://doi.org/10.1126/sciadv.aav2189.

Tapponnier, P., Zhiqin, X., Roger, F., Meyer, B., Arnaud, N., Wittlinger, G., & Jingsui, Y. (2001). Oblique, stepwise rise and growth of the Tibet Plateau. *Science, 294,* 1671–1677.

Tarduno, J. A., Blackman, E. G., & Mamajek, E. E. (2014). Detecting the oldest geodynamo and attendant shielding from the solar wind: Implications for habitability. *Physics of the Earth and Planetary Interiors, 233,* 68–87.

Tarduno, J. A., Cottrell, R. D., Davis, W. J., Nimmo, F., & Bono, R. K. (2015). A Hadean to Paleoarchean geodynamo recorded by single zircon crystals. *Science, 349,* 521–524.

Torsvik, T. H., Rousse, S., Smethurst, M. A., (2009). A new scheme for the opening of the South Atlantic Ocean and the dissection of an Aptian salt basin. *Geophysical Journal International, 183(1),* 29–34.

Torsvik, T. H., Van der Voo, R., Preeden, U., Mac Niocaill, C., Steinberger, B., Doubrovine, P. V., et al. (2012). Phanerozoic polar wander, palaeogeography and dynamics. *Earth Science Reviews, 114,* 325–368.

U.S. Geological Survey Geologic Names Committee. (2018). Divisions of geologic time—Major chronostratigraphic and geochronologic units. U.S. Geological Survey Fact Sheet 2018–3054, 2 p. https://doi.org/10.3133/fs20183054.

Vail, P. R. (1977). Seismic stratigraphy and global changes of sea level. *Bulletin American Association of Petroleum Geologists Memoir, 26,* 49–212.

Vérard, C., Hochard, C., Baumgartner, P. O., Stampfli, G. M., & Liu, M. (2015). 3D palaeogeographic reconstructions of the Phanerozoic versus sea-level and Sr-ratio variations. *Journal of Palaeogeography, 4(1),* 64–84.

Villeneuve, M. (2008). Review of the orogenic belts on the western side of the West African craton: The Bassarides, Rokelides and Mauritanides. In N. Ennih & J.-P. Liégeois (Eds.), *The boundaries of the West African Craton* (Vol. 297, pp. 169–201). Geological Society, London, Special Publications.

Vine, F. J., & Matthews, D. H. (1963). Magnetic anomalies over oceanic ridges. *Nature, 199,* 947.

Yount, L. (2009). *Alfred Wegener: Creator of the continental drift theory.* New York: Infobase Publishing.

Zhao, G., Cawood, P. A., Wilde, S. A., & Sun, M. (2002). Review of global 2.1-1.8 Ga orogens: Implications for a pre-Rodinia supercontinent. *Earth-Science Reviews, 59,* 125–162.

Introduction to Geochronology

Hervé Guillou

Accurate knowledge of climate variations in the past is an essential preamble to any realistic modeling of future climate, and requires an understanding of the mechanisms that govern the natural dynamics of the climate, and especially of its rapid changes. One of the current major concerns for this research is the quantification of the phase shift in climate between different regions of the globe. This requires having reliable, precise and comprehensive chrono-stratigraphic tools in order to temporally locate and to synchronize the various archives.

Establishing a common time frame for all climate archives remains a major challenge. Research on long time scales emphasizes the climate system's response to external forcing, but the study of rapid and abrupt changes in climate allows the internal variability of the climate system and the interactions between its various components to be investigated. The last glacial period was characterized by a succession of very rapid changes in the climate in the North Atlantic, which resulted in massive reorganization of the climate system on a global scale and was manifested in particular by the massive discharge of icebergs into the ocean, known as Heinrich events. These events mainly occurred during glacial periods but some occurred as soon as ice sheets developed on land in the northern hemisphere at the end of the last interglacial, and also at the beginning of the Holocene, even though interglacial periods seem to have been much more stable. In order to understand and model the mechanisms involved, it is important to know the precise chronology of all these events.

Another purpose of geochronological studies is to enable the comparison of climate records on a common and absolute time scale. It is only through this approach that the phase shifts between the hemispheres or between low and high latitudes can be understood and explained. By harmonizing the time scales for different types of records, both from the land and ocean, isotopic stratigraphy plays an essential role in providing a better understanding of the chronology and dynamics of the mechanisms responsible for climate variations.

Variations in climate are caused by many factors with characteristic durations ranging from hundreds of millions of years for the evolution of the Sun to a few years for internal reorganization of the climate system. In addition, records of climate signals that can potentially be dated are represented in very different substrates (sediments, ice, coral, cave concretions known as speleothems). The choice of the best adapted geochronological tools to date them will depend on the nature of the records, their age, the time span of the phenomena to be dated and the desired level of accuracy.

The dating methods most commonly used in paleoclimatology are: dendrochronology, ^{14}C, the Uranium/Thorium relationship, Potassium-Argon ($^{40}K/^{40}Ar$) and its variant $^{40}Ar/^{39}Ar$ (isotopic methods), and magnetic stratigraphy (indirect method of dating). Often, in order to provide an accurate geo-chronological framework, two or more of these methods need to be compared.

In the following chapters, we will present the absolute dating methods implemented to provide a time scale independent of astronomical parameters and to refine the stratigraphic scales commonly used in paleoclimatology and in paleo-oceanography, very often based on variations in the $^{18}O/^{16}O$ relationship in ice and benthic foraminifera related to the orbital signal. The principles of the methods mentioned above, their field of application, their implementation in the laboratory, their accuracy and limitations will also be presented. The scope of each method will be illustrated with a concrete example.

Geochronology plays an essential role for both geologists and paleoclimatologists. For geologists, it has allowed traditional stratigraphy to be linked to a time scale covering the full history of the Earth and to estimate the time constants of the great geological phenomena (plate tectonics, uplift of mountains, renewal of ocean basins, long-term

H. Guillou (✉)
Laboratoire des Sciences du Climat et de l'Environnement, LSCE/IPSL, CEA-CNRS-UVSQ, Université Paris-Saclay, 91190 Gif-sur-Yvette, France
e-mail: herve.guillou@lsce.ipsl.fr

global climate changes). The latest version of the chronostratigraphic scale is available on the website of the International Commission on Stratigraphy: http://www.stratigraphy.org/.

For paleoclimatologists, geochronology has provided a time scale entirely independent of astronomical parameters. This has allowed the Milankovitch theory to be validated, a work that has been a focus of attention of geochemists over the last forty years. Currently, a major experimental effort is underway to try to improve the dating accuracy and to detect the phase shifts that accompany the response of various components of climate system to the insolation forcing.

Carbon-14

Martine Paterne, Élisabeth Michel,
and Christine Hatté et Jean-Claude Dutay

Seventy years after its discovery by W.B.F. Libby and collaborators (Arnold and Libby 1949; Libby 1952), the radiocarbon (^{14}C) method of dating is still of great interest in many scientific fields in biology, earth science, climate, environment and archeology. Libby received the Nobel Prize in Chemistry in 1960 and the chairman of the Nobel Committee highlighted the importance of this discovery in these terms: "*Seldom has a single discovery in chemistry had such an impact on the thinking of so many fields of human endeavor. Seldom has a single discovery generated such wide public interest*". The history of the ^{14}C method is an excellent example of the fruitful exchanges among different scientific fields and of the complementarity between scientific advances and technological innovations. Since its discovery, more than a hundred and fifty laboratories in the world are now dedicated to ^{14}C dating. In the 1980s, new technologies, notably the accelerator mass spectrometry, allowed the use of samples of increasingly reduced sizes and a better precision of the ^{14}C ages. Now, the physical and chemical processes involved in biological and environmental changes may be analyzed at the molecular scale.

The method and techniques of ^{14}C dating have been the subject of several web and journal publications to which readers may refer (Libby 1981; Taylor 1987; Taylor et al. 1992; Currie 2004).

Principles of the Radiocarbon Method

Discovery of the Method

Kamen (1963) reported the history of the ^{14}C discovery and Libby's meeting with ^{14}C in 1939 at Berkeley. He attributed the *physical* prediction of the existence of this isotope to the physicist Kurie (Kamen 1963), who studied neutron-induced disintegration of light elements such as nitrogen (^{14}N). During these experiments, Kurie observed infrequent and *abnormal* long thin traces in a cloud chamber filled with air. He attributed them to the emission of protons following the reaction ^{14}N (n, ^{1}H) ^{14}C although other reactions such as ^{14}N (n, ^{2}H) ^{12}C and ^{14}N (n, ^{3}H) ^{12}C could also have been possible. In 1936, Burcham and Goldhaber demonstrated that no α-particles were emitted in the slow neutron disintegration of ^{14}N and only the reaction ^{14}N(n, ^{1}H)^{14}C was possible with proton emission and formation of ^{14}C noted ^{14}N(n, p)^{14}C (Kamen 1963).

The evidence of the *chemical* existence of ^{14}C is due to Ruben, a chemist and Libby's student, and to Kamen, a radiochemist of the Lawrence Livermore Radiation Laboratory at Berkeley. They investigated the assimilation processes of CO_2 during photosynthesis by incubating plant species with the radioactive isotope ^{11}C, which was produced in the Livermore cyclotron (Ruben et al. 1949). Labeled intermediate solutions were deposited on a blotting paper, and, once dried, the paper was protected by a plastic film and wrapped inside a screen-wall counter. The use of ^{11}C in biology was however very difficult due to long separation phases of various photosynthetic pigments by ultracentrifugation and a half-life of 21 min. Furthermore, it was not very competitive with the ^{13}C labeling of plants. At the request of Lawrence, who invented the cyclotron and received the Nobel Prize in Physics in 1939, the existence or not of long-lived radioactive isotopes was systematically sought for each element of the first column of the periodic table (H, C, N, O), and thus the search for the *chemical* existence of ^{14}C (Kamen 1963). Kamen submitted a graphite target to a deuteron beam in the cyclotron overnight. After burning the graphite, Ruben precipitated the CO_2 into a carbonate. This precipitate was furnished to their colleague in the chemistry department, W.F. Libby, who developed proportional counters to measure the radioactivity of elements such as neodymium, samarium, rubidium and lutetium, to determine their period (Libby 1934). The detection

M. Paterne (✉) · É. Michel · C. H. et Jean-Claude Dutay
Laboratoire des Sciences du Climat et de l'Environnement,
LSCE/IPSL, CEA-CNRS-UVSQ, Université Paris-Saclay,
91190 Gif-sur-Yvette, France
e-mail: martine.paterne@lsce.ipsl.fr

of a weak activity was the first proof of the existence of artificially created ^{14}C; its half-life was estimated to be between 10^3 years and 10^5 years (Kamen 1963).

This research was interrupted in the early 1940s as Libby joined Harold Urey's team in Chicago to develop techniques of isotopic enrichment of uranium for nuclear weapons in the frame of the Manhattan Project. He would later use these enrichment techniques in natural samples for ^{14}C studies. While the ^{14}C was artificially created, the existence of a natural production of ^{14}C still needed to be proven and thus the existence of slow neutrons in the Earth's atmosphere in order to assess the feasibility of ^{14}C dating.

The method of ^{14}C dating is linked to the discovery of cosmic radiations, later termed cosmic rays, by Victor Hess in 1912 by using electroscopes aboard a balloon (Libby 1964; Rossi 1952). This discovery was the beginning of numerous studies, which investigated the composition, intensity, origin, and effect of the cosmic rays on the Earth's atmosphere. Cosmic rays caused nuclear reactions in the atmosphere, which were suspected by Grosse (Libby 1981). These reactions were first evidenced by Blau in 1932, who pioneered the technique of photographic plates covered by thick nuclear emulsions to separate the α-particles from proton tracks. Such plates were exposed in the Austrian Alps revealing disintegration stars in the emulsion (Rossi 1952). Rumbaugh and Locher determined the nature of this radiation by sending photographic plates into the stratosphere to an altitude of about 20 km in the gondola of a balloon (Rossi 1952). Some plates were covered with different materials about 1 cm thick and the others were free of materials. The control plates showed no traces, while those covered by paraffin, for example, showed four times more traces than those covered by lead or carbon. Due to the absence of traces on the control plates, the traces could be only protons and not α particles, and these protons could have been emitted only from the different materials during collision of atoms with neutrons. Korff and colleagues then performed new experiments using proportional counters aboard balloon, some filled with boron trifluoride (boron is a neutron absorber and emits a α particle upon collision with a neutron), and others filled with a mixture of hydrogen, methane and carbon monoxide (sensitive to fast neutrons) (Rossi 1952; Simpson 2000). These counters permitted precise measurement of the density of the neutrons and their energy spectrum. These authors have thus shown that the density of slow neutrons reached a maximum at an altitude between 12 and 16 km, and then decreased towards the sea level. When entering the atmosphere, the protons, which compose about 90% of the cosmic rays, collide with atoms and molecules (mainly nitrogen and oxygen). The products of their disintegration are protons and neutrons, which collide with other atoms while neutrons lose some energy on each collision. This explains the increase in neutron density at around 16 km and the decrease towards the sea-level.

During collisions, the neutrons slow down, and Korff suggested that these secondary slow neutrons were captured by nitrogen nuclei to form the cosmogenic isotope ^{14}C following the reaction ^{14}N (n, p) ^{14}C (Korff 1951). When receiving the Nobel Prize in 1960, Libby (1964) indicated that the idea of the ^{14}C dating method was inspired by Korff's results. Later, Simpson (2000) showed that the density of neutrons varied with the latitude, as a function of the lines of the Earth's magnetic field that deflect the (electrically charged) particles of cosmic rays. The average production of ^{14}C is in the range of 2.25 ± 0.1 atoms of $^{14}C/cm^2/s$. It varies from one to six between the equator and the poles, and at the poles, it can vary by a factor of almost four depending on solar activity. Of minor importance, other reactions on ^{16}O, ^{17}O, ^{13}C, contribute also to the formation of atmospheric ^{14}C.

Principle of the ^{14}C Dating Method

Libby postulated in 1946 (Arnold and Libby 1949) that the production of ^{14}C atoms and their decay as ^{14}N by emitting a β particle would be in equilibrium at steady-state conditions (Fig. 4.1). As the cosmic rays continuously bombard the Earth and as the Earth age is much higher than the estimated period of 10^3–10^5 years, the distribution of ^{14}C would be in equilibrium within all the reservoirs of exchangeable carbon (atmosphere, ocean, biosphere). Estimating the production of neutrons per cm^2 and per second based on the distribution of neutrons observed by Korff (1951) and the amount of exchangeable carbon between the reservoirs, Libby wrote that the specific activity of exchangeable carbon could be easily calculated taking into account the balance between production and decay:

$$\frac{d^{14}C}{dt} = Q - \lambda^{14}C = 0$$

where Q is the production of a ^{14}C atom per second and λ, the decay constant, is equal to $\ln(2)/T_{1/2}$, where $T_{1/2}$ is the half-life (half of the radioactive atoms have decayed).

$$\frac{d\left(\frac{^{14}C}{^{12}C}\right)}{dt} = 0 = \frac{1}{(^{12}C)^2}\left[^{12}C\frac{d^{14}C}{dt} - ^{14}C\frac{d^{12}C}{dt}\right]$$
$$= \frac{^{12}C}{(^{12}C)^2}\left[Q - \lambda^{14}C\right]$$

because

$$^{14}C\frac{d^{12}C}{dt} = 0$$

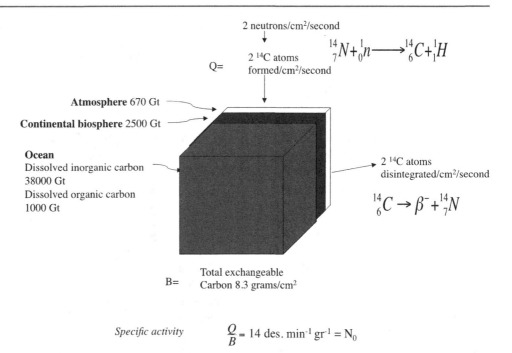

Fig. 4.1 Diagram of the ^{14}C formation and mixing in the different terrestrial reservoirs (Libby 1964)

The carbon stocks are expressed in Gt (Gigatons) and the number of neutrons per cm^2 per second corresponds to pre-1950 estimates (modified according to 10):

$$\lambda \frac{^{14}C}{^{12}\{C\}} = \frac{Q}{^{12}C}$$

The predicted activity should have been between 1 and 10 decays per minute per gram of carbon, given the uncertainties, in all living matter. ^{14}C activities of 10.5 disintegrations per minute per gram were measured from isotopically enriched samples of biomethane in 1947, in good agreement with the prediction. Libby and collaborators measured then the specific activities of natural tree samples from different continents by reducing background counting with lead and iron shielding. They found worldwide homogeneous ^{14}C activities at around 12.5 disintegrations/min/g of carbon[1] (Libby 1981). A precise measurement of the ^{14}C half-life, and the specific activity of samples of known ages were further undertaken to validate the ^{14}C dating method.

Estimation of the Half-Life and the First ^{14}C Dating

The half-life of ^{14}C was determined through many experiments (Engelkemeir et al. 1949; Olsson et al. 1962). One of them consisted of measuring, by mass spectrometry, the isotopic ratio $^{14}C/^{12}C$ of highly enriched barium carbonates [BaCO$_3$] (up to 6%) produced by submitting to neutrons beams, solutions of ammonium nitrate in a cyclotron (Engelkemeir et al. 1949). Proportional counters were then filled with the CO_2 released by BaCO$_3$ hydrolysis. By measuring the number of disintegrations per minute and per gram of carbon (dN/dt) and knowing the number of atoms of ^{14}C (N) in the samples, the half-life was estimated at 5720 ± 47 years. Averaging all the published estimates, Libby estimated the half-life at 5568 years. Redeterminations have led to a value of 5730 ± 40 years (Godwin 1962). The latter value has recently been debated (Chiu et al. 2007).

Libby and collaborators then undertook the dating of samples of known ages, mostly from the tombs of the Egyptian kingdoms, and published them as the Curve of Knowns (Libby 1964).

Principle of the Method

Carbon-14 is formed in the upper atmosphere, where it is rapidly oxidized to form $^{14}CO_2$ molecules. All living matter contains carbon, and, thus a very small proportion of ^{14}C. The ^{14}C abundance is of some 1.2×10^{-10}% (or 1.2×10^{-12} g of ^{14}C per g of carbon), while those of the isotopes ^{13}C and ^{12}C are respectively 1.108% and 98.892%. The ^{14}C exchanges between the living material and their environment cease at the death of the animals or plants. The time (t) since the death can be measured by comparing the residual specific activity in dead organisms to that of the atmosphere.

---

[1] The specific activity is now determined at 13.56 ± 0.07 disintegrations/min/g of carbon.

The radiocarbon age (t) is calculated using the radioactive exponential decay:

$$\frac{^{14}C}{^{12}C} = \left(\frac{^{14}C}{^{12}C}\right)_0 e^{-\lambda t}$$

where $(^{14}C/^{12}C)_0$ is the atmospheric ratio and $\lambda = \ln(2)/T_{1/2}$ the decay constant. By convention, the Libby's half-life at 5568 years is used to calculate the ^{14}C ages. The mean lifetime of the ^{14}C atoms before decay is $T_{1/2}/\ln(2)$.

The calculation of age becomes:

$$t = \frac{1}{\lambda} \ln\left(\frac{\frac{^{14}C}{^{12}C}}{\left(\frac{^{14}C}{^{12}C}\right)_0}\right)$$

The ^{14}C dating method is based on the hypothesis of a constant radioactive equilibrium between the ^{14}C formation and its disintegration in ^{14}N. If we look at Libby's diagram (Fig. 4.1), we may observe that this is true if the production of ^{14}C, the size of the various reservoirs of carbon (atmosphere, oceans, land and marine biosphere) and their carbon content remain constant over time, as well as the fluxes between the various reservoirs. In addition, the physicochemical integrity of the dated fossils must have been preserved after the death of organisms. For example, no isotopic exchange or secondary crystallization should have occurred. Finally, the samples should have not moved from their burying sites to date precisely any events.

Validity of the Assumptions and Definition of a Reference Standard for the Atmosphere

The first offsets between the ^{14}C and known ages appeared very quickly, notably with the major contribution of dendrochronology, a counting method of the annual tree-ring growth. In 1955, Suess demonstrated (Taylor 1987; Damon et al. 1978) that the ^{14}C content in the atmosphere varied in the last hundred years and decreased from 1890 AD to 1950 AD (Fig. 4.2). He suggested that the decrease was the result of the CO_2 release into the atmosphere from the domestic and industrial combustion of ^{14}C-depleted fossil fuels (coal, oil). These annual emissions, approximately 150Gt of C as CO_2 until 1950 AD, were responsible for a ^{14}C aging of the atmosphere of about 160 years between 1890 AD and 1950 AD, the so-called '*Suess effect*'. In 1957, Rafter and Fergusson observed a rapid atmospheric ^{14}C increase that they attributed to the ^{14}C production during the aerial atomic bomb tests. These peaked between 1960 and 1961 and doubled the ^{14}C concentration in atmospheric CO_2. This atmospheric ^{14}C spike led later to life-size experiments to monitor carbon exchanges between the various earth reservoirs.

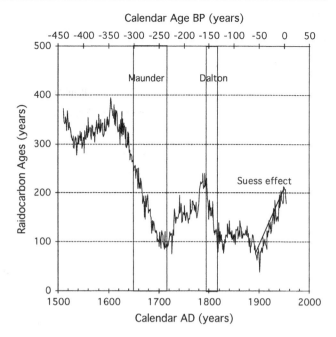

Fig. 4.2 Variations of the ^{14}C ages as a function of calendar ages in tree-rings. The time interval of the '*Suess effect*', the Maunder and Dalton minima of solar magnetic activity are shown. The last two coincide with a rapid decrease of the atmospheric ^{14}C ages due to lower filtering of cosmic protons by the solar magnetic field. The 'Suess effect' corresponds to the increase of the ^{14}C ages due to the dilution of atmospheric $^{14}CO_2$ by the industrial and domestic emissions of CO_2 into the atmosphere from the ^{14}C-free fossil fuels

Besides the anthropogenic changes of ^{14}C, the natural variations of ^{14}C in the different carbon reservoirs were then identified by comparing the ^{14}C and dendrochronological ages over the past millennia (Damon et al. 1978; Stuiver et al. 1991). They are attributed to changes in ^{14}C production in the upper atmosphere and to variations in the natural carbon cycle linked to the size of the carbon reservoirs, their composition and the carbon exchange fluxes.

De Vries observed rapid fluctuations (wiggles) in the atmospheric ^{14}C from tree-ring analyses (Damon et al. 1978; Stuiver et al. 1991). He attributed these fluctuations to changes in climate and solar activity, as both the Little Ice Age (about 1560 AD to 1830 AD) and the Maunder and Dalton minima, two time intervals of few sunspots, occurred over this time.

The charged particles released from the Sun, known as the solar wind, create a magnetic field in the interplanetary space. These eruptions are a manifestation of the magnetic activity of the Sun, with a minimum dipolar magnetic field corresponding to the maximum equatorial activity, and their intensity is reversed every 11 years. The solar magnetic field varies with a cycle of 22 years, reversing its polarity every 11 years at every change of solar activity. The particles of the galactic cosmic rays are deflected by the solar wind. The higher the solar magnetic activity, the fewer particles penetrate into the upper atmosphere. As a result, the ^{14}C production decreases as the solar activity increases, and vice versa. Similarly, the cosmic ray intensity in the upper atmosphere is modulated by the changes in the Earth's magnetic field, which acts as a shield against the electrically-charged cosmic protons. The greater the intensity of the geomagnetic field, the fewer the number of cosmic protons arriving in the upper atmosphere, and therefore the production of ^{14}C is lower in the atmosphere (Damon et al. 1978; Stuiver et al. 1991). Changes in the Earth's magnetic field could explain about 50% of the ^{14}C variation in the atmosphere between the last glacial and the Holocene, and the ^{14}C aging of the atmosphere during the last glacial related to a high magnetic field intensity (Lal and Charles 2007).

The dendrochronological record of the atmospheric ^{14}C content over the last 10,000 years has, in turn, given the opportunity to evaluate the changes of solar activity beyond the first observations in the seventeenth century. Spectral analyses of the atmospheric ^{14}C emphasize long cycles of 88 years, 208 years and 2050 years attributed to changes in solar activity (Damon and Peristykh 2000). During the last 70 years, the solar activity was at an exceptionally high level and many other periods of high activity, although shorter, occurred in the past, such as, for example, at the beginning of the Holocene. In addition, some authors have thought that the unusual solar activity may have contributed, in a small part, to the recent climate change observed in the late twentieth century (Muscheler et al. 2005).

Climate changes modify the size of the carbon reservoirs and the CO_2 fluxes between them and therefore the atmospheric $^{14}CO_2$ as noted by de Vries (Stuiver et al. 1991). Libby (1952) had already estimated the impact of climate changes on the ^{14}C ages. He assumed that the sea level lowering ($\sim$100 m) and the temperature decrease during the last ice age reduced the oceanic carbon inventory, resulting in an increase of the specific activity of ^{14}C (Fig. 4.1). Assuming that the ^{14}C activity in the reservoir exchange changed by 10%, Libby calculated that the glacial ^{14}C ages would be too young by some 800 years. We know now that the ^{14}C ages are too young by about 2000 years with respect to 'true' ages during the last glacial maximum (Reimer et al. 2013). The ocean circulation and the carbon cycle were deeply modified during the last ice age that modulated the concentration of atmospheric CO_2. Between the last glacial maximum and the Holocene, the atmospheric CO_2 concentration increased from 190 to 280 ppm as recorded in the Antarctic ice cores.

The growing evidence of the atmospheric ^{14}C variations through time led to two consequences: the establishment of a standard, i.e. a reference value of ^{14}C for the atmosphere (N_0), and the calibration process which precisely quantifies the difference between the true or 'absolute' ages and the ^{14}C ages.

As requested by Arnold in 1956, the National Bureau of Standards (Washington) prepared a standard (NBS-I) of 450 kg of oxalic acid (HOOC-COOH), an organic compound extracted from a French crop of sugar beet in 1955 (Arnold and Libby 1949). It provided the reference activity of year zero, from which the ^{14}C age of a sample is calculated. The reference activity was taken at 95% that of NBS-I to account for the Suess and nuclear bomb effects and the reference year has been set arbitrarily at 1950. Since that time, other standards were prepared (NBS-II, sucrose). The ^{14}C ages are expressed in years BP (Before Present, present being equal to 1950 AD). In archeology, the terms Anno Domini (AD) and Before Christ (BC) are used. Lately appears the term Common Era (CE) which is equivalent to AD and BCE to BC.

Before measuring the ^{14}C activity in a sample and calculating its ^{14}C age, the fractionation of the stable isotopes of carbon must be considered. Craig in 1953 demonstrated that the $^{13}C/^{12}C$ ratios ($\delta^{13}C$) vary in contemporary materials as a function of the reservoir's $\delta^{13}C$ in which they form. The amount of enrichment/depletion of ^{14}C due to biological and physicochemical fractionation processes is approximately two times that measured by $\delta^{13}C$ in the same sample. The $\delta^{13}C$ value, expressed in ‰ compared to the standard PDB (Pee Dee Belemnite), is about −6.5‰ for the pre-industrial atmospheric CO_2. It varies between −2‰ and +3‰ in the carbonates of seashells and between −20 and +3‰ for those of lake shells. The $\delta^{13}C$ values of plants vary between −27 and −14‰ due to photosynthesis processes. Processes of assimilation of atmospheric CO_2 by plants are carried out mainly according to two cycles of transformation of organic compounds, called the Calvin cycle and the Hatch and Slack cycle, or also as C3 and C4, the second with a lower isotope discrimination than the first. As a result of the great variability of the $\delta^{13}C$ values, the ^{14}C activities have been normalized to a common reference of $\delta^{13}C$ set at −25‰ (see below). The complete procedure for calculation of age is explained in Box 1.

Box 1
As in most definition of isotopes, the ^{14}C content is expressed by a δ in ‰, which defines the difference between a sample and a standard, which may be the NBS-I standard, the $\delta^{13}C$ of which is equal to −19‰ relative to the PDB (Broecker and Olson 1959; Olsson

and Osadebe 1974; Stuiver and Polach 1977; Stenström et al. 2011).

The $\delta^{14}C$ may be written as follows:

$$\delta^{14}C = \left(\frac{A_s}{A_{Ox}} - 1\right) \times 1000 \quad (4.1)$$

With A_s and A_{Ox}, the respective activities of the sample and of the standard NBS-I of oxalic acid.

$$A_{ON} = 0.95 \times A_{Ox}\left[1 - \frac{2 \times (19 + \delta^{13}C_{Ox})}{1000}\right] \quad (4.2)$$

The ^{14}C activity in the atmosphere in 1950 (A_{ON}) is equal to 95% of the NBS-I standard activity, corrected for $\delta^{13}C$ (Eq. 4.2). The isotopic fractionation of $\delta^{13}C$, which affects the abundance of the mass 14, is noted by the number 2. It indicates that the fractionation between the mass 14 and mass 12 is double that between the masses 13 and 12. The negative sign assigned to fractionation $\delta^{13}C$ means that the ^{14}C activity measured in organisms with a negative $\delta^{13}C$, that is to say with less affinity for the mass 13 than the PDB standard, must increase to compensate for the loss caused in ^{14}C by fractionation (Broecker and Olson 1959; Olsson and Osadebe 1974; Stuiver and Polach 1977; Stenström et al. 2011).

$d^{14}C$ is defined as:

$$d^{14}C = \left(\frac{A_s}{A_{ON}} - 1\right) \times 1000 \quad (4.3)$$

To take into account the variability in the fractionation of isotopic $\delta^{13}C$ measured in the samples, $\delta^{14}C$ is normalized to a common value, $\delta^{13}C$, set at $-25‰$ versus PDB, regardless of the sample type (carbonate, dissolved inorganic carbon or organic matter). This value was obtained by averaging the measurements of the isotopic ratio $\delta^{13}C$ of several pieces of wood with an age less than 1890 AD (Broecker and Olson 1959; Olsson and Osadebe 1974).

We define:

$$D^{14}C(‰) = \left(\frac{A_{SN}}{A_{ON}} - 1\right) \times 1000 \quad (4.4)$$

with

$$A_{SN} = A_S\left(1 - \frac{2 \times (25 + \delta^{13}C_S)}{1000}\right) \quad (4.5)$$

$$D^{14}C(‰) = d^{14}C - 2(\delta^{13}C_S + 25)\left(1 + \frac{d^{14}C}{1000}\right) \quad (4.6)$$

The calculation of the ^{14}C age is then:

$$t(years) = \frac{1}{\lambda} \times \ln\left(\frac{1}{1 + D^{14}C/1000}\right)$$

In the case of precise measurements of activity performed in oceanography or to calibrate the ^{14}C ages, it is necessary to consider the ^{14}C decay between the age of the sample (x) and its measurement (y) using the period T of 5730 years. Equations 4.1 and 4.6 then become:

$$\delta_n^{14}C = \left(\frac{A_{SN}e^{\lambda(x-y)}}{A_{OxN}e^{\lambda(y-1950)}} - 1\right) \times 1000$$

since the ^{14}C activity of the sample decreases at the same rate as the standard

$$\delta_n^{14}C = \left(\frac{A_{SN}e^{\lambda(1950-x)}}{A_{OxN}} - 1\right) \times 1000$$

and

$$\Delta^{14}C(‰) = \delta^{14}C - 2(\delta^{13}C_S + 25)\left(1 + \frac{\delta^{14}C}{1000}\right)$$

As the activity of a sample in 1950 is the same than that of the standard, then $\Delta^{14}C$ is zero in 1950 AD.

Now, the activity of a sample is expressed as fraction modern (F) that represents the $^{14}C/^{12}C$ ratio in a sample such that:

$$F^{14}C = \frac{A_{SN}}{A_{ON}}$$

In oceanography, the 14C concentration is expressed as $\Delta^{14}C$ or Δ in ‰:

$$\Delta^{14}C = \left(\frac{A_{SN}}{A_{ABS}} - 1\right) \times 1000 \text{(without age correction)}$$

$$\Delta = \left(\frac{A_{SN}e^{\lambda(y-x)}}{A_{ABS}} - 1\right) \times 1000 \text{(with age correction)}$$

with AABS is the absolute age, y the year of measurement, x the year of growth, and $\lambda = \ln(2)/5730$. The notation Δ is often written as $\Delta^{14}C$-age corrected.

Calibration of the ^{14}C Ages

It is now well-established that the ^{14}C ages are not equal to the absolute ages because (i) the ^{14}C ages are calculated with the 5568 year half-life, (ii) the ^{14}C concentration in the atmosphere varies as a function of time due to changes in the production rate by cosmic rays which are modulated by the solar and earth magnetic fields and (iii) of the changes in the carbon cycle. The ^{14}C calibration consists of the precise measurement of the difference between an absolute (calendar) age and a ^{14}C age. An international group of scientists led by Stuiver and Reimer joins efforts to iterate the calibration datasets. The last product is the IntCal13 calibration, which extends over the past 50,000 years (Reimer et al. 2013).

Methods and Results

The calibration procedure consists in measuring the ^{14}C age of a sample while the absolute ages are determined by three methods with the best possible accuracy. The first is based either on the counting of annual tree-ring growth (dendrochronology) or on the counting of the annual laminae (varves) deposited in marine or lake sediments. The second consists in the precise U-Th dating of carbonates (warm water corals or speleothems) by mass spectrometry (Chap. 6). The third consists in synchronizing the variations of climate proxies in marine sediments and in speleothems, the last two being dated by the U-Th method.

In the eighties, the ^{14}C and U-Th dating benefitted from new techniques in mass spectrometry coupled to an accelerator or by thermal ionization (TIMS), respectively—which allowed their precision to be greatly improved and the size of samples to be reduced by a factor of 1000 (from one gram to a few milligrams). Bard had first proposed to extend the calibration from 10,000 to 22,000 cal BP through paired ^{14}C and Th-U dating of corals from the Barbados islands (Fig. 4.3) (Bard et al. 1990; Reimer et al. 2013). The calibration record IntCal13 is based on the ^{14}C dating of the terrestrial and marine samples referenced onto an 'absolute' timescale. It extends up to 13,900 cal BP by means of dendrochronology and up to ∼50,000 cal BP from terrestrial plant species deposited in the varved sediments of Lake Suigetsu. It is completed by U-Th-dated samples and has corrected ^{14}C ages (see below) of the Hulu speleothem and of Atlantic and Pacific corals. Finally, corrected ^{14}C ages of foraminifera in the varved and unvarved sediments from the Cariaco basin linked to the Hulu speleothem by coeval climatic fluctuations are also included. The marine calibration record similarly covers the past 50,000 cal BP and it is based on the dating of corals from the Atlantic and Pacific oceans and of planktonic foraminifera in the Cariaco sediments in the Caribbean Sea (Fig. 4.4). The sea surface ^{14}C is not in equilibrium with that of the atmosphere. The comparison of the atmospheric and marine ^{14}C dating at a same calendar age between 0 and 13,900 cal BP allowed the atmosphere-ocean ^{14}C difference (the marine reservoir age-see below) to be quantified and the marine calibration to be anchored to the atmospheric record by subtracting the reservoir age to the marine ^{14}C ages. In IntCal13, the variability of the reservoir ages during the glacial period were taken into account by augmenting the value by 200 years prior to 13,900 cal BP. Similarly, the ^{14}C in speleothems is not equal to that of the atmosphere, and the difference (dead carbon fraction—see below) is measured between the time interval 0–13,900 cal BP and then considered constant prior to 13,900 cal BP.

As initiated with IntCal04, the mathematical approach that defines the envelope of IntCal13 takes into account both uncertainties on ^{14}C dating and those associated with the absolute ages. In addition, it also takes into account the diversity of records and the representativeness of atmospheric ^{14}C by assigning different statistical weights to them before returning the most likely fit between measured ^{14}C activity and absolute age. The IntCal13 error envelope represents the best fit between the included datasets and it is much smoother than those of the previous calibration records. It does not take into account the scattering of the ^{14}C ages in the different datasets, which do not permit 'real ^{14}C variations' to be discriminated from measurement noise. Those real variations are likely to be attributable to the complex history of each carbon reservoir in either the terrestrial (atmospheric) or marine environments over the past 50,000 cal year due to climate and oceanic circulation fluctuations, and to production changes through the variations of the Earth and solar magnetic field.

Examples of Precise Calibration of ^{14}C Ages

The Dating of the Eruption of Santorini

During the eruption of Santorini in Cyclades in the Aegean Sea, huge amounts of volcanic ash were emitted into the atmosphere and spread eastward and southward, covering much of the Middle East. The ash bed is a useful chronological marker for the whole region, allowing the chronology of human connections to be refined throughout the eastern basin of the Mediterranean Sea during the second millennium BC. Archaeologists, for example, linked the apogee of the Minoan civilization to that of the New Kingdom in Egypt in the sixteenth century BC, on the basis of the elegance of the decorations of objects found in the ash layer in Santorini. The date of this eruption could not be defined to within 60 years, despite hundreds of ^{14}C dating of charcoals found

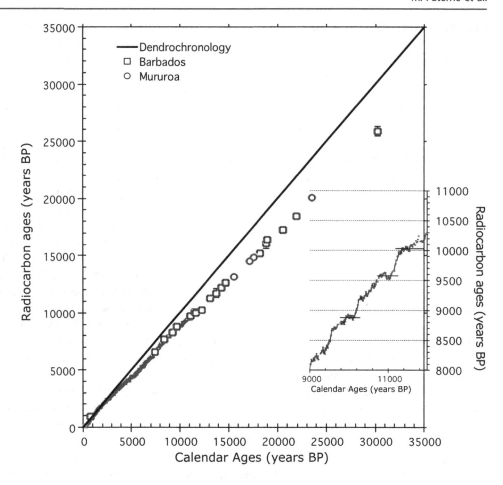

Fig. 4.3 Variations between ^{14}C and calendar ages (Reimer et al. 2013). The calendar ages are based on tree ring counting back to 11,950 cal BP and on the U-Th dating of corals from Barbados and Tahiti (in Reimer et al. 2013). The ^{14}C age plateau are shown in the inset between 9000 cal BP and 12,000 cal BP

in the ash-bed (Friedrich et al. 2006). The discovery of an olive branch with leaves buried in the ash bed enabled a very precise dating of the eruption. Friedrich et al. (2006) dated the annual tree-rings of a piece of branch (Fig. 4.5). The variation (wiggle) of the ^{14}C ages in the tree-ring matched precisely that of the calibration record, the number of calendar years being equal to that of the tree-rings.

The calendar age of the eruption of Santorini was thus estimated to be between 1627 and 1600 BC, with a confidence level of 95%. The apogee of the Minoan civilization would then be earlier than that of the New Kingdom, and would be contemporary to the time when foreign kings—Hyksos—occupied the country (Friedrich et al. 2006).

The Bipolar Seesaw or the North-South Heat Transfer

Rapid fluctuations of climate, referred to as Heinrich events, Younger Dryas, and Dansgard-Oeschger events, punctuated the last glacial period and the deglaciation.

They were attributed to variations in the thermohaline oceanic circulation (see Chap. 21). The mechanisms of these rapid fluctuations were clarified by the analysis of the ^{14}C variations in the varved deposits in the Cariaco Basin (Hughen et al. 1998; Reimer et al. 2013). The atmospheric $^{14}CO_2$ concentration is very sensitive to the rate of formation of deep water, because the CO_2 in the atmosphere is transferred to the deeper layers of the ocean via the oceanic circulation. This rate also regulates the heat transfer to the northern high latitudes. During the cold event of the Younger Dryas, which lasted approximately 1200 calendar years, the atmospheric $\Delta^{14}C$ increased (younger ^{14}C ages) during the first 200 years and then decreased during the following 1000 years (Fig. 4.6a). The increase of $\Delta^{14}C$ is in agreement with a reduction in the deepwater formation in the northern North Atlantic, contributing to a cold climate, but how may the further $\Delta^{14}C$ decrease be explained when the climate is still cold in the northern latitudes?

Glaciologists have shown that the abrupt cooling observed in the northern hemisphere was accompanied by a warming in Antarctica, and vice versa (Blunier et al. 1998). Comparing these two results, Broecker (1998) proposed a seesaw pattern of the north-south heat transfer via the ocean or the "thermal bipolar seesaw" (Fig. 4.6b): a reduction in deep water formation in the North Atlantic Ocean led to the installation of a cold climate in the northern hemisphere and an increase in the atmospheric ^{14}C. Deep waters further form in the Southern Ocean contributing to the atmospheric $\Delta^{14}C$ decrease.

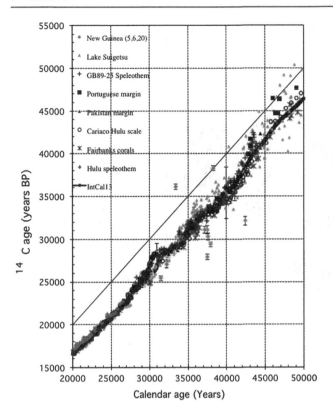

Fig. 4.4 Variations between ^{14}C and calendar ages between 20,000 and 50,000 cal BP from the calibration record IntCal 13 (Reimer et al. 2013). The straight line represents the 1:1 relation

Apparent Ages

Oceanic Environments: The Reservoir Ages

Carbon exists in several forms: carbon monoxide CO and carbon dioxide CO_2, methane CH_4, many complex organic molecules (cellulose, etc.). The dominant forms are the bicarbonate and carbonate ions (CO_3^{2-}, HCO_3^-, H_2CO_3) dissolved in water, and calcium, barium and magnesium carbonates ($CaCO_3$, $BaCO_3$, $MgCO_3$). Most of the Earth's living species directly assimilate the $^{14}CO_2$ from the atmosphere. In the ocean, the organisms incorporate ^{14}C, which is not in equilibrium with the atmospheric $^{14}CO_2$. The benthic foraminifera are calcareous organisms living on the surface of marine sediments. When collected alive at a water-depth of 2000 m in the North Atlantic and Pacific oceans, they will have ^{14}C ages of about 600 years and 1400 years, respectively (before the nuclear bomb tests). These ^{14}C ages are apparent ages that include the age of the organism and that of the water body in which the organisms formed their tests from the dissolved ions according to:

$$CO_2 + H_2O \leftrightarrow H_2CO_3 \leftrightarrow H^+ + HCO_3^-$$
$$HCO_3^- \leftrightarrow H^+ + CO_3^{2-}$$
$$(Ca^{2+} + 2HCO_3^- \rightarrow CaCO_3 + H_2O + CO_2).$$

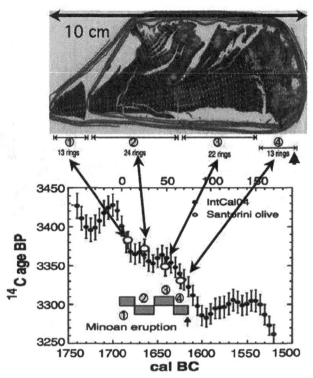

Fig. 4.5 ^{14}C and calendar ages of the Santorini eruption (modified from Friedrich et al. 2006). Four groups of rings counted (grey rectangles) from an olive branch of an approximately ten centimeters length were dated. The total number of rings equals the number of calendar years. The ^{14}C ages match the ^{14}C variations of the calibration record (Reimer et al. 2013). The calendar age of the eruption range between 1621 and 1605 BC at the 1σ confidence level (68.2%) and between 1627 and 1600 BC at the 2σ confidence level (95.4%)

The ^{14}C aging of deep waters is the result of the conveyor (thermohaline) circulation. In the modern ocean, wind-driven warm saline surface waters from the North Atlantic low latitudes flow poleward where they cool and sink to form the North Atlantic Deep Water (NADW). NADW flows at depth to the Indian and Pacific Oceans where it mixes and upwells then flowing back to the North Atlantic. All the deep waters flow to the Southern Ocean where they upwell and downwell to form a return flow to the north through the intermediate waters and the Antarctic bottom waters. Once isolated from the atmosphere, the dissolved $^{14}CO_2$ in the intermediate and deep waters starts to decrease. Although in contact with the atmosphere, the sea surface ^{14}C is not equal to the atmospheric ^{14}C, because of the mixing between the sea surface and the underlying sub-surface waters. The difference between the apparent ^{14}C ages of living organisms in the ocean and the atmospheric ^{14}C ages for the same absolute age is termed the reservoir age R. The sea surface values of R as well as those of the intermediate or deep waters differ in each one of the ocean basins and regionally within the basins.

Fig. 4.6 a Variations in the atmospheric ^{14}C content ($\Delta^{14}C$ in ‰) (triangle) and grey scale (line) in marine varved sediments from the Cariaco Basin as a function of calendar ages by varve counting (Hughen et al. 1998; Reimer et al. 2013). **b** Variations of $\delta^{18}O$ trapped in CO_2 air bubbles in the Greenland ice core GISP2 sites in Greenland (line) and in the Byrd ice core in Antarctica (diamond) as a function of synchronized time from analyses of CH_4 concentrations in the two sites (Blunier et al. 1998)

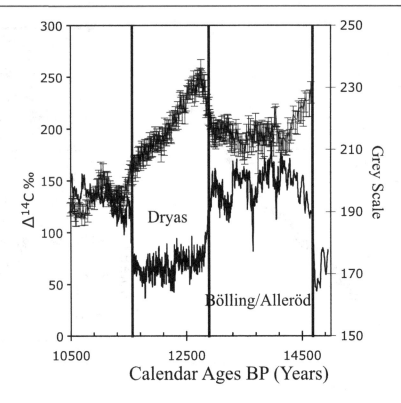

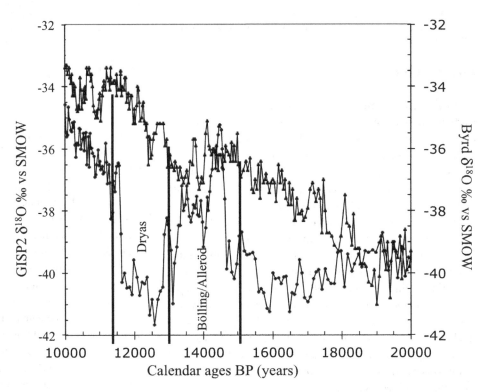

The modern values of R are quantified by the ^{14}C dating of carbonates (mollusk shells, corals) of known ages from the historic collections of museums (Bard 1988; Siani et al. 2000; Tisnérat-Laborde et al. 2010) and, for older ages than those of these collections, by subtracting the ^{14}C age of a marine organism from that of the atmospheric calibration record at the same absolute age (Reimer et al. 2013). The mean value of R is about 300–400 years in the subtropical sea surface, and it may be some 1100 years around the Antarctic continent. Values of approximately 700 years were measured in oceanic regions close to upwelling of intermediate and deep waters in the North Indian and Pacific Oceans. ΔR is the local departure from the global mean value of R. The latter is calculated from the modeled global sea surface ^{14}C age, using a global box-diffusion carbon model, which accounts for the ^{14}C production and carbon cycle (Reimer et al. 2013).

In the past, the oceanic circulation changed as the reservoir ages did. Changes of the differences between the marine and atmospheric ^{14}C ages may be estimated by dating the charcoals and foraminifera from widespread volcanic ash layers in both marine sediments and on land (Bard et al. 1994; Siani et al. 2001; Austin et al. 2011; Thornalley et al. 2011; Siani et al. 2013; Sikes and Guilderson 2016), as well as by dating paired charcoals and mollusk shells found in the same deposits (Bondevik et al. 2006; Ascough et al. 2009). These studies demonstrated that sea surface R changed during the first step of the deglaciation from the modern value of 400 to 2000 years in the North Atlantic (Bard et al. 1994; Bondevik et al. 2006; Austin et al. 2011; Thornalley et al. 2011) and to 800 years in the Mediterranean Sea (Siani et al. 2013). In the South Pacific, the subtropical R augmented from ~ 300 to ~ 700 years (Sikes and Guilderson 2016), while those of the sub Antarctic surface waters increased from ~ 800 to 1400 years and to 3200 years (Siani et al. 2013; Sikes and Guilderson 2016).

Past increases in the sea ice extent at high latitudes during the glacial periods prevents the atmosphere-ocean $^{14}CO_2$ exchanges contributing to the increase of the sea surface R. In the IntCal13 calibration record, the modern value of R was subtracted to the ^{14}C ages of the marine samples to be compared to the atmospheric ^{14}C ages. To account for changes in the oceanic circulation, the modern value of R was augmented by 200 years from $\sim 14,000$ cal BP and then considered as constant down to 50,000 cal BP (Reimer et al. 2013).

Continental Environments: The Hard Water and Dead Carbon Effects

Modern lake vegetation and calcium carbonates in lakes in calcareous regions exhibit older ^{14}C ages than those of the atmosphere. The dissolved inorganic carbon $^{14}CO_2$ in lakes, used during photosynthetic processes and during the precipitation of calcium carbonates originates from the dissolution of ^{14}C-free carbonates of geological age (dead carbon) and from the mineralization of old organic matter enclosed in lake sediments. Impact of the later on the apparent ^{14}C ages of lacustrine plants is much higher in lake surrounded by peats and in artificial lakes implemented by soils flooding. The resulting aging of ^{14}C ages is called '*hard water*' effect. In addition, water stratification in lakes or the presence of an ice cover in high altitude and polar lakes tend to prevent the atmospheric $^{14}CO_2$ input to the lake waters, that tends to increase the ^{14}C age in lake waters relatively to that of the atmosphere.

The speleothems are composed of calcium carbonates, formed by the dissolution of ^{14}C-free geological carbonates by slightly acidified waters by CO_2 from the atmosphere and from the degradation of the soil organic matter. In Fig. 4.7 the procedure to estimate the fraction of dead carbon (DCF) in a Bahamas speleothem, located in the western North Atlantic, is shown (Beck et al. 2001). To an U-Th age corresponds an atmospheric ^{14}C age in the calibration record. The DCF is calculated by subtracting the ^{14}C age of the speleothem to that of the contemporaneous atmosphere. In this example, the ^{14}C ages of the speleothem are older than those of the atmosphere at an average of about 1450 years between 11,000 cal. BP and 15,000 cal. BP, which corresponds to a DCF of 16%. This aging is not constant as a function of time and varies between 1000 years and 2000 years. These variations appear to be closely correlated to the climatic fluctuations recorded either in the Greenland ice-core GISP2 or in the marine varved sediments of the Cariaco Basin. About 30% of the variability of the DCF in this speleothem may be explained by such fluctuations by the way of changes of the local temperature and rainfall patterns.

The ^{14}C Exchanges in the Carbon Reservoirs

Radiocarbon is commonly used to test numerical simulations of the oceanic circulation in Ocean General Circulation Models (O-GCM) (Toggweiler et al. 1989; Key et al. 2004). The systematic measurements of the ^{14}C content of the dissolved inorganic carbon (DIC) in the different world ocean basins started with the international oceanographic campaigns GEOSECS (1972-1978), followed by WOCE (World Ocean Experiment, 1990–2002), and others (Broecker et al. 1995; Key et al. 1996). In addition to ^{14}C analyses, the physicochemical properties of the worldwide basin water masses were measured along depth profiles. Both the natural and anthropogenic components of the ^{14}C concentration in the oceans offer the opportunity to validate the general ocean circulation simulated by numerical models (Toggweiler et al. 1989; Key et al. 2004). The natural component tests the circulation of deep waters, while the anthropogenic component, resulting from the thermonuclear tests in the 1960s, allows the analysis of physical processes with time constants of a few decades, such as the formation of deep and intermediate waters and the ventilation of the thermocline (transition zone between the cold intermediate and deep waters and the warm surface waters). Because of radioactive decay, the ^{14}C content of an ocean water mass decreases during the oceanic transport, once the water mass is isolated from exchanges with the atmosphere.

Fig. 4.7 Variations of the ^{14}C ages as a function of U-Th calendar ages from a Bahamas speleothem (Beck et al. 2001, Reimer et al. 2013). **a** The black squares represent the ^{14}C ages from the speleothem and the white squares to the ^{14}C ages of the speleothem adjusted onto the atmospheric calibration record (grey squares). The mean dead carbon age is 1450 years. **b** Comparison of the variations of dead carbon (bold line) (Paterne, unpublished) and of GISP 2 $\delta^{18}O$ (Blunier et al. 1998)

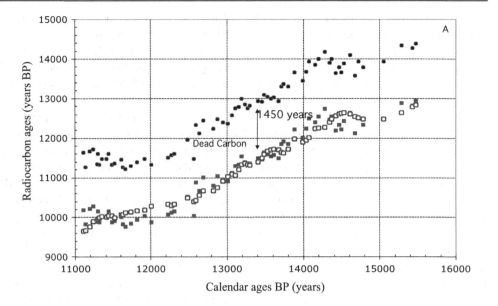

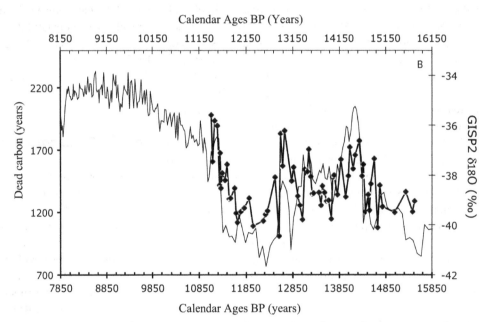

The ^{14}C provides independent and additional constraints for ocean circulation models validation beside conventional tracers of the water masses (temperature, salinity), by allowing estimates of the time constants of the ocean ventilation.

Examples of Simulation of Modern Oceanic Circulation

Figure 4.8 represents the ^{14}C distribution simulated with the NEMO global ocean circulation model. The simulation is evaluated using the GEOSECS data (Broecker et al. 1995; Key et al. 1996). The model reproduces a realistic structure of the ventilation of the deep ocean producing concentrations similar to observations. In particular, the ventilation signature of the deep ocean associated with the Antarctic deepwater formation coming from the Southern Ocean (Antarctic Bottom Water AABW) is reproduced, characterized by high values of $\Delta^{14}C$ at the ocean floor.

The eastern Pacific is characterized by a thick water mass between 2000 m and 3500 m of almost homogeneous ^{14}C-depleted values at around −200 and −240‰. These deep waters correspond to the oldest deep waters of the world ocean with ^{14}C ages of 1790 years and 2200 years respectively.

The temporal changes of the sea surface ^{14}C may be measured from the annual growth bands of recent corals over

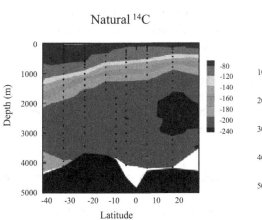
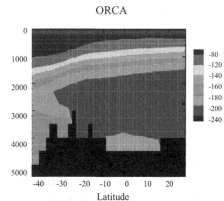

Fig. 4.8 Observed and simulated ^{14}C concentration (in ‰) in the eastern Pacific Ocean. The negative value indicates an aging of the water bodies compared with the age of the atmosphere. A decrease of 10‰ is equivalent to an aging of about 80 years. A mass of nearly homogeneous water between −200 and −240‰ occupies the eastern Pacific Ocean from depths between 2000 and 3500 m. At about 40 °S, the presence of a more recent water mass, between −180 and −200‰ at 4000 m depth, formed on the edge of the Antarctic continent should be noted. It corresponds to the Antarctic Bottom Water (AABW)

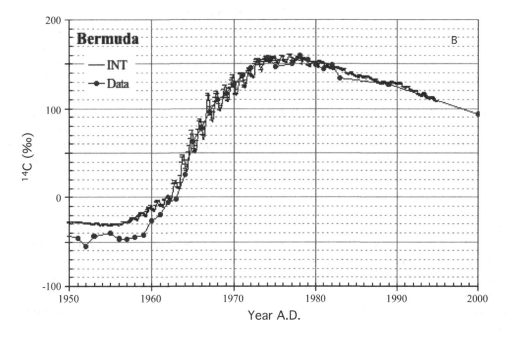

Fig. 4.9 Variations in Δ^{14}C in ‰ (circles) in a banded coral collected near Bermuda (Druffel 1989), in the subtropical surface waters of the North Atlantic between 1950 and 1990. Δ^{14}C increased rapidly between 1960 AD and 1965 AD due to the ^{14}C input into the atmosphere during the aerial nuclear bomb tests. Results of the interannual OPA simulation (INT) (8.1) are represented. This simulation highlights the importance of the winds in the atmosphere-ocean exchanges in the subtropical North Atlantic (Tisnérat, Dutay, personal communication)

several decades. These data provide additional validation of general circulation models, which, in turn, allow a better identification of the causes and mechanisms of the temporal variability of ^{14}C. Figure 4.9 represents the ORCA/IPSL model simulation of ^{14}C data measured in one coral from Bermuda shore (Druffel 1989).

Both the ^{14}C coral data and the modeled ^{14}C at the sea surface show coeval changes, indicating that the model correctly reproduced the transfer of the tracer at the air-sea interface, as well as its penetration below the sea surface into the ocean.

Oceanic Paleocirculation

In the meantime, from the systematic ^{14}C measurements in the world ocean layers, the idea quickly emerged to recover past changes in ocean circulation from ^{14}C analyses of marine biocarbonates. The ^{14}C concentration of large volume samples such as corals and mollusks shells were first measured by radioactive decay measurements (Stuiver et al. 1986; Druffel 1989). This could be achieved for small volume samples with the new AMS ^{14}C technique. The planktonic and poorly abundant benthic foraminifera contemporaneously deposited in deep-sea sediment cores from different water-depths could henceforth be easily dated allowing the paleo-ventilation of the paleocean layers to be recovered to the limit of the ^{14}C dating (theoretically 10 times the half-life). Meantime the changes of the Earth's orbital parameters, and thus changes of the insolation, the atmospheric and oceanic circulation changes modulate the Earth's climate. Considerable efforts, compiled in (Zhao et al. 2018), have been made to measure the ^{14}C differences in the deep to surface ocean water masses in the past from benthic-planktonic foraminifera. Benthic corals have provided additional estimates of the deep Δ^{14}C as a function of calendar ages as they can be dated by both the U-Th and ^{14}C methods (Adkins et al. 1998; Goldstein et al. 2001; Robinson et al. 2005; Burke and Robinson 2012; Chen et al. 2015). Opposite results in the estimates of the paleo-ventilation of the ocean during the deglaciation emerged when using either the ^{14}C dating of paired benthic-planktonic foraminifera or paired ^{14}C/U-Th dating of benthic corals (Adkins et al. 1998; Goldstein et al. 2001; Robinson et al. 2005; Burke and Robinson 2012; Chen et al. 2015) or the paired atmospheric-marine ^{14}C dating from ash-layers (Sikes et al. 2000; Ikehara et al. 2011; Siani et al. 2013; Ezat et al. 2017). Values of Δ^{14}C are obtained from the absolute (calendar) age and the ^{14}C age (see Box 1). Thus the conflicting results are very likely related to the estimated Δ^{14}C from paired benthic-planktonic foraminifera due to (i) an incorrect estimate of the sea surface reservoir age subtracted to marine ^{14}C ages (in order to be referenced to calendar ages), (ii) the use of different atmospheric ^{14}C calibration records, notably the much smoother variations of ^{14}C in IntCal13 than those estimated in previous records (Reimer et al. 2013), (iii) the use of ^{14}C values of foraminifera picked in deep-sea sediment cores with a low sedimentation rate (see below: the bioturbation effects), and (iv) few planktonic foraminifera spend their entire life at the sea surface, and their ^{14}C content represents that of the upper ten to hundreds of meters of the water column in which they lived. More constraints on the changes of the sea surface reservoir ages allowing robust calculation of Δ^{14}C from benthic foraminifera and more ^{14}C analyses from deep sea corals will be very useful for model simulations of the oceanic circulation during the last glacial maximum and the deglaciation using three-dimensional models (Tagliabue et al. 2009). In the study by Zhao et al., (2018), the box model resolution, which should at least take into account the changing geometry of water masses in the last 25 kyr (Michel et al. 1995), is a limitation to estimate regional changes in paleoventilation.

Mineralization of Organic Matter in Soil

The contribution of soils and their role as sinks and sources in the global carbon cycle remain misunderstood until now. The stock of soil organic matter is defined as a balance between the input of organic matter through vegetation and the loss through microbial decomposition. The balance can be disrupted by changes in agricultural practices and climate variations. For instance, a temperature increase may clearly increase the activity of soil microorganisms and the subsequent soil organic matter mineralization. No consensus has however been reached on the relative importance of the various climatic factors that affect soil organic matter dynamics, such as temperature, aridity, land use. To better evaluate the effect of these disturbances on the global carbon cycle, it is essential not only to characterize soil carbon stocks but also soil carbon dynamics. To do so, ^{14}C is a powerful tool as it can be considered as a clock that registers the carbon residence time in the soil organic mixture (Scharpenseel and Shiffmann 1971; Balesdent and Guillet 1982).

Conceptual views of soil organic carbon dynamics have greatly evolved with time. Carbon sequestration was considered to be related to the chemical structure of the components (lignin having a longer mean residence time than sugar), to the accessibility of organic matter in aggregates (the higher the pore, the more labile the organic matter), to the affinity between organic matter and mineral surface (the stronger the bond, the more refractory the components) (Six et al. 2002). The conceptual view is still developing (Kleber et al. 2007) and ^{14}C brings powerful elements. ^{14}C measurement is thus done on bulk organic matter, on density fractions, on granulometric fractions, on molecular fractions and on molecules according to the process or to the turnover to be characterized.

Isotopic methods, such as dating by carbon-14, natural (percentage of plants in C3 and in C4) and artificial carbon-13 labeling are very powerful tools in so far as they make it possible to estimate the residence time of natural organic matter in the soil. Instrumented monitoring of the ^{13}C content in experimental soil plots can be used to estimate the mean residence time from a few years to some decades. Many studies have been conducted since the beginning of the ^{14}C method (Balesdent and Guillet 1982; Gaudinsky et al. 2000). At first, they were carried out by radioactive counting measurements (a few grams of carbon), and then by mass spectrometry coupled to an accelerator (AMS), which did not allow the targeted molecular level to be reached, because a few milligrams of carbon were still required. Nevertheless, their scope was large, and they identified kinetic pools among the various elements of the soil's organic matter, in other words, compartments that can be defined by a specific carbon residence time. It has been shown that the residence time of different carbon compartments in soil can range from one to several decades or even to a few thousand years for the stable fraction.

Using the contamination of plant species by ^{14}C nuclear explosions in 1960 AD, Gaudinsky et al. (2000) modeled over time the ^{14}C activity of the different compartments of the soil receiving a constant organic input every year, according to their residence time (Fig. 4.10). The pool with a residence time of 10 years reached a maximum activity of ^{14}C in 1972, eight years after that recorded in the atmosphere. The one with a 50-year residence time recorded a maximum activity in 1985. Using the results of this modeling to interpret the measurements conducted on the separate fractions from the same soil, the residence time of the different fractions can be estimated. These, along with the relative weights of the different fractions, can then be used to characterize the dynamics of carbon of the soil and to assess carbon stocks over time.

It is important to keep in mind that whatever the fraction and since soil is the result of balance between input and output, the dated sample is always a mixture of components of different ^{14}C ages. The resulting ^{14}C is only a mean age of all components of different ^{14}C age. Likewise, it is important to remind that soil is alive and any molecule is recycled. It might reach the soil as vegetal sugar and as a source of energy for microbial life be bio-assimilated and metabolized into a microbial lipid. It will however keep the same ^{14}C signature of the original vegetal sugar. That's why a microorganism molecule can give its old ^{14}C age even if the microorganism is still alive. We definitively characterize the mean age of carbon and not the mean age of the molecule we analyzed.

Treatment of Samples and Calculations of ^{14}C Ages

Radiocarbon dating is based on either decay-counting from gas (CO_2, C_2H_2, C_6H_6) in proportional or liquid scintillation counters or atom-counting of ^{14}C, ^{13}C and ^{12}C from graphite targets by mass spectrometry (Accelerator mass spectrometry: AMS). The greatest advantage of the AMS technique is the very small sample size required, nowadays as small as a few tens of micrograms. At the end of the seventies, the

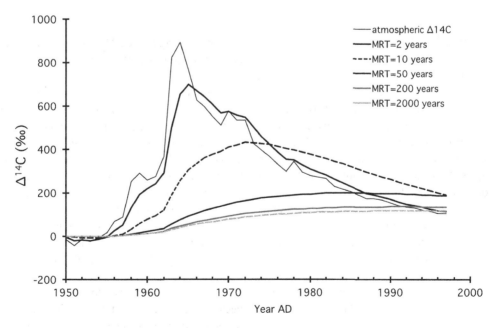

Fig. 4.10 ^{14}C activity in a soil with a constant annual supply of carbon over time, expressed as $\Delta^{14}C$ in ‰ (Gaudinsky et al. 2000). The thin black line represents the atmospheric ^{14}C activity in the northern hemisphere; the bold line represents that of the soil fractions with different mean residence time (MRT). We note that the peak of bomb detonations is clearly reflected in the more recent fractions (<10 years), while the effect is diluted in the fractions with MRT of a few decades. It is virtually nonexistent in older fractions

extracted ions from the graphite by a Cesium gun, were accelerated to a voltage difference of 2 MeV while the acceleration voltage is nowadays reduced at 0.5 MeV. The reader is referred to various publications and web descriptions on the equipment and measurement methods (Taylor et al. 1992).

Physical-Chemical Treatment

The samples are first cleaned by physicochemical treatments to eliminate contaminants. Vegetation samples collected in soils may include old carbonates and living rootlets, and the ^{14}C ages would thus increase or decrease respectively. After a visual examination, the organic remains (seeds, coal, plants) undergo the classic Acid-Alkali-Acid chemical treatment (noted AAA). The aim is to eliminate contaminants from bacterial decomposition of organic matter since the burial of the sample. Finally, CO_2 is obtained by burning the sample into an evacuated quartz sealed tube filled with copper oxide at about 800 °C. Carbonate samples (foraminifera, pteropods, corals, speleothems, earthworm granules) are examined under the microscope to check their homogeneity. Corals and mollusk shells are pre-cleaned by sand blasting to eliminate secondary calcite precipitation, which may lower the ^{14}C ages. All the carbonates are leached in a weak acid to remove surface contaminants before hydrolysis in a vacuum device. For AMS ^{14}C measurement, the obtained CO_2 is converted into graphite by metal catalysis (Tisnérat-Laborde et al. 2001; Hatté et al. 2003).

Determination of a ^{14}C Age

The ^{14}C age is obtained by comparing the activity of a sample to that of a standard reference, representative of the atmospheric ^{14}C content in 1950 (NBS-I; NBS-II; sucrose) regardless of the dating techniques. Chemicals as well as vacuum lines and counting devices are contaminated by modern $^{14}CO_2$ which has a $^{14}C/^{12}C$ ratio of 1.2×10^{-12} in 1950 AD. The effect of such a contamination on the ^{14}C ages of a 1 mg carbon sample may be estimated using a simple mixing equation. A ^{14}C age of 40,000 years ($^{14}C/C = 6.88 \times 10^{-15}$) would be lowered by 4300 years to 7200 years, due to a modern contamination of 5 µg to 10 µg, respectively. This emphasizes the importance of a careful cleaning of the sample. The same amounts of ^{14}C-free contaminants will have little effect: about 40 years and 80 years respectively regardless of the ^{14}C age of the sample.

The first ^{14}C dating of cave paintings revealed an age of about 31,000 years BP that deeply modified our understanding of how human art evolved (Valladas et al. 1992; Cuzange et al. 2007). Clearly, such old ages cannot be due to a contamination by the carbonates precipitate on cave walls, as was frequently hypothesized. To increase the ^{14}C age of a painting of 1 mg carbon and 15,000 years old, for example, to an age of about 33,000 years would require some 900 µg of ^{14}C-free contaminants, which is almost the entire sample and this would have been seen through visual examination.

The *internal* contamination of samples, due to the cleaning treatments, is carefully assessed by measuring the ^{14}C activity of an old (^{14}C-free) sample (blank), cleaned in a similar manner to samples of unknown ages. During a ^{14}C-AMS run of measurements, standard references, blank samples and the samples to date are inserted. The "blank" or "background" activity is subtracted from the activity of the sample of unknown age. The smaller the sample, the greater the effect of the *internal* contamination on the ^{14}C age. This contamination varies depending on the nature of the samples. Even if the blank activity is very low, its variability determines the accuracy of the ^{14}C age and the ^{14}C age limit. Consider two samples of carbonate and charcoal, with a $^{14}C/C$ activity of 6.88×10^{-15} (an age of 40,000 years BP) and an absolute error of 5%. The blank variability accounts for 25% and the subtracted blank value for the carbonate and charcoal is 5×10^{-16} and 19×10^{-16}, respectively. If the blank variability increases by a factor of 2, the ^{14}C age uncertainties will increase by 100 years for the carbonate and by 700 years for the charcoal, respectively. The precision of a ^{14}C age older than 30,000 years BP may be appreciably affected by a few thousand years when the blank activity is not well-estimated.

Some Examples of Post-depositional Disturbances of the ^{14}C Ages

The validity of the ^{14}C dating of climatic, archaeological or geological events depends, in many cases, of mechanical or biological disturbances of sedimentary deposits that may occur after the death and the burial of organisms. In archaeological sites, stratigraphic inversions of ^{14}C ages may be due to soils disturbances linked either to successive occupations by human and animals, or to wind and water effects, or both. The increasingly small size of ^{14}C-dated samples, which tend to migrate through the sedimentary deposits, favors such ^{14}C age inversions.

Archaeological sites contain numerous ^{14}C datable remains, such as mollusk shells, charcoals and bones. Researchers give priority to the ^{14}C dating of charcoals, because the ^{14}C dating of marine shells requires a correction of the reservoir ages to be compared with those of bone and vegetation samples. In a seaside Peruvian site, paired marine mollusk shells and charcoals were associated in several sediment layers to allow the sea surface reservoirs ages

Fig. 4.11 Variations in the $\delta^{18}O$ (bold line and diamonds) and in the number of foraminifera per cm^2 and per gram of sediment (thin line): **a** for the subpolar planktonic foraminifera *Globigerina bulloides*; **b** for the polar planktonic foraminifera *Neogloboquadrina pachyderma sinistral* as a function of depth in a North Atlantic deep-sea sediment core. The ^{14}C ages at 1 sigma (in BP years) at the different depths (arrows) are shown (modified from Bard et al. 1987)

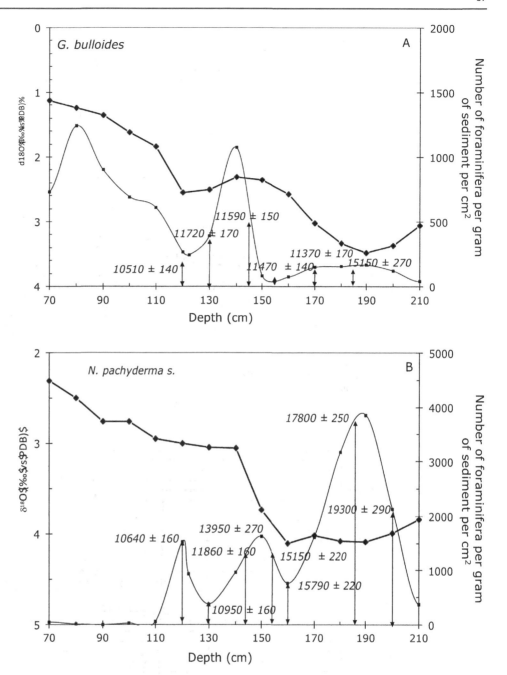

(R) during the Holocene (Kennett et al. 2002) to be quantified. While the modern value of R varies from 540 years to 970 years depending of the intensity of the Peruvian upwelling (the rise to the surface of ^{14}C-depleted intermediate and deep waters), it was slightly lower at 430 years between 5000 and 5300 cal BP and much lower at only 30 years between 5600 and 5900 cal BP. The much lower value of R is very unlikely, as charcoals and associated mollusk shells would have the same ^{14}C ages. This may be due to two phenomena: the charcoals originate from the burning of old trees at the time of burial (living over some hundred years) or the small charcoals migrated downwards in the archaeological layers. Only statistical analyses of the ^{14}C ages of charcoals and shells would enable a conclusion to be reached. This Peruvian study clearly emphasizes that charcoals are not always the best candidates in radiocarbon dating.

A second example is related to the ^{14}C dating of two species of planktonic foraminifera during the deglaciation in a North Atlantic deep-sea sediment core (Bard et al. 1987). The species are characteristic of the cold polar waters (*Neogloboquadrina pachyderma sinistral*) and of the warmer subpolar waters (*Globigerina bulloides*). It can be seen from Fig. 4.11 that the ^{14}C ages of *G. bulloides* are constant

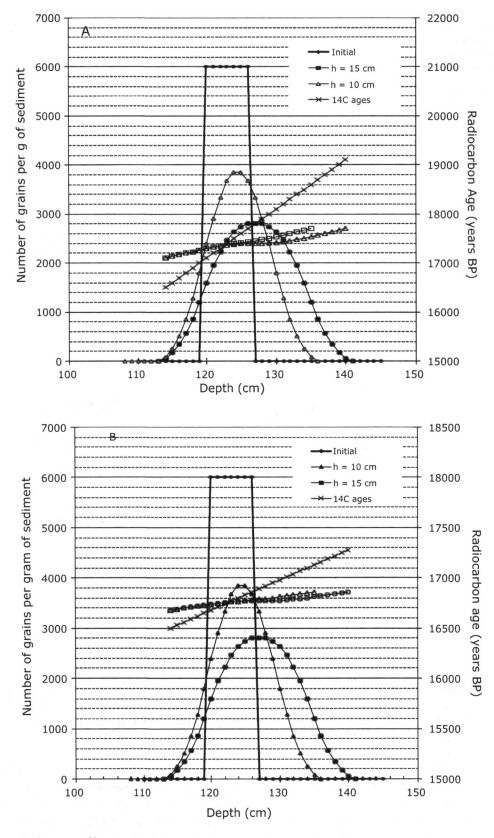

Fig. 4.12 Bioturbation effects on ^{14}C age of an event, observed in a deep sea sediment core with sedimentation rates of 10 cm/1000 years (**a**) and 30 cm/1000 years (**b**) using bioturbation depth (h) of 10 cm (black triangles) and 15 cm (black squares). The initial ^{14}C ages are plotted by crosses, and the resulting deviations from a bioturbation depth of 10 cm (white triangles) and 15 cm (white squares). Note the change of the ^{14}C age scales in **a** and **b**

between 170 cm and 140 cm, while those of *N. pachyderma* decreased. Moreover, the latter are much older than the subpolar species. The abundance of these species varies along core depth and peaks of abundance spreads over several tens of cm. The ^{14}C age distribution of the two species is due to bioturbation, a mixing process of sediment grains by benthic organisms, which move up and down through the sedimentary deposits. Applying a simple mixing filter of a constant thickness (h) over time, we can observe that the estimated abundance is lower than at the time of deposition. The abundance peak is no longer at its initial position in the sediment, and its position depends of the mixing depth (h) (Fig. 4.12). In the examples, the bioturbation effects on ^{14}C ages led to lower the ^{14}C age with respect to the 'real' ^{14}C age at the onset of the event. The effect is 900 and 1400 years depending on the h value with a sedimentation rate of 10 cm per 1000 years. A much higher sedimentation rate of 30 cm per 1000 years tends to lower the bioturbation effects as the ^{14}C ages decrease by 270 years and 420 years, with h equal to 10 cm and 15 cm respectively. The ^{14}C age of the end of the event increased by 600 years using a sedimentation rate of 10 cm per 1000 years. The higher the sedimentation rates, the lower the impact of bioturbation on the ^{14}C ages.

Accurate marine ^{14}C ages are obtained from the ^{14}C dating of foraminifera picked at the depth of maximum abundance in deep-sea cores with a high sedimentation rate. The example on Fig. 4.11 also emphasizes that the selected species must be in adequacy with the climatic period to be dated. The variability in the compiled deep and surface Δ^{14}C values in Zhao et al. (2018) may be partly explained by bioturbation and inadequately dated species.

References

Adkins, J. F., Cheng, H., Boyle, E. A., Druffel, E. R., & Edwards, R. L. (1998). Deep-sea coral evidence for rapid change in ventilation of the deep North Atlantic 15,400 years ago. *Science, 280,* 725–728.
Arnold, J. R., & Libby, W. F. (1949). Age determinations by radiocarbon content: Checks with samples of known Age. *Science, 110,* 678–680.
Ascough, P. L., Cook, G. T., & Dugmore, A. J. (2009). North Atlantic marine ^{14}C reservoir effects: Implications for late-Holocene chronological studies Quaternary. *Geochronology, 4,* 171–180.
Austin, W. E. N., Telford, R. J., Ninnemann, U. S., Brown, L., Wilson, L. J., Small, D. P., & Bryant, C. L. (2011). North Atlantic reservoir ages linked to high Younger Dryas atmospheric radiocarbon concentrations. *Global and Planetary Change, 79,* 226–233.
Balesdent, J., & Guillet, B. (1982). Les datations par le ^{14}C des matières organiques des sols. *Science du sol, 2,* 93–112.
Bard, E. (1988). Correction of accelerator mass spectrometry ^{14}C ages measured in planktonic foraminifera: Paleoceanographic implications. *Paleoceanography, 3,* 635–645.
Bard, E., Arnold, M., Duprat, J., Moyes, J., & Duplessy, J. C. (1987). Reconstruction of the last deglaciation: Deconvolved records of δ^{18}O profiles, micropaleontological variations and accelerator mass spectrometric ^{14}C dating. *Climate Dynamics, 1,* 101–112.
Bard, E., Arnold, M., Mangerud, J., Paterne, M., Labeyrie, L., Duprat, J., et al. (1994). The North Atlantic atmosphere-sea surface ^{14}C gradient during the Younger Dryas climatic event. *Earth and Planetary Science Letters, 126,* 275–287.
Bard, E., Hamelin, B., Fairbanks, R. G., & Zindler, A. (1990). Calibration of the ^{14}C timescale over the past 30,000 years using mass spectrometric U-Th ages from Barbados corals. *Nature, 345,* 405–410.
Beck, J. W., Richards, D. A., Edwards, R. L., Bernard, W., Silverman, B. W., Smart, P. L., et al. (2001). Extremely large variations of atmospheric C-14 concentration during the last glacial period. *Science, 292,* 2453–2458.
Blunier, T., Chappellaz, J., Schwander, J., Dällenbach, A., Stauffer, B., Stocker, T., et al. (1998). Asynchrony of Antarctic and Greenland climate change during the last glacial period. *Nature, 394,* 739–743.
Bondevik, S., Mangerud, J., Birks, H. H., Gulliksen, S., & Reimer, P. (2006). Changes in North Atlantic radiocarbon reservoir ages during the Allerød and Younger Dryas. *Science, 312,* 1514–1517.
Broecker, W. B. F. (1998). Paleocean Circulation During the Last Deglaciation: A Bipolar Seesaw. *Paleoceanography, 13,* 119–121.
Broecker, W. S., & Olson, E. A. (1959). Lamont radiocarbon measurements VI. *American Journal of Science. Radiocarbon Supplement, 1,* 111–132.
Broecker, W. S., Sutherland, S., Smethie, W., Peng, T. S., & Ostlund, G. (1995). Oceanic radiocarbon: Separation of the natural and bomb component. *Global Biogeochemical Cycles, 9,* 263–288.
Burke, A., & Robinson, L. F. (2012). The Southern Ocean's role in carbon exchange during the last deglaciation. *Science, 335,* 557–561.
Chen, T., Robinson, L. F., Burke, A., Southon, J., Spooner, P., Morris, P. J., et al. (2015). Synchronous centennial abrupt events in the ocean and atmosphere during the last deglaciation. *Science, 349* (6255), 1537–1541.
Chiu, T. C., Fairbanks, R. G., Cao, L., & Mortlock, R. A. (2007). Analysis of the atmospheric C-14 record spanning the past 50 000 years derived from high-precision Th-230/U-234/U-238, Pa-231/U-235 and C-14 dates on fossil corals. *Quaternary Science Review, 26,* 18–36.
Currie, L. A. (2004). The remarkable metrological history of radiocarbon dating. *Journal of Research of the National Institute of Standards and Technology, 109,* 185–217.
Cuzange, M. T., Delque-Kolic, E., Goslar, T., Grootes, P. M., Higham, T., Kaltnecker, E., et al. (2007). Radiocarbon intercomparison program for the chauvet cave. *Radiocarbon, 49,* 339–347.
Damon, P. E., Lerman, J. C., & Long, A. (1978). Temporal fluctuations of atmospheric ^{14}C: Causal factors and implications. *The Annual Review of Earth and Planetary Sciences, 6,* 457–494.
Damon, P. E., & Peristykh, A. N. (2000). Radiocarbon calibration and application to geophysics, solar physics, and astrophysics. *Radiocarbon, 42,* 137–150.
Druffel, E. R. M. (1989). Decade time scale variability of ventilation in the North Atlantic: High-precision measurements of bomb radiocarbon in banded corals. *Journal of Geophysical Research, 94,* 3271–3285.
Engelkemeir, A., Hamill, W. H., Inghram, M. G., & Libby, W. F. (1949). The half-life of radiocarbon (^{14}C). *The Physical Review, 75,* 1825–1833.
Ezat, M. M., Rasmussen, T. L., Thornalley, D. J. R., Olsen, J., Skinner, L. C., Hönisch, B., et al. (2017). Ventilation history of Nordic Seas over-flows during the last (de)glacial period revealed by species-specific benthic foraminiferal ^{14}C dates. *Paleoceanography, 32,* 172–181.

Friedrich, W. L., Kromer, B., Friedrich, M., Heinemeier, J., Pfeiffer, T., & Talamo, S. (2006). Santorini eruption radiocarbon dated to 1627–1600 BC. *Science, 312*, 548.

Gaudinsky, J. B., Trumbore, S., Davidson, E. A., & Zheng, S. (2000). Soil carbon cycling in a temperate forest: Radiocarbon-based estimates of residence times, sequestration rates and partitioning of fluxes. *Biogeochemistry, 51*, 33–69.

Godwin, H. (1962). Half-life of radiocarbon. *Nature, 195*, 984.

Goldstein, S. J., Lea, D. W., Chakraborty, S., Kashgarian, M., & Murrell, M. T. (2001). Uranium-series and radiocarbon geochronology of deep-sea corals: Implications for Southern Ocean ventilation rates and the oceanic carbon cycle. *Earth and Planetary Science Letters, 193*, 167–182.

Hatté, C., Poupeau, J. J., Tannau, J. F., & Paterne, M. (2003). Development of an automated system for preparation of organic samples. *Radiocarbon, 45*(3), 421–430.

Hughen, K. A., Overpeck, J. T., Lehman, S. J., Kashgarian, M., Southon, J., Peterson, L. C., et al. (1998). Deglacial changes in ocean circulation from an extended radiocarbon calibration. *Nature, 391*, 65–68.

Ikehara, K., Danhara, T., Yamashita, T., Tanahashi, M., Morita, S., & Ohkushia, K. (2011). Paleoceanographic control on a large marine reservoir effect offshore of Tokai, south of Japan, NW Pacific, during the last glacial maximum-deglaciation. *Quaternary International, 246*, 213–221.

Kamen, M. D. (1963). Early history of carbon-14. *Science, 140*, 584–590.

Kennett, D. J., Ingramm, L. B., Southon, J. R., & Wise, K. (2002). Differences in ^{14}C age between stratigraphically associated charcoal and marine shell from the archaic period site of kilometer 4, Southern Peru: Old wood or old water? *Radiocarbon, 44*, 53–58.

Key, R. M., Kozyr, A., Sabine, C. L., Lee, K., Wanninkhof, R., Bullister, J. L., Feely, R. A., Millero, F. J., Mordy, C., & Peng, T.H. (2004). A global ocean carbon climatology: Results from Global Data Analysis Project (GLODAP). *Global Biogeochemical Cycles, 18*, GB4031. https://doi.org/10.1029/2004gb002247.

Key, R. M., Quay, P. D., Jones, G. A., McNichol, A. P., von Reden, K. F., & Schneider, R. J. (1996). WOCE AMS radiocarbon I: Pacific ocean results; P6, P16 & P17. *Radiocarbon, 38*, 425–518.

Kleber, M., Sollins, P., & Sutton, R. (2007). A conceptual model of organo-mineral interactions in soils: self-assembly of organic molecular fragments into zonal structures on mineral surfaces. *Biogeochemistry, 85*, 9–24.

Korff, S. A. (1951). Cosmic-ray neutrons. *American Journal of Physics, 19*, 226–229.

Lal, D., & Charles, C. (2007). Deconvolution of the atmospheric radiocarbon record in the last 50,000 years. *Earth and Planetary Science Letters, 258*, 550–560.

Libby, W. F. (1934). Radioactivity of neodymium and samarium. *Physical Review, 45*, 196–204.

Libby, W. F. (1952). *Radiocarbon dating* (Vol. 37, 124 p.). Chicago, IL: The University of Chicago Press.

Libby, W. F. (1964). *Radiocarbon dating*. Nobel Lecture 12 December 1960. Nobel Lectures, Chemistry 1942–1962, Amsterdam: Elsevier Publishing Company.

Libby, W. F. (1981). In R.Berger & L. M. Libby (Eds.), *Radiocarbon and tritium* (Vol. I, 300 p.). Ocean Way Santa Monica, CA: 129 Geo Science Analytical Inc., 90402.

Michel, E., Labeyrie, L. D., Duplessy, J.-C., Gorfti, N., Labracherie, M., & Turon, J.-L. (1995). Could deep Subantarctic convection feed the world deep basins during the Last Glacial Maximum? *Paleoceanography, 10*, 927–942.

Muscheler, R., Joos, F., Müller, S. A., & Snowball, I. (2005). Climate: How unusual is today's solar activity? *Nature, 436*, E3–E4.

Olsson, I., Kaplen, I., Turnbull, A. H., & Prosser, N. J. D. (1962). Determination of the half-life of ^{14}C with a proportional counter. *Arkiv för fysik, 14*, 237–255.

Olsson, I., & Osadebe, F. (1974). Carbon isotope variations and fractionation corrections in ^{14}C dating. *Boreas, 3*, 139–146.

Reimer, P. J., Bard, E., Bayliss, A., Beck, J. W., Blackwell, P. G., Bronk Ramsey, C., et al. (2013). IntCal13 and Marine13 radiocarbon age calibration curves 0–50,000 years cal BP. *Radiocarbon, 55*, 1869–1887.

Robinson, L. F., Adkins, J. F., Keigwin, L. D., Southon, J., Fernandez, D. P., Wang, S., et al. (2005). Radiocarbon variability in the western North Atlantic during the last deglaciation. *Science, 310*, 1469–1473.

Rossi, B. (1952). *High Energy Particles* (p. 268). Englewood cliffs, NJ: Prentice Hall Physics series Inc.

Ruben, S., Kamen, M. D., & Hassid, W. (1949). Photosynthesis with radioactive carbon: II. Chemical properties of the intermediates. *Journal of the American Chemical Society, 62*, 3443–3450.

Scharpenseel, H. W., & Shiffmann, H. (1971). Radiocarbon dating of soils, a review. *Zeitschrift für Planzenernährung, Düngung, Bodenkunde, 140*, 159–174.

Siani, G., Michel, E., De Pol-Holz, R., Tim DeVries, T., Lamy, F., Carel, M., et al. (2013). Carbon isotope records reveal precise timing of enhanced Southern Ocean upwelling during the last deglaciation. *Nature Communications, 4*, 2758. https://doi.org/10.1038/ncomms3758.

Siani, G., Paterne, M., Arnold, M., Bard, E., Métivier, B., Tisnerat, N., et al. (2000). Radiocarbon reservoir ages in the Mediterranean Sea and Black Sea. *Radiocarbon, 42*(2), 271–280.

Siani, G., Paterne, M., Michel, E., Sulpizio, R., Sbrana, A., Arnold, M., & Haddad, G. (2001). Mediterranean Sea surface radiocarbon reservoir age changes since the last glacial maximum. *Science, 294*, 1917–1920.

Sikes, E. L., & Guilderson, T. P. (2016). Southwest Pacific Ocean surface reservoir ages since the last glaciation: Circulation insights from multiple-core studies. *Paleoceanography, 31*, 298–310.

Sikes, E. L., Samson, C. R., Guilderson, T. P., & Howard, W. R. (2000). Old radiocarbon ages in the southwest Pacific Ocean during last glacial period and deglaciation. *Nature, 405*, 555–559.

Simpson, J. A. (2000). The cosmic ray nucleonic component: The invention and scientific uses of the neutron monitor. *Space Science Reviews, 93*, 11–32.

Six, J., Conant, R. T., Paul, A., & Paustian, K. (2002). Stabilization mechanisms of soil organic matter: Implications for C-saturation of soils. *Plant and Soil, 241*, 155–176.

Stenström, K. E., Skog, G., Georgiadou, E., Genberg, J., & Johansson, A. (2011). A guide to radiocarbon units and calculations. International Report LUNFD6 (NFFR-3111), Lund University, 1–17.

Stuiver, M., Braziunas, T. F., Becker, B., & Kromer, B. (1991). Climatic, Solar, Oceanic and Geomagnetic Influences on Late-Glacial and Holocene Atmospheric $^{14}C/^{12}C$ Change. *Quaternary Research, 35*, 1–24.

Stuiver, M., Pearson, G. W., & Braziunas, T. (1986). Radiocarbon age calibration of marine samples back to 9000 cal yr bp. *Radiocarbon, 28*, 980–1021.

Stuiver, M., & Polach, H. A. (1977). Reporting of ^{14}C. *Radiocarbon, 19*, 355–363.

Tagliabue, A., Bopp, L., Roche, D. M., Bouttes, N., Dutay, J.-C., Alkama, R., et al. (2009). Quantifying the roles of ocean circulation and biogeochemistry in governing ocean carbon-13 and atmospheric carbon dioxide at the last glacial maximum. *Climate of the Past, 5*, 695–706.

Taylor, R. E. (1987). *Radiocarbon dating: An archeological perspective* (212 p.). London: Academic Press Inc. Ltd.

Taylor, R. E., Long, A., & Kra, R. S. (1992). *Radiocarbon after four decades. An interdisciplinary perspective* (596 p.). New York: Springer.

Thornalley, D. J. R., McCave, N., & Elderfield, H. (2011). Tephra in deglacial ocean sediments south of Iceland: Stratigraphy, geochemistry and oceanic reservoir ages. *Journal of Quaternary Science, 26,* 190–198.

Tisnérat-Laborde, N., Paterne, M., Métivier, B., Arnold, M., Yiou, P., Blamart, D., et al. (2010). Variability of the northeast Atlantic sea surface $\Delta^{14}C$ and marine reservoir age and the North Atlantic Oscillation (NAO). *Quaternary Science Reviews, 29,* 2633–2646.

Tisnérat-Laborde, N., Poupeau, J. J., Tannau, J. F., & Paterne, M. (2001). Development of a semi-automated system for routine preparation of carbonate sample. *Radiocarbon, 43,* 299–304.

Toggweiler, J. R., Dixon, K., & Bryan, K. (1989). Simulations of radiocarbon in a coarse-resolution world ocean model. 1. Steady state prebomb distributions. *Journal of Geophysical Research, 94,* 8217–8242.

Valladas, H., Cachier, H., Maurice, P., De Quirost, F. B., Clottes, J., Valdes, V. C., et al. (1992). Direct radiocarbon dates for prehistoric paintings at the Altamira, El Castillo and Niaux caves. *Nature, 357,* 68–70.

Zhao, N., Marchal, O., Keigwin, L., Amrhein, D., & Gebbie, G. (2018). A synthesis of deglacial deep-sea radiocarbon records and their (in) consistency with modern ocean ventilation. *Paleoceanography and Paleoclimatology, 33,* 128–151.

The ^{40}K/^{40}Ar and ^{40}Ar/^{39}Ar Methods

Hervé Guillou, Sébastien Nomade, and Vincent Scao

The ^{40}K/^{40}Ar method and its variant, ^{40}Ar/^{39}Ar, are based on the natural radioactive decay of ^{40}K, one of the isotopes of potassium, in ^{40}Ar, one of the isotopes of argon. ^{40}K decreases in ^{40}Ar* (the * symbol indicates that this is a radiogenic isotope) with a period of 1.25×10^9 years, according to the law of radioactive decay $N = N_0 \, e^{-\lambda t}$. In other words, if we consider a closed system, containing at an initial time (t_0) N_0 atoms of ^{40}K, then $N_0/2$ atoms of ^{40}K will remain in the system after 1.25×10^9 years. This gives us an indication of the geochronological application. If, in a geological sample, both the number of parent atoms remaining (^{40}K) and the number of daughter atoms formed (^{40}Ar*) can be measured, then it is possible to calculate the age of formation of this sample. The relatively high abundance of the isotope ^{40}K (K is the seventh most abundant element on Earth), combined with a low decay rate, makes the ^{40}K/^{40}Ar method and its variant ^{40}Ar/^{39}Ar two of the most widely used geochronological tools in Earth Sciences. They are applicable to various geological materials and cover a wide range of ages, given the long period of ^{40}K.

Already in 1921, Aston, using a mass spectrograph, proved the existence of two isotopes of potassium (^{39}K and ^{41}K). In 1935, Klemperer, and also Neuman and Walker, experimentally demonstrated the natural radioactive decay of ^{40}K to ^{40}Ca and ^{40}Ar*. In 1948, Aldrich and Nier confirmed the radiogenic origin of argon ^{40}Ar*. They experimentally determined the ^{40}Ar/^{36}Ar ratio of several potassic minerals and compared it to that of the atmosphere, assuming it to be constant and equal to 298.56 (this updated value, determined by Lee et al. (2006), replaces the previous one of 295.5 established by Steiger and Jäger (1977)). As the ratios obtained were superior to that of the atmosphere, the source of argon ^{40}Ar* by radioactive decay of ^{40}K was demonstrated. In parallel, understanding of the decay constant of ^{40}K became more accurate. Aldrich and Nier could therefore see the potential of the ^{40}K/^{40}Ar pair for the dating of rocks.

The K-Ar clock is based on the principles of radioactive decay and the accumulation of a daughter isotope. However, studies subsequent to Aldrich and Nier's work showed that there were many causes of disturbance in the K/Ar clock. Among these are the inability of certain rocks or minerals to retain all of the radiogenic argon-40 (^{40}Ar*), or the presence of 'excess argon' in some samples. Consequently, for a K-Ar age to be accepted as correct, the following must be true:

1. when starting the clock (at time zero t_0), the ^{40}Ar/^{36}Ar ratio in the sample is the same as that of the atmosphere (298.56), in other words that ^{40}Ar$^* = 0$;
2. and between t_0 and the moment the sample is dated, it behaves as a closed system with regard to ^{40}K and ^{40}Ar.

The conventional K/Ar method does not allow verification of these two major assumptions. To remedy this, the ^{40}Ar/^{39}Ar variant was developed by Wänke and König (1959) and Merrihue (1965), which showed that the ^{40}K/^{40}Ar ages can be obtained by irradiating samples of rocks or minerals. When a sample is subjected to a neutron flux in a reactor, some ^{39}K becomes ^{39}Ar. The measurement of the ^{39}Ar content by a counting method calculates the number of parent atoms (^{40}K) remaining in the sample, since we know the relative abundances of different isotopes of potassium (see below). Using the same counting method, the number of ^{40}Ar isotopes present in the sample can also be measured. However, this approach is unsatisfactory because it does not allow for necessary corrections related to the process of irradiation to be made, as we shall see later, and the precision achieved on the different concentrations of argon isotopes is insufficient. The crucial work for the ^{40}Ar/^{39}Ar method was carried out by Merrihue in 1965. He showed that the argon ^{39}Ar generated in a nuclear reactor from the ^{39}K of a sample can be measured precisely by mass spectrometry. This ^{39}Ar, derived from the ^{39}K, is annotated as ^{39}Ar$_K$. In addition, the other isotopes of argon, ^{40}Ar and

H. Guillou (✉) · S. Nomade · V. Scao
Laboratoire des Sciences du Climat et de l'Environnement, LSCE/IPSL, CEA-CNRS-UVSQ, Université Paris-Saclay, 91190 Gif-sur-Yvette, France
e-mail: herve.guillou@lsce.ipsl.fr

^{36}Ar (essential for the atmospheric correction and ^{40}Ar* content calculation) can be measured in the same way. Thus, in a single measurement, one can calculate the proportions of ^{40}K and ^{40}Ar* present in a sample, and from this, calculate its age. In their 1966 article, Merrihue and Turner established the fundamentals in terms of approach and concepts for the ^{40}Ar/^{39}Ar method. In particular, they showed that the relative proportions of radioactive parents and radiogenic daughters can be calculated accurately from a measurement by mass spectrometry.

In addition, since isotopic ratios can be measured more precisely than the concentrations of K and Ar, this method improves the accuracy of the ages and can be used for dating smaller samples than the ^{40}K/^{40}Ar method does. This same work laid the groundwork for the application of isochrones and age spectra (concepts that will be discussed below) to the ^{40}Ar/^{39}Ar method.

The ^{40}K/^{40}Ar and ^{40}Ar/^{39}Ar methods became very popular in geology as they are applicable to different terrestrial geological materials, such as terrestrial magmatic rocks (volcanic, plutonic, metamorphic) and extraterrestrial (meteorites, moon samples) ones. For some measurements, the ^{40}K/^{40}Ar method is also well suited to the dating of clay minerals. The range of application of these isotopic age dating methods has an upper limit of 3 billion years and a lower limit of 10,000 years.

These two methods were used to date major events in the history of the Earth (fauna and flora of the Mesozoic and Cenozoic, mass extinctions, origin and evolution of hominids, major volcanic eruptions, genesis and evolution of the large mountain chains, etc.). They were, and still are, used to establish and calibrate the geological time scale, including the time scale of the reversals of Earth's magnetic field, very useful tie-points in paleoclimatology.

In the following, we present the main principles and areas of application of these two methods. For further details, the reader may refer to the works of Dalrymple and Lanphere (1969) and McDougall and Harrison (1988).

Principles of the K-Ar Method

Diagram of Radioactive Decay in ^{40}K

The principle of the ^{40}K/^{40}Ar method is based on the natural radioactive decay of ^{40}K in ^{40}Ar (Fig. 5.1). The decay of ^{40}K is complex. At 88.8%, the ^{40}K decays to ^{40}Ca by emitting β^-. At 11.2% it decays to ^{40}Ar*, either by emitting β^+ (0.01%), or by direct electronic capture (0.16%) or by electron capture followed by a γ emission (11%). This last mechanism is the most common. An electron from the atom is captured, resulting in the formation of a neutron at the expense of a proton. The ^{40}Ar atom thus produced is in an excited state. It then returns quickly to its ground state by emitting gamma radiation.

The Age Equation

As with the other isotopic clocks, the fundamental law of radioactive decay applies:

$$N = N_0 e^{-\lambda t} \quad (5.1)$$

N: number of radioactive parent atoms (^{40}K) at time t, N_0: number of radioactive parent atoms at t_0, λ: decay constant.

From Eq. (5.1), we can calculate the number of daughter atoms ($D^* = {}^{40}$K + ^{40}Ca) formed over time t:

$$N_o = N e^{\lambda t}$$

$$D^* = N_o - N = N e^{\lambda t} - N = N(e^{\lambda t} - 1) \quad (5.2)$$

The constants and isotopic abundances required for the age calculation are listed in Table 5.1.

The age equation is established from Eq. (5.2):

$$^{40}\text{Ar}^* = \frac{\lambda_\epsilon}{\lambda} {}^{40}\text{K}(e^{\lambda t} - 1) \quad (5.3)$$

where ^{40}Ar* is the isotope of argon produced from the in situ decay of ^{40}K, λ the total radioactive decay constant of ^{40}K equal to $\lambda_\epsilon + \lambda_\beta$. The ratio of proportionality $\frac{\lambda_\epsilon}{\lambda}$ corresponds to the fraction of the decay leading to the formation of ^{40}Ar* (and not of ^{40}Ca*).

From Eq. (5.3) we get:

$$t = \frac{1}{\lambda} \ln\left(\frac{^{40}Ar^*}{^{40}K} \frac{\lambda}{\lambda_\epsilon} + 1\right) \quad (5.4)$$

with t expressed in years.

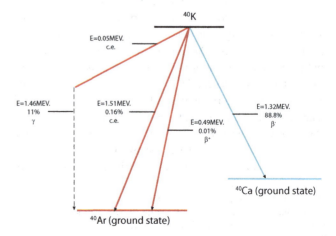

Fig. 5.1 Diagram of radioactive decay of ^{40}K

Table 5.1 Decay constants of ^{40}K and isotopic abundances of K and Ar. The values attributed to the constants were determined by Min et al. (2000), and for isotopic ratios by Garner et al. (1975) and Nier (1950)

Constant	Value
$\lambda_\varepsilon = \lambda^{40}Ar$	$(5.80 \pm 0.014) \times 10^{-11}$ a^{-1}
$\lambda_\beta = \lambda^{40}Ca$	$(4.884 \pm 0.099) \times 10^{-10}$ a^{-1}
$\lambda = \lambda_\varepsilon + \lambda_\beta$	$(5.464 \pm 0.107) \times 10^{-10}$ a^{-1}
^{39}K	93.2581%
^{40}K	0.01167%
^{41}K	6.7302%
^{40}Ar	99.600%
^{38}Ar	0.0632%
^{36}Ar	0.3364%

The half-life period ($N = N_0/2$) T is calculated from (5.1):

$$T = \frac{\ln 2}{\lambda} = 1.25 \times 10^9 \text{ years.}$$

Operation of the Potassium-Argon Clock

The radioactive clock ^{40}K/^{40}Ar is based on the process of accumulation. K is one of the components of magma. When this is in liquid form, the argon ^{40}Ar* formed from the decay of ^{40}K escapes from the system. During a volcanic eruption, the magma that reaches the surface cools very quickly. Thus, argon ^{40}Ar* is trapped in the solidified lava and accumulates in the crystalline lattice. The radiogenic argon (^{40}Ar*) thus trapped can only escape if the rock or mineral are either melted or recrystallized, or heated to temperatures generally greater than or equal to 200 °C, in such a way that the argon can diffuse through the crystal lattice. Dalrymple and Lanphere (1969) illustrated the operation of the ^{40}K/^{40}Ar clock (Fig. 5.2) in a diagram, taking the crystallization of magma as an example. Ideally, there are three distinct stages. During the first stage, at high temperatures, the phenomenon of diffusion prevails. ^{40}Ar* is not retained in the lattice. The second stage corresponds to a start of cooling and partial accumulation of the argon ^{40}Ar*. The last step corresponds to the rapid cooling of the surface of the silicate or magmatic melt. At this point, ^{40}Ar* is retained entirely within the crystal lattice.

From this evolution came the basic assumptions for the application of the K-Ar clock which are detailed below.

1. The parent isotope ^{40}K decays at a constant rate, independently of the physical conditions of the system (P and T). The constants used are those in Table 5.1.
2. The ^{40}K/K$_{total}$ ratio is constant in natural materials. This condition is important because it is not ^{40}K that is directly measured, but the total K (K-Ar) or ^{39}Ar$_K$ (^{40}Ar/^{39}Ar). ^{40}K is deduced from the isotopic composition of K. This ratio has changed over time due to radioactive decay, but this term is not included in the age equation. At a given t, this ratio is constant in all materials because these isotopes do not fractionate as a result of the geological processes.
3. We consider that at $t = 0$, the moment of formation of the sample, it is devoid of radiogenic argon (^{40}Ar* = 0); otherwise, ages obtained would be marred by an error of excess argon. In geochronology, this is the same as assuming that at $t = 0$, the ^{40}Ar/^{36}Ar ratio of the sample, called the initial ratio, is considered to be equal to that of the atmosphere, or 298.56. There are some deviations from this principle. These are cases of excess argon or inherited argon, which cannot be directly detected by the K-Ar method, but can be detected more easily by the ^{40}Ar/^{39}Ar method. These excesses of argon show up as an overestimation of the calculated ages and are an important limitation of the K-Ar method.
4. It is also necessary that the formation time of the system be negligible compared to the age of the sample. Therefore, volcanic rocks that form by very rapid cooling provide the most suitable samples for this method of dating.
5. It is essential to assume that the sample evolved within a closed system with regard to K and Ar ever since the geological event to be dated. This condition involves rigorously selecting unaltered samples, in order to avoid any disruption (re-opening subsequent to formation) in the isotopic system.

Datable Materials and Age Ranges

The main materials suitable for testing by the K-Ar and ^{39}Ar/^{40}Ar methods as well as the age ranges are listed below in Fig. 5.3:

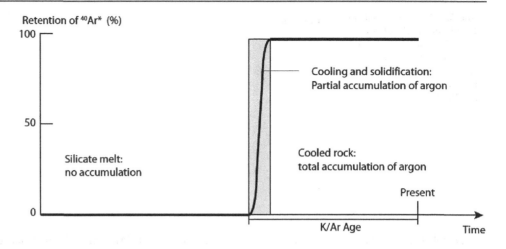

Fig. 5.2 Principle of the $^{40}K/^{40}Ar$ clock in the case of a simple-story magmatic rock (from Dalrymple and Lanphere 1969)

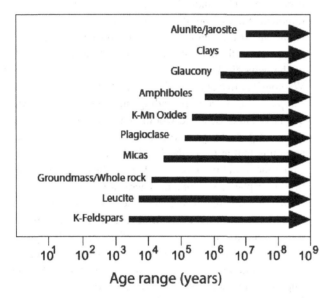

Fig. 5.3 Applicable age range of K-Ar and $^{40}Ar/^{39}Ar$ dating for various materials (from Renne 2000)

- volcanic rocks (calibration of time scales, stratigraphic studies); minerals: sanidine (≥ 2 ka) anorthoclase (≥ 5 ka), plagioclase (≥ 200 ka), amphibole (≥ 1 Ma), leucite, nepheline;
- rocks (lava and tephra): mesostasis (groundmass) of all types of unaltered rock ($\geq 3 \, 4$ ka), non-hydrated glass (≥ 300 ka);
- plutonic rocks, metamorphic minerals; potassic feldspar, plagioclase, biotite, amphibole, muscovite, phengite, alunite, adularia;
- sediments; neoformed clays, glauconite, evaporites, detrital minerals rich in K, alunite/jarosite.

The Unspiked K-Ar Method

Selection and Preparation of the Samples

The stratigraphic and geographic location of the sample needs to be noted and established precisely in the field. This will enable external limits to be placed on the ages obtained. Only unaltered samples are retained, mainly to comply with the condition of evolution within a closed system, as discussed above. Freshness is controlled by macroscopic and microscopic observations, and by chemical analysis which measures the loss on ignition (L.O.I. almost equivalent to the water content), a reliable marker of the degree of alteration of the sample. For example, a basalt is considered unaltered if its L.O.I. value is less than 1%. The phase of the sample (mineral, glass, etc.) must be representative of the event to be dated. In the case of volcanic rocks, the selected phase for lava is the microcrystalline groundmass because it is formed by rapid cooling when the magma arrives at the surface; for tephra, phenocrysts (feldspar, mica and amphibole) synchronous with the eruption are selected. After sampling in the field, the rocks are cut, ground and sieved to a 0.250–0.125 mm size. The granulate obtained is washed in an ultrasonic bath of acetic acid (1 N) at 50 °C. The sample is then rinsed thoroughly with deionized water. Then, the separation of the mineralogical phases is done in the laboratory, by magnetic sorting and densitometry or picking under microscope. In the case of tephra, the preparatory procedure is identical. However, given the nature of the deposit, the grinding phase is bypassed.

Determining $^{40}Ar^*$

To calculate an age from Eq. (5.4), only the measurement of two variables is required: the content of radiogenic argon 40 ($^{40}Ar^*$) and of potassium 40 (^{40}K). The K-Ar method of dating used by the LSCE is the unspiked (with no tracer) K-Ar technique developed by Cassignol et al. (1978), Cassignol and Gillot (1982).

A schematic representation of the ultra-high vacuum line connected to the K-Ar mass spectrometer is given in Fig. 5.4. Prior to the dosing of argon, several stages, such as extraction and purification of the argon are required. Groundmass splits (0.5–2.0 g) of samples are wrapped into 99.5% copper foil packets, loaded in the sample holder, which has been turbo-molecular pumped for about 20 h (Fig. 5.4). During the last two hours of that stage, the molybdemium (Mo) crucible is degassed at about 1500 °C until the pressure decreases to 10^{-9} Torr. The sample is then dropped into the Mo crucible and becomes molten at full power of the induction furnace. During the melting stage (i.e. 20 min), the extracted gas is adsorbed by the first active charcoal finger at liquid nitrogen temperature.

After the melting, the gas is released by heating the charcoal to 110 °C and purified via the mutual action of a titanium sublimation pump and a SAES 10 GP-MK3 Zr-Al getter operated at 400 °C. This first step of gas clean-up (i.e. elimination of active gases) generally lasts 30 min (Fig. 5.5) and is followed by three consecutive exposures of five minutes each of the gas to SAES 10 GP-MK3 Zr-Al getters also operated at 400 °C. The remaining gas, mostly argon, is then adsorbed 5 min by a second active charcoal maintained at liquid nitrogen temperature.

The argon is then freed from the active charcoal finger n° 2 by bringing it to room temperature. After a rapid cryo-pumping of the spectrometer, the argon is introduced into the mass spectrometer. The argon, an inert gas, is ionised in the mass spectrometer, under the effect of an electronic source. ^{40}Ar becomes $^{40}Ar^+$ and ^{36}Ar becomes $^{36}Ar^+$. The atoms thus charged are accelerated under the influence of a difference in potential (about 620 volts). They then pass through a magnetic field. At this point, their trajectory becomes circular. In a chamber with a high vacuum, these ions with a mass of 40 and 36 and charged e, animated at speed due to a difference in potential (620 V), trace a trajectory of radius R, as they pass through a magnetic field H (3600 Gauss), according to the equation:

$$R = \frac{1439}{H} \left[\frac{m}{e} V\right]^{\frac{1}{2}}$$

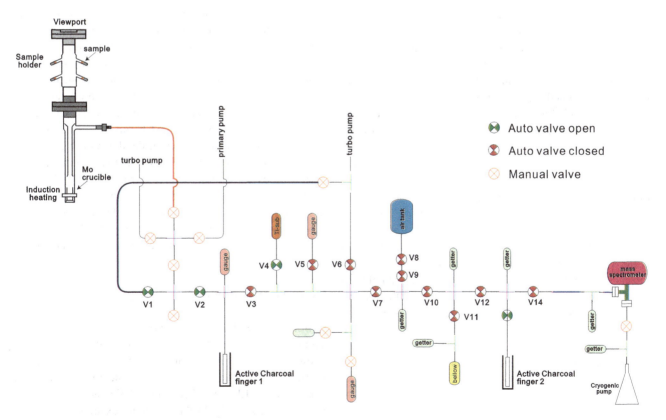

Fig. 5.4 Schematic representation of the ultra-high vacuum line connected to the K-Ar mass spectrometer. Ti-Sub: Titanium sublimation pump, V1-14: Ultra-high vacuum all metal valves. NB: the diagram is at a different scale between the preparation line and the sample holder

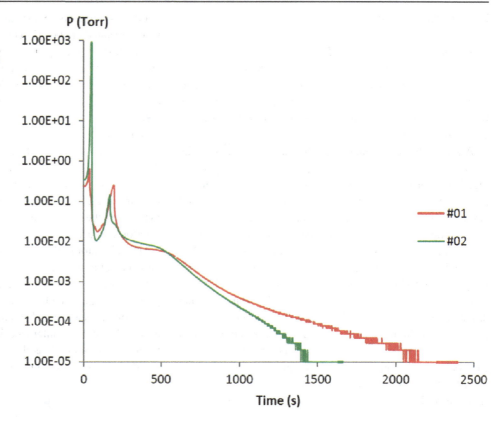

Fig. 5.5 Evolution of pressure versus time during the gas clean-up by means of a Titanium sublimation pump and a SAES 10 GP-MK3 Zr-Al getter. #01 and #02 are the reference numbers for the experiment

with R: radius in mm, H: magnetic field measured in Gauss, m: atomic mass of the ion, e: number of elemental charges carried by the ion, V: difference in potential in volts.

Thus, for a spectrometer configured with $H = 3600$ Gauss and $V = 620$ volts, the following is obtained: $R_{40} = 62.9$ mm and $R_{36} = 59.7$ mm.

The ^{40}Ar and ^{36}Ar isotopes of the sample are measured simultaneously on a dual manifold comprised of two Faraday cups, set at $m/e = 40$ and $m/e = 36$. The signals are integrated over a period of 100 s. Once the analysis of the gas is made, the sample is removed from the mass spectrometer by cryo-pumping. The reference atmospheric argon (at.) is taken by means of a double valve from a cylinder containing air from the laboratory (designed as Ref.Atm.). It is then introduced into the mass spectrometer and measured at the same pressure conditions as the sample. It allows the direct comparison of the two aliquots of gas (the sample and atmospheric reference) and the determination of the relative content of radiogenic argon (Fig. 5.6). This is done by varying the analysis volume via a variable volume (VV) connected to the mass-spectrometer cell.

The content of ^{40}Ar* is given by the equation:

$$^{40}Ar^* = \frac{\frac{^{40}Ar_s}{^{36}Ar_s} - \frac{^{40}Ar_{at}}{^{36}A_{at}}}{\frac{^{40}Ar_s}{^{36}Ar_s}}$$

The third measurement of an aliquot of gas is performed for calibration, in other words, the conversion of an electrical signal into a number of atoms. A known number of argon ^{40}Ar is fed from a calibrated cylinder into the mass spectrometer. The number of atoms is deduced from the measurement of standard minerals of a known age. The procedure for establishing the equation of the calibration curve (Fig. 5.7) is described in Charbit et al., (1998). The three previous steps are shown in the diagram in Fig. 5.6.

The characteristics of the standard minerals commonly used in K-Ar for calibration are shown in Table 5.2.

Potassium analysis is done by flame spectrometry (atomic absorption and emission), on several selections taken independently to ensure sample homogeneity. As this method is a conventional one, we will not enter into the details here.

Example of a Calculation of Age

Calibration of the Mass Spectrometer

Calibration requires an analysis of mineral standards of a known age. First, the ^{40}Ar and ^{36}Ar content of the standard is measured by mass spectrometry. The mineral standard studied here is LP-6. Its K content is 8.37% and that of ^{40}Ar* is 1.158×10^{15} at./g. In the following example, 0.04442 g of this standard was molten. Then, three aliquots of air were

Fig. 5.6 Principle of the K-Ar method with no tracer. Two isotopes of argon (^{40}Ar and ^{36}Ar) are measured by mass spectrometer. The sample gas is composed of atmospheric argon ^{40}Ar$_{at}$ and ^{36}Ar$_{at}$, as well as radiogenic argon ^{40}Ar*, resulting from the decay of ^{40}K.
s = sample; at = atmospheric, cd = calibrated dose

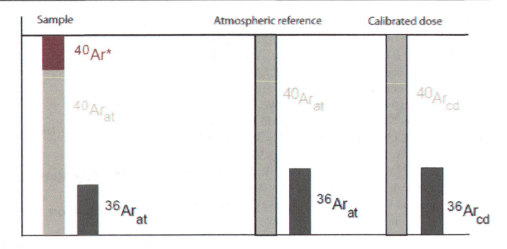

Fig. 5.7 Calibration curve of the mass-spectrometer on the 19th of November 2001

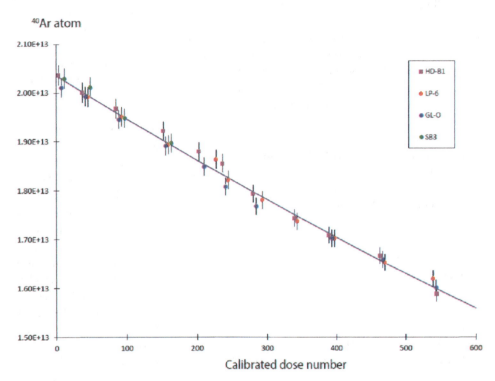

Table 5.2 Values of mineral standards

Standard	K (%)	^{40}Ar* (10^{-9} mol/g)	Age (Ma)	References
HD-B1	7.987	0.335	24.03 ± 0.41	Fuhrmann et al. (1987)
GL-O	6.56	1.093	93.60 ± 0.90	Charbit et al. (1998)
LP-6	8.37	1.923	127.8 ± 1.4	Odin (1982)
SB-3	7.483	2.213	162.9 ± 0.9	Lanphere and Dalrymple (2000)

taken up using a double valve system (pipette) from a calibration canister, and the ^{40}Ar and ^{36}Ar contents are measured.

The results of the analysis by mass spectrometry are shown in the following table.

	^{40}Ar in volts	^{36}Ar in millivolts
Sample (LP-6)	8.921	3.110
Atmospheric reference	7.637	28.252
Calibrated dose (3 doses)	7.637	

From these values, the equivalent ^{40}Ar/^{36}Ar ratios can be calculated as follows: $(^{40}$Ar/^{36}Ar$)_{sample}$ = $R.I._s$ = 2 868.49; $(^{40}$Ar/^{36}Ar$)_{tam}$ = $R.I._{atm}$ = 270.32; and the concentration of ^{40}Ar* as: ^{40}Ar*, hence: $(R.I._s - R.I._{atm})/(R.I._s)$ = 90.58%.

We know the content of ^{40}Ar* (1.158 × 10^{15} at/g) and the melted weight (0.04442 g) of the standard. We can calculate the number of argon ^{40}Ar* atoms introduced in the mass spectrometer: N_{at} = 1.158 × 10^{15} × 0.044 42 = 5.144 × 10^{13} at.

Out of the ^{40}Ar signal of the sample (8.921 V), 90.58% corresponds to ^{40}Ar* that is to say 8080 V. From this, we deduce that 8080 V corresponds to 5.144 × 10^{13} at. of ^{40}Ar*.

Three aliquots of air produce a signal of 7.637 V, this therefore corresponds to (7.637 × 5.144 × 10^{13})/8.080 = 4.862 × 10^{13} at. of argon ^{40}Ar*.

An aliquot of air taken from the calibration canister is therefore equivalent to 4.862 × 10^{13}/3 = 1.621 × 10^{13} at. of ^{40}Ar. This canister can then be used to calibrate the mass spectrometer for the measurement of ordinary samples. Obviously, for each calibration measurement, the content of ^{40}Ar decreases in the cylinder. This change is monitored by a regular measurement of the mineral standard. This is the curve shown in Fig. 5.7.

Measuring a Sample of Unknown Age

Experimental data: melted weight: 1.0669 g; potassium content of the analyzed rock K % = 0.643; 1 calibrated dose = 1.608 2 × 10^{13} atoms.

	Sample	Atmospheric reference	Calibrated dose
^{40}Ar (V)	1.196	1.323	3.912
^{36}Ar (mV)	4.225	4.884	14.474

The calculation of R.I. isotope ratios (^{40}Ar/^{36}Ar) gives: $R.I._s$ = 283.08; $R.I._{atm}$ = 270.88; $R.I._{cd}$ = 270.28.

The level of radiogenic argon is calculated by:

$$^{40}\text{Ar}^*\% = \frac{R.I._s - R.I._{atm}}{R.I._s} = 4.31\%$$

The number of argon ^{40}Ar atoms in the sample is calculated using the calibration data. We know from the calibration curve that 3.912 volts correspond to 1.608 2 × 10^{13} atoms. Therefore, 1.196 V (^{40}Ar sample) is equivalent to 0.492 × 10^{13} atoms. The concentration of atoms per gram is this value divided by the weight of the melted sample (1.0669 g) so 0.461 × 10^{13} at./g. 4.31% of the measured ^{40}Ar is radiogenic. This gives ^{40}Ar* at./g = 0.461 × 10^{13} × 0.0431 × 1.987 = 10^{13}.

The number of ^{40}K atoms is calculated using:

$$^{40}\text{K} = \left(\frac{K \times 0.01}{39.098304} \times 0.0001167\right) \times 6.023 \times 10^{23}$$
$$= 1.156 \times 10^{16}$$

The age is obtained from the equation:

$$t = \frac{1}{\lambda} \times \ln\left[\frac{^{40}\text{Ar}^*}{^{40}\text{K}}\left(\frac{\lambda}{\lambda\varepsilon}\right) + 1\right] = 296000 \text{ years}$$

(λ = 5.543 × 10^{-10} and λ_ε = 0.581 × 10^{-10}).

The ^{40}Ar/^{39}Ar Method: General Principles

The Age Equation

This method is a variant of the K-Ar method. Firstly, the samples undergo neutron activation. This activation under fast neutron flux within a nuclear reactor is intended to transform the isotope ^{39}K to ^{39}Ar. The amount of argon ^{39}Ar thus generated is proportional to the number of ^{39}K atoms and therefore of ^{40}K (parent atoms) present in the sample, the ^{40}K/^{39}K ratio being (supposedly) constant in nature. To do this, the samples are placed with samples of known age (standards) in aluminum discs themselves stacked in an aluminum tube (shuttle). This shuttle is then subjected to a fast neutron flux, for a period of between a few minutes and 24 h depending on the age and nature of the samples.

The irradiation causes the formation of an artificial argon isotope, ^{39}Ar, according to the reaction $^{39}_{18}$K$(n,p)^{39}_{18}$Ar (capture band of 80 to 100 mbarn, Mitchell 1968; Roddick 1983). ^{39}Ar is radioactive. Its disintegration period is 265 years. As the analysis by spectrometer is performed less than one year after irradiation, the error margin on its estimation is negligible. The advantage of producing argon ^{39}Ar in proportion to the parent element (^{40}K) is that this transformation replaces the measurement of the ^{40}K/^{40}Ar ratio by two different methods (atomic absorption for ^{40}K and mass spectrometry for ^{40}Ar) with the direct measurement of the ^{40}Ar/^{39}Ar ratio (by mass spectrometry).

The precise knowledge of ^{39}Ar production yield is obtained by referring to known age standards. These standards are irradiated in the same shuttles as the samples. The radiation yield is calculated according to the equation established by Mitchell (1968).

$$^{39}\text{Ar}_s = {}^{39}\text{K}\Delta T \int \Phi_E \sigma_E d_E = {}^{39}\text{K}\Delta T I \quad (5.5)$$

$$I = \int_0^\infty \Phi_E \sigma_E d_E$$

with: ^{39}K being the number of atoms of ^{39}K in the standard sample; ^{39}Ar$_s$ the number of atoms of ^{39}Ar produced in the standard sample; Φ_E, the energy flux; σ_E, zone of efficient capture of the reaction $^{39}K \rightarrow {}^{39}Ar$ at energy E; ΔT, the irradiation duration.

The amount of $^{40}Ar^*$ produced by the disintegration of ^{40}K follows the equation:

$$^{40}Ar^* = \frac{\lambda \varepsilon}{\lambda} {}^{40}K\left(e^{\lambda t_s} - 1\right) \quad (5.6)$$

with t_s, known age of the standard.

Combining (5.5) and (5.6), we obtain:

$$\frac{^{40}Ar^*}{^{39}Ar^*} = \frac{^{40}K}{^{39}K} \frac{\lambda \varepsilon}{\lambda} \frac{1}{\Delta T} \frac{(e^{\lambda t_s} - 1)}{I} \quad (5.7)$$

Equation (5.7) is simplified by defining the J parameter which is the radiation flux actually received by the sample:

$$J = \frac{^{39}K}{^{40}K} \frac{\lambda}{\lambda \varepsilon} \Delta T I \quad (5.8)$$

From (5.7), J becomes,

$$J = \frac{e^{\lambda t_s} - 1}{^{40}Ar^*/^{39}Ar} \quad (5.9)$$

It is then possible to resolve the age equation:

$$t_e = \frac{1}{\lambda} \ln\left[1 + \frac{(^{40}Ar^*/^{39}Ar)_e}{(^{40}Ar^*/^{39}Ar)_s}\left(e^{\lambda t_s} - 1\right)\right] \quad (5.10)$$

with s = standard and e = sample.

The following table shows the age of the main standard (or flux) minerals for $^{40}K/^{40}Ar$ and $^{40}Ar/^{39}Ar$ methods.

Name	Mineral	Age (Ma)	References
Hb-3gr	Hornblende	1072 ± 11	Turner et al. (1971)
MMhb-1	Hornblende	520.4 ± 1.7	Samson and Alexander (1987)
LP-6	Biotite	127.9 ± 1.1	Odin (1982)
SB-2	Biotite	162.1 ± 2.0	Dalrymple et al. (1981)
GA-1550	Biotite	97.9 ± 0.9	McDougall and Roksandic (1974)
B4M	Muscovite	18.6 ± 0.4	Flish (1982)
B4B	Biotite	17.3 ± 0.2	Flish (1982)
FCTs	Sanidine	28.187 ± 0.019	Phillips et al. (2017)
ACRs	Sanidine	1.18404 ± 0.00068	Phillips et al. (2017)

The age assigned to the standards may vary depending on the authors. With regard to FCTs and ACRs, the following references may be consulted among others: Kuiper et al. (2008), Renne et al. (2010), Jicha et al. (2016), Niespolo et al. (2017).

Corrections for Atmospheric Argon and Interference of Mass

As for the $^{40}K/^{40}Ar$ method, the correction for atmospheric argon is essential in order to determine the proportion of $^{40}Ar^*$. This adjustment is done by repeated mass spectrometric measurements of aliquots of air. This defines the instrumental $^{40}Ar/^{36}Ar$ atmospheric ratio. Most mass spectrometers give values slightly different from 298.56. It is therefore through the repeated measurements of aliquots of air that the bias in the measuring apparatus can be calculated. As a first approximation and for samples without calcium, the determination of the percentage of radiogenic argon can be done by simply comparing the $^{40}Ar/^{36}Ar$ ratio of the sample and the instrumental $^{40}Ar/^{36}Ar$ ratio of the atmosphere.

However, during irradiation, secondary reactions occur from Ca, K and Cl isotopes which also produce artificial isotopes of argon (Fig. 5.8):

- $^{40}Ca(n, n\alpha)$ ^{36}Ar
- $^{42}Ca(n, \alpha)$ ^{39}Ar
- $^{40}K(n, p)$ ^{40}Ar
- $^{35}Cl(n, \gamma)$ $^{36}Cl - \beta^- \rightarrow {}^{36}Ar$ $t_{1/2} = 300 \times 10^3$ years
- $^{37}Cl(n, \gamma)$ $^{38}Cl - \beta^- \rightarrow {}^{38}Ar$ $t_{1/2} = 37.3$ min

The correction for interference of the masses 40, 39 and 36 due to Ca is possible because of an additional reaction:

- $^{40}Ca(n, \alpha)$ ^{37}Ar $t_{1/2} = 35.1$ days

To know the $^{39}Ar_{Ca}$ and $^{36}Ar_{Ca}$ contents, it is necessary to irradiate a pure calcium salt, such as CaF_2, and measure its 39/37 and 40/37 ratios with the mass spectrometer. The initial value of argon ^{37}Ar of the irradiated sample, when it is taken from the reactor, also needs to be calculated. This is done by applying the law of radioactive decay:

$$^{37}Ar_0 = {}^{37}Ar_m e^{\lambda_{37} t_i} \lambda_{37} t_i / \left(1 - e^{\lambda_{37} t_i}\right)$$

with $^{37}Ar_0$ = amount of the isotope 37 produced at the end of the irradiation; $^{37}Ar_m$ = amount of the isotope 37 measured on the day of analysis; t = duration of the irradiation; t_i = time interval between irradiation and analysis; $\lambda_{37} = 0.0197\ 4\ j^{-1}$.

The correction factors $(^{39}Ar/^{37}Ar)_{Ca}$ and $(^{36}Ar/^{37}Ar)_{Ca}$, which depend on the yield from the irradiation on the salts, are thus well defined.

Fig. 5.8 Mass spectrum of an irradiated sample, deciphering the origin of the different isotopes (in red) generated from the neutronic activation

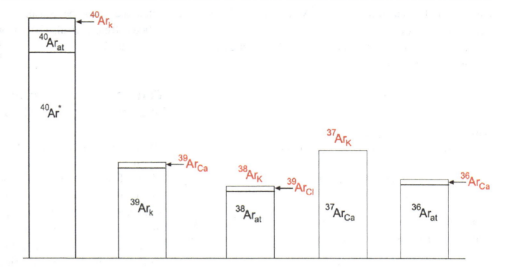

In order to know the correction factor associated with the production of ^{40}Ar from ^{40}K, a pure salt of K (K$_2$SO$_4$ or KF) is also irradiated. This allows the ratio $(^{40}\text{Ar}/^{39}\text{Ar})_K$ to be defined, which is then used to calculate the share of argon ^{40}Ar resulting from irradiation of ^{40}K, which will be cut back from the total argon ^{40}Ar.

In summary, the masses corrected for interferences related to radiation and to the atmospheric component can be written:

$$^{40}\text{Ar}^* = {}^{40}\text{Ar}_m - \left({}^{40}\text{Ar}_{at} + {}^{40}\text{Ar}_K\right)$$
$$^{39}\text{Ar} = {}^{39}\text{Ar}_m - {}^{39}\text{Ar}_{Ca}$$
$$^{36}\text{Ar} = {}^{36}\text{Ar}_{at} - {}^{36}\text{Ar}_{Ca}$$

The Age Spectra

The ^{40}Ar/^{39}Ar method allows for the collection of more comprehensive information on the behavior of the radioisotopic clock than the ^{40}K/^{40}Ar method. In the experimental approach known as 'step-heating' (Turner et al. 1966), the sample is gradually heated in steps of increasing temperature (for example, by steps of 60°C). At each step, the isotopic composition of argon in the extracted and purified gas is measured by mass spectrometry. An apparent age can thus be calculated for each step. In the end, this results in an age spectrum. The general appearance of these spectra shows whether the sample and, consequently, the ^{40}K/^{40}Ar clock, were disrupted or not.

In the case of an undisturbed sample (Fig. 5.9), which evolved in a closed system, the K is homogeneously distributed in the crystal lattice. This is also true for ^{40}Ar* and ^{39}Ar. When a sample is subjected to degassing in increasing temperature steps, the ^{40}Ar* and ^{39}Ar isotopes will be extracted at a constant ratio. As a consequence, a similar apparent age, within error margins, will be obtained for all

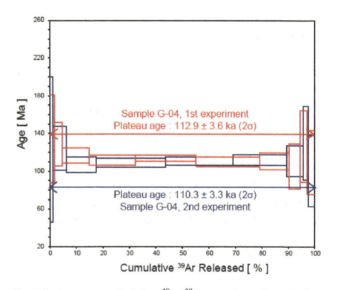

Fig. 5.9 Age spectra depicting ^{40}Ar/^{39}Ar experimental results from duplicated measurements of sample G-04, a lava from the Gelso section of Vulcano (Aeolian Islands, Italy). Uncertainties are ±2σ. In red, the first step-heating experiment; in blue, the second step-heating experiment

steps. The result of such experiments will be a consistent 'age spectrum' in horizontal form which defines the plateau age. Several definitions have been proposed for a plateau age (Dalrymple and Lanphere 1974; Berger and York 1981; McDougall and Harrison 1988). It is generally considered that a plateau is composed of at least three successive steps with at least 60% of the ^{39}Ar$_k$ released and whose apparent ages are consistent within ±2 sigma (i.e. within the respective analytical error bars at 95.6% confidence interval).

Cases of inconsistent age spectra are relatively common and the reasons for this discrepancy are varied. One of the most frequently cited reasons is related to the 'recoil effect'. The transformation of ^{39}K into ^{39}Ar by the reaction ^{39}K(n, p)^{39}Ar can be accompanied by a loss of ^{39}Ar. In other words, the ^{39}Ar atom can be founder-distributed, due to this reaction, in a different crystallographic site than that occupied by its parent, ^{39}K. The distance of this backwards movement is proportional, firstly, to the energy employed during the neutron activation, and secondly, to the sample density. Thus, this recoil effect can lead to a redistribution of ^{39}Ar$_K$ in the samples with grains of a size less than 5–10 μm. This redistribution of ^{39}Ar$_K$ can cause an over or underestimation of apparent ages at different temperature steps. The loss of ^{39}Ar$_K$, related to the recoil effect, is seen in an overestimation of the ages obtained.

Other natural phenomena are also the cause of inconsistent age spectra. Alteration, metamorphism, hydrothermalism are all processes that might disrupt the K/Ar clock. These processes are the source of migration, loss or gain of argon and potassium isotopes.

The Single Grain Method

Technological developments of recent decades (increasing sensitivity of mass spectrometers, laser fusion system) make it possible to work on samples of smaller and smaller size. It is thus possible to get down to the level of the crystal. This approach is used in particular for dating tephra. Within a given tephra layer, several crystals of the same mineral type are selected. After irradiation, each crystal is individually melted by laser, the gas is extracted and it is then analyzed by mass spectrometer. An age is obtained for each constituent crystal at the tephra level. It is thus possible to establish spectra of age probability for a given tephra (Deino and Potts 1992). The analysis of these spectra allows homogeneity at the stratigraphic level to be estimated and the most statistically probable age to be defined (Fig. 5.10).

The Isochrones

The ^{40}Ar/^{39}Ar method is particularly suited to data processing by isochrones. In the inverse isochron diagram ^{40}Ar/^{36}Ar versus ^{39}Ar/^{36}Ar (Fig. 5.11), the slope is equivalent to the ^{40}Ar*/^{39}Ar$_K$ ratio which is itself proportional to the age, and the *intercept* on the y axis corresponds to the (^{40}Ar/^{36}Ar)$_I$ ratio. This last ratio indicates the proportion of ^{40}Ar and of ^{36}Ar at $t = 0$, in other words, at the moment the system closed. This value is directly comparable to the ^{40}Ar/^{36}Ar atmospheric ratio. It is thus possible, from the inverse isochron diagram, to highlight the presence or absence of excess argon. This information is particularly important because it allows one of the basic assumptions for application of the clock to be checked, namely that an age is considered correct if, at $t = 0$, ^{40}Ar* = 0 and (^{40}Ar/^{36}Ar)$_{initial}$ = (^{40}Ar/^{36}Ar)$_{atmospheric}$.

This analysis by isochron is particularly useful for the dating of tephra. Indeed, ideally, all the minerals from the same layer of a tephra should be on the same isochron, as they have, in principle, the same age. Furthermore, the value (^{40}Ar/^{36}Ar)$_i$ must be equivalent to the atmospheric value. If some experimental points are not on this isochron, we can deduce that the corresponding crystals are xenocrysts i.e. older crystals remobilized during the eruptive event at the origin of the tephra (Fig. 5.11).

Selection and Preparation of Samples

The procedure is the same as that followed for the K/Ar method.

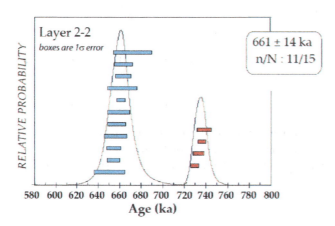

Fig. 5.10 Probability diagram obtained for the stratigraphic level N° 2-2 (Notarchirico archaeological site, Basilicata, Italy). 15 crystals were analyzed. Experiments in blue (11 crystals) define an age of 661 ± 14 ka. Red boxes are xenocrysts (4 crystals out of the 15 analyzed) and as a consequence, eliminated from the age calculation. Data are from Pereira et al. (2017)

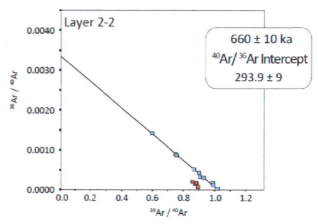

Fig. 5.11 Inverse isochron diagram obtained for the stratigraphic level N° 2-2 (Notarchirico archaeological site, Basilicata, Italy). 15 crystals were analyzed. Experiments in blue define an age of 660 ± 10 ka. Red boxes are xenocrysts, eliminated from the age calculation. Data are from Pereira et al. (2017)

Mass-Spectrometric Analysis

- **Stage 1: pumping—pre-degassing.** The sample (crystal or microcrystalline groundmass) is placed in a vacuum chamber including a crucible in the case of fusion in a furnace, and a viewport in the case of fusion by Laser CO_2), then placed under secondary high vacuum by means of turbo-molecular pumps. The sample is then heated to about 500 °C. This first gas extracted is eliminated by pumping.
- **Stage 2: fusion.** The sample is melted either by a CO_2 laser or in a double vacuum resistance furnace. In the case of fusion by laser, it applies essentially to the analysis of single crystals and to small groups of crystals (about 5–15 grains). The furnace is used for the analysis of microcrystalline groundmass. As seen before, this allows a step by step fusion, and hence a gradual degassing of the sample required for the 'step-heating' method.
- **Stage 3: purification.** The gas extracted from the sample is purified by the combined effect of Getters pumps and a titanium sublimation pump.
- **Stage 4: measurement by mass spectrometer.** After purification, the gas is introduced into the mass spectrometer. The purified gas is measured using a high-sensitivity noble gas GV5400 instrument operated in ion-counting mode. One analytical run consists of 20 peak scans of each argon isotope with integration times of 1 s (^{40}Ar, ^{39}Ar) or 10 s (^{36}Ar, ^{37}Ar, ^{38}Ar, baseline), first preceded by a peak centering routine on the five Ar isotopes, upon admission of the sample into the mass-spectrometer. The precision and accuracy of the mass discrimination correction is monitored by periodical measurements of air argon. This monitoring is performed using a dedicated air-calibration system featuring a 6 L tank filled with purified atmospheric argon. This tank is connected to the mass spectrometer vacuum line via two pneumatically- actuated air pipettes of approximately 0.1 and 1.0 cc. This system allows for a 1 cc (e.g. 600,000 counts s^{-1} (cps on ^{40}Ar)) and a 0.1 cc (e.g. 70 000 cps on ^{40}Ar) atmospheric aliquots to be delivered into the mass spectrometer and permits a careful monitoring of the mass discrimination over a wide dynamic range. The mineral standards used to calculate the flux are analyzed in the same way as the ordinary samples.

Calculation of Age

Determination of the J factor (Neutron flux received during irradiation)

J is determined for each sample. See the following analysis of a ACR-2 standard grain (1.194 Ma) subjected to a fast neutron flux for 30 min (Osiris reactor, CEA Saclay):

	^{40}Ar	^{39}Ar	^{38}Ar	^{37}Ar	^{36}Ar
Measured (mV)	9.246×10^{-3}	2.321×10^{-3}	3.576×10^{-5}	1.188×10^{-6}	1.375×10^{-6}
Blank (mV)	2.351×10^{-5}	2.586×10^{-7}	2.238×10^{-8}	6.672×10^{-7}	2.383×10^{-7}
Corrected measurement	2.312×10^{-3}	7.967×10^{-4}	1.234×10^{-5}	3.161×10^{-7}	2.981×10^{-6}

Using Eq. (5.9), J can be calculated by setting the $^{40}Ar^*/^{39}Ar_K$ ratio or R_e as follows:

$$R_e = \frac{[^{40}Ar/^{39}Ar]_m - [^{40}Ar/^{36}Ar]_A [^{36}Ar/^{39}Ar]_m + [^{40}Ar/^{36}Ar]_A [^{36}Ar/^{37}Ar]_{Ca}[^{37}Ar/^{39}Ar]_m}{1 - [^{39}Ar/^{37}Ar]_{Ca}[^{37}Ar/^{39}Ar]_m}$$
$$- \left[\frac{^{40}Ar}{^{39}Ar}\right]_K$$

with m: measured ratios (see table above).

$[^{40}Ar/^{36}Ar]_A$ = atmospheric reference ratio = 292.8 (for this sample);

$[^{36}Ar/^{37}Ar]_{Ca}$ = (given by calcium salt) 5.60×10^{-4};

$[^{39}Ar/^{37}Ar]_{Ca}$ = (given by calcium salt) $6.95 \ 10^{-4}$;

$[^{40}Ar/^{39}Ar]_K$ = (given by potassium salt) $3.52 \ 10^{-3}$;

$$R_e = \frac{3.974 - 292.8 \times 4.898 \times 10^{-4} + 292.8 \times 5.60 \times 10^{-4} \times 5.09 \times 10^{-2}}{1 - 6.95 \times 10^{-4} \times 5.09 \times 10^{-2}}$$
$$- 3.52 \times 10^{-3} = \underline{3.8355}$$

If t_s = 1.194 Ma and $\lambda = 5.543 \times 10^{-10}$; J is then calculated as follows:

$$J = \left(e^{1.194 \times 0.0000000005543} - 1\right)/3.8355 = \underline{1.726 \times 10^{-4}}$$

The measurement is repeated on at least three grains to quantify any possible external errors originating from the heterogeneity of the age standard (~1%). A weighted average is then calculated from these measurements and used to calculate the sample value.

Age Calculation of a Sample

Here are the measurements obtained for 1 sanidine crystal irradiated for 90 min in the Osiris reactor (CEA Saclay):

	^{40}Ar	^{39}Ar	^{38}Ar	^{37}Ar	^{36}Ar
Measured (mV)	2.491×10^{-03}	7.969×10^{-04}	1.241×10^{-05}	4.612×10^{-07}	3.136×10^{-06}
Blank (mV)	1.791×10^{-04}	1.515×10^{-07}	7.158×10^{-08}	1.451×10^{-07}	1.554×10^{-07}
Corrected measurement	2.312×10^{-03}	7.967×10^{-04}	1.234×10^{-05}	3.161×10^{-07}	2.981×10^{-06}

By combining the expression of J (Eq. 5.9) with Eq. 5.10, the age calculation gives:

$$t_e = \frac{1}{\lambda} \ln\left[1 + J \frac{^{40}Ar^*}{^{39}Ar_K}\right]$$

with $J = 6.530 \times 10^{-4}$ (calculated for this sample);

$\lambda = 5.543 \times 10^{-10}$ (constant for total decay of ^{40}K);

$R_e = 1.790\ 8$ (see calculation of R_e above with $[^{40}Ar/^{36}Ar]_A = 296.1$);

$t_e = 1/(5.543 \times 10^{-10}) \times \ln(1 + 6.530 \times 10^{-4} \times 1.790\ 8) = \underline{2.108\ Ma}$.

Advantages and Limitations of the $^{40}K/^{40}Ar$ and $^{40}Ar/^{39}Ar$ Methods

The table below summarizes the advantages and limitations of both methods.

Method	$^{40}K/^{40}Ar$	$^{40}Ar/^{39}Ar$
Advantages	• Rapid implementation • No need for prior irradiation of samples • Precise measurement of low amounts of $^{40}Ar^*$ (well-suited to young basalts of mid-oceanic ridges)	• Basic assumptions can be verified (age spectrum, isochrons) • Dating possible on very small sample sizes (grain by grain dating well suited to tephra)
Limitations	• The basic assumptions (initial $^{40}Ar/^{36}Ar = 298.56$, evolution in a closed system) for application of the clock are not verified • Large weight (>1 g) of sample required These two points prohibit the dating of tephra by the K-Ar method	• Pre-irradiation leads to corrections (interference of masses) • The recoil effect makes dating of very fine-grained samples or ones with a glassy texture complicated

Application: Example of the Dating of the Laschamp Event

In the chapter on magnetic stratigraphy (Chap. 7), the importance of the dating of geomagnetic events is discussed. Here, we will show how the Laschamp excursion could be correctly dated with a high degree of precision.

The dating of this excursion was obtained by a geochronological study combining the $^{40}K/^{40}Ar$ and $^{40}Ar/^{39}Ar$ methods applied to two lavas from the Massif Central (Guillou et al. 2004). Prior to this study, estimates of ages were imprecise and inconsistent with the ages deduced from other means of dating, such as astronomical calibration.

Two lava flows have been subjected to paleomagnetic and geochronological study. One of these comes from the 'Puy de Laschamp' part of the *Chaine des Puys*, located in the French Massif Central. The second, called 'the Olby flow' comes from the 'Puy de Barme' also part of the Chaine des Puys.

For each sample, the K-Ar ages (unspiked method) are calculated from two independent measures of potassium and three, also independent, measures of argon. The ages obtained for the Laschamps lava flow (41.5 ± 1.9 ka) and that of Olby (41.4 ± 1.9 ka) are identical at the two sigma level. The weighted average of these two values gives an age of 41.4 ± 1.4 ka. As for $^{40}Ar/^{39}Ar$ ages (Fig. 5.12), seven experiments out of thirteen give consistent age spectra, for which 100% of the extracted gas could be used to define a plateau age. For the other six experiments, between 76 and 96% of the extracted gas was used to define a plateau age. Furthermore, the *intercept* values calculated from the inverse isochron diagrams are equivalent to the atmospheric ratio. This indicates that the age determinations are not marred by error due to either a loss or gain of argon. The weighted average for isochron ages for the two sampling sites for the Laschamps lava flow were 39.4 ± 2.6 ka and 38.3 ± 2.6 ka. The Olby lava flow has an age of 39.2 ± 4.9 ka. The combination of these three ages give a weighted average of 38.9 ± 1.7 ka.

The K-Ar and $^{40}Ar/^{39}Ar$ measurements are compatible at the two sigma level. As these two flows recorded the same paleomagnetic excursion, this is dated to 40.4 ± 1.1 ka. Note that if we take the uncertainty (2.4%) on the potassium decay constant into account, the error in age goes from 1.1 to 2.0 ka. Thus, the age retained for the excursion is 40.4 ± 2.0 ka. This age is comparable to that obtained by independent chronological methods (see Chap. 7). A new study (Laj et al. 2014), combining K-Ar and $^{40}Ar/^{39}Ar$ dating, associated with paleomagnetism, was applied to a larger number of volcanoes from the Chaine des Puys, and has since made it possible to narrow down the age of the Laschamps excursion to 41.2 ± 1.6 ka.

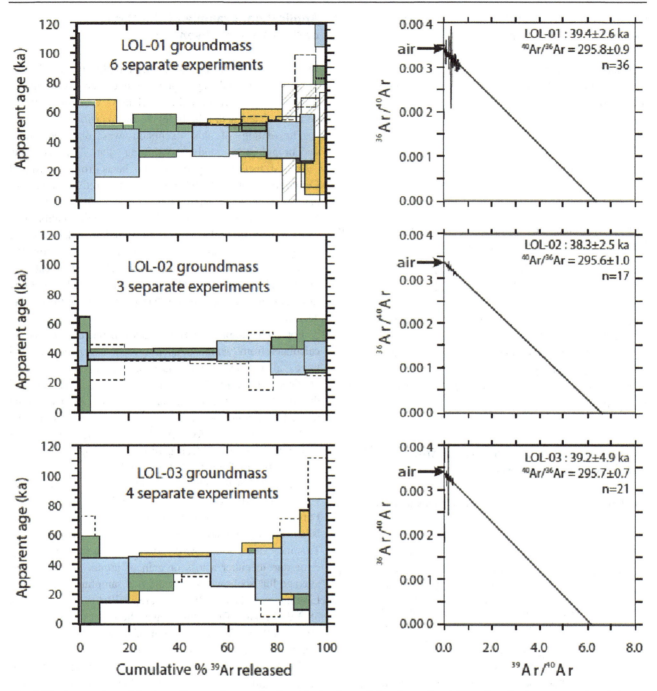

Fig. 5.12 Age spectra and isochron diagrams for samples from the Laschamp-1, Laschamp-2 and Olby sites. Isochron ages and statistics are weighted mean values from experiments on all the subsamples, (in Guillou et al. 2004). n = number of steps retained in the age calculation

This example highlights the potential of the K-Ar dating clock to date recent events of the Quaternary period. One of the applications of these chronological data to calibrate time scales, with various other useful applications in the earth sciences, archeology and paleoclimatology.

References

Berger, G. W., & York, D. (1981). Geothermometry from $^{40}Ar/^{39}Ar$ dating experiments. *Geochimica et Cosmochimica Acta, 45*, 795–811.

Cassignol, C., Cornette, Y., David, B., & Gillot, P. Y. (1978). Technologie potassium-argon. C.E.N., Saclay, Rapport CEA R-4802, 37 p.

Cassignol, C., & Gillot, P.-Y. (1982). Range and effectiveness of unspiked potassium-argon dating: Experimental groundwork and examples. In G. S. Odin (Ed.), *Numerical dating in stratigraphy* (pp. 159–179). Chichester: Wiley.

Charbit, S., Guillou, H., & Turpin, L. (1998). Cross calibration of K-Ar standard minerals using an unspiked Ar measurement technique. *Chemical Geology, 150*, 147–159.

Dalrymple, G. B., Alexander, E. C. Lanphere, M., & Kraker, G. P. (1981). Irradiation of samples for $^{40}Ar/^{39}Ar$ dating using the geological survey TRIGA reactor. U.S. Geological Survey Professional Paper, 1176.

Dalrymple, G. B., & Lanphere, M. A. (1969). In J. Gilluly & A. O. Woodford (Eds.), *Potassium-argon dating. Principles, techniques and applications to geochronology* (251 p.). San Francisco, CA: W. H. Freeman and Company.

Dalrymple, G. B., & Lanphere, M. A. (1974). $^{40}Ar/^{39}Ar$ age spectra of some undisturbed terrestrial samples. *Geochimica et Cosmochimica Acta, 38*, 715–738.

Deino, A. L., & Potts, R. (1992). Age-probability spectra for examination of single crystal $^{40}Ar/^{39}Ar$ dating results: Examples from Olorgesailie, Southern Kenya Rift. *Quaternary International, 13*(14), 47–53.

Flish, M. (1982). Potassium-argon analysis. In G. S. Odin (Ed.), *Numerical dating in stratigraphy* (pp. 151–158). New York: Wiley.

Fuhrmann, U., Lippolt, H., & Hess, C. J. (1987). HD-B1 biotite reference material for K-Ar Chronometry. *Chemical Geology, 66*, 41–51.

Garner, E. L., Murphy, T. J., Gramlich, J. W., Paulsen, P. J., & Barnes, I. L. (1975). Absolute isotopic abundance ratios and the atomic weight of a reference sample of potassium. *Journal of Research of the National Bureau of Standards, 79A*, 713–725.

Guillou, H., Singer, B. S., Laj, C., Kissel, C., Scaillet, S., & Jicha, B. R. (2004). On the age of the Laschamp geomagnetic excursion. *Earth and Planetary Science Letters, 227*(3–4), 331–343.

Jicha, B. R., Singer, B. S., & Sobol, P. (2016). Re-evaluation of the ages of $^{40}Ar/^{39}Ar$ sanidine standards and super-eruptions in the western US using a Noblesse multi-collector mass spectrometer. *Chemical Geology, 431*, 54–66.

Kuiper, K. F., Deino, A., Hilgen, F. J., Krijgsman, W., Renne, P. R., & Wijbrans, J. R. (2008). Synchronizing rock clocks of Earth history. *Science, 320*, 500–505.

Laj, C., Guillou, H., & Kissel, C. (2014). Dynamics of the earth magnetic field in the 10-75 kyr period comprising the Laschamp and Mono Lake excursions: New results from the French Chaîne des Puys in a global perspective. *Earth and Planetary Science Letters, 387*, 184–197.

Lanphere, M. A., & Dalrymple, G. B. (2000). First principles calibration of ^{38}Ar Tracers. U.S. Geological Survey Professional Paper, 1621.

Lee, J. Y., Marti, K., Severinghaus, K., Kawamura, K., Yoo, H. S., Lee, J. B., et al. (2006). A redetermination of the isotopic abundances of atmospheric Ar. *Geochimica Cosmochimica Acta, 70*, 4507–4512.

McDougall, I., & Harrison, T. M. (1988). *Geochronology and thermochronology by the $^{40}Ar/^{39}Ar$ method* (p. 212). New York: Oxford University Press.

McDougall, I., & Roksandic, Z. (1974). Total fusion $^{40}Ar/^{39}Ar$ ages using HIFAR reactor. *Geological Society of Australia, 21*, 81–89.

Merrihue, C. (1965). Trace element determinations and potassium argon dating by mass spectroscopy of neutron irradiated samples. *Transactions American Geophysical Union, 46*, 125.

Merrihue, C., & Turner, G. (1966). Potassium-argon dating by activation with fast neutrons. *Journal of Geophysical Research, 71*, 2852–2857.

Min, K. W., Mundil, R., Renne, P. R., & Ludwig, K. R. (2000). A test for systematic errors in $^{40}Ar/^{39}Ar$ geochronology through comparison with U-Pb analysis of a 1.1 Ga rhyolite. *Geochimica Cosmochimica Acta, 64*, 73–98.

Mitchell, J. G. (1968). The Argon-40/Argon-39 dating in coesite-bearing and associated units of the Dora Maira Massif, Western Alps. *European Journal of Mineralogy, 3*, 239–262.

Nier, A. O. (1950). A redetermination of the relative abundances of the isotopes of carbon, nitrogen, oxygen, argon and potassium. *Physical Review, 77*, 789–793.

Niespolo, E. M., Rutte, D., Deino, A. L., & Renne, P. R. (2017). Intercalibration and age of the Alder Creek sanidine $^{40}Ar/^{39}Ar$ standard. *Quaternary Geochronology, 39*, 205–213.

Odin, G. S. (1982). Interlaboratory standards for dating purposes. In G. S. Odin (Ed.), *Numerical dating in stratigraphy* (pp. 123–158). New York: Wiley.

Pereira, A., Nomade, S., Bahain, J.-J., & Piperno, M. (2017). Datation par $^{40}Ar/^{39}Ar$ sur monocristaux de feldspaths potassiques: exemple d'application sur le site pléistocène moyen ancien de Notarchirico (Basilicate, Italie). *Quaternaire, 28*(2), 149–154.

Phillips, D., Matchan, E. L., Honda, M., & Kuiper, K. F. (2017). Astronomical calibration of $^{40}Ar/^{39}Ar$ reference minerals using high-precision, multi-collector (ARGUS VI) mass spectrometry. *Geochimica Cosmochimica Acta, 196*, 351–369.

Renne, P. R. (2000). In J. S. Noller, J. M. Sowers, & W. R. Lettis (Eds.), *K-Ar and $^{40}Ar/^{39}Ar$ dating, in quaternary geochronology: Methods and applications*. Washington, D.C.: American Geophysical Union. https://doi.org/10.1029/rf004p0077.

Renne, P. R., Mundil, R., Balco, G., Min, K., & Ludwig, K. R. (2010). Joint determination of 40 K decay constants and $^{40}Ar^*/^{40}K$ for the Fish Canyon sanidine standard and improved accuracy for $^{40}Ar/^{39}Ar$ geochronology. *Geochemica et Cosmochimica Acta, 74*, 5349–5367.

Roddick, J. C. (1983). High precision intercalibration of $^{40}Ar-^{39}Ar$ standards. *Geochemical and Cosmochimical Acta, 47*, 887–898.

Samson, S. D., & Alexander, E. C., Jr. (1987). Calibration of interlaboratory $^{40}Ar-^{39}Ar$ Dating Standard, MMhb-1. *Chemical Geology, 66*, 27–34.

Steiger, R. H., & Jäger, E. (1977). Subcommission on geochronology: Convention on the use of decay constants in geo- and cosmochronology. *Earth and Planetary Science Letters, 5*, 320–324.

Turner, G., Huneke, J. C., Podosek, F. A., & Wasserburg, G. J. (1971). $^{40}Ar-^{39}Ar$ ages and cosmic ray exposure rays of apollo 14 samples. *Earth and Planetary Science Letters, 12*, 19–35.

Turner, G., Miller, J. A., & Grasty, R. L. (1966). The thermal history of the bruderheim meteorite. *Earth and Planetary Science Letters, 1*, 155–157.

Wänke, H., & König, H. (1959). Eine neue Methode zür Kalium-Argon Alterbestimmung und ihre Anwendung auf Steinmeteorite. *Zeitschrift für Naturforschung A, 14a*, 860–866.

Dating of Corals and Other Geological Samples via the Radioactive Disequilibrium of Uranium and Thorium Isotopes

Norbert Frank and Freya Hemsing

Abstract

U/Th dating methods have become cornerstone tools for the determination of the age of climate change recorded in marine and continental carbonates. Here we describe the theoretical principles and analytical methods along the example of U/Th dating the aragonite skeletons of tropical corals. We demonstrate that a precision limiting factor is built in the dating principle, known as U-series open system behavior. Moreover, above all the quality of the samples is crucial for a successful and accurate age determination. When using well preserved fossil coral fragments ages provide measures of past sea level, contribute to the calibration of the radiocarbon time scale, and allow for the reconstruction of reef accumulation rates, tectonic subsidence, or uplift. We finally emphasize that U/Th dating also works for secondary carbonates such as stalagmites, calcareous tuff and travertine, but the boundary conditions regarding the U concentration, and the initial U and Th isotopic composition vary wildly and need careful consideration. When doing so, highest precision and accuracy can Is feasible.

Dating methods based on the radioactive disequilibrium in the uranium decay series were developed over the last fifty years. They can be applied to minerals that, at the time of their formation, incorporate uranium into their crystal lattice, but not thorium, whose isotope, ^{230}Th, is a daughter of ^{234}U (Fig. 6.1). This is the case for corals which form their aragonitic, calcareous skeletons from elements present in the seawater. They are used as an example throughout this chapter. The basic concept is that, in seawater, there is dissolved uranium but very little thorium, an insoluble element. Therefore, each crystal of aragonite formed by the coral incorporates only uranium and not its first-generation daughters which are all thorium isotopes. Uranium concentration in seawater is very stable and homogeneous with 3.3 μg of U per liter of water. The activity ratio (^{234}U/^{238}U) is also very stable in the ocean and is slightly above the radioactive equilibrium with a value of 1.1468 ± 0.0001 in the open ocean (Andersen et al. 2010). This slight excess of ^{234}U is due to preferential leaching of this isotope from rocks during weathering of the continental crust (Ivanovich and Harmon 1992). Henderson (2002) proposed that the seawater ratio (^{234}U/^{238}U) has remained constant for at least the past 800,000 years. However, moderate variations of ± 0.01 in this ratio are possible, due to changes in weathering of the continental crust, sea level changes, and variations in freshwater runoff from rivers resulting from climate changes (Esat and Yokoyama 2006). Very recently, it has been suggested that this ratio may also be dependent on the ocean circulation (Chen et al. 2016).

Let us go back to the coral. If the skeleton of a coral remains a chemically closed system after its formation (in other words, without any exchange of uranium or thorium with its sedimentary environment) the ^{230}Th from the decay of ^{234}U accumulates progressively over time, while the excess of ^{234}U decreases.

The state of this radioactive disequilibrium allows for a very precise determination of the coral age depending on the measurement technique used. When this dating method was developed, the isotopes of uranium and thorium were measured by their radioactivity, either directly by α spectrometry or indirectly by γ spectrometry, enabling age determination ranging from a few thousand years to about 300,000 years. Now, we are able to measure the abundance of these isotopes by characterizing them according to their masses, giving a much better accuracy. The age range measurable by thermal ionization mass spectrometry or even multi-collector inductively coupled plasma source mass spectrometry now

N. Frank
Laboratoire des Sciences du Climat et de L'Environnement, LSCE/IPSL, CEA-CNRS-UVSQ, Université Paris-Saclay, Gif-Sur-Yvette, 91190, France

N. Frank (✉) · F. Hemsing
Institute of Environmental Physics, Heidelberg University, Im Neuenheimer Feld 229, 69120 Heidelberg, Germany
e-mail: Norbert.Frank@iup.uni-heidelberg.de

© Springer Nature Switzerland AG 2021
G. Ramstein et al. (eds.), *Paleoclimatology*, Frontiers in Earth Sciences,
https://doi.org/10.1007/978-3-030-24982-3_6

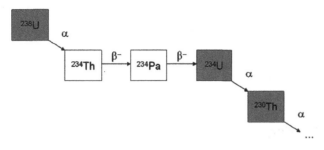

Fig. 6.1 Start of the radioactive decay chain ^{238}U

allows the dating range to be extended from a few years to a few hundred thousand years (>500,000 years). As a result, U/Th dating has now become an indispensable tool for precision geochronology and research on the environmental and climate changes in the late Quaternary.

Undoubtedly, one of the greatest successes of this chronometer is the precise reconstruction of sea levels throughout the climate cycles from growth series of corals (Thompson and Goldstein 2006). This geochronological technique also permits the determination of subsidence or elevation of reefs caused by tectonic movements over time (Frank et al. 2006). Significant progress has also been made on the calibration of ^{14}C ages by combined ^{230}Th/U and ^{14}C analysis (Reimer et al. 2013).

Over the past decade, it has become clear that the mineral system of a coral is not a completely closed system but sometimes uranium and thorium exchanges with its sedimentary environment occur. This conclusion was reached from the observation that uranium ^{234}U and ^{230}Th are often in excess of the theoretical levels of the activity ratios (^{234}U/^{238}U and ^{230}Th/^{238}U) that would be expected from radioactive decay. These inconsistencies are the result of recrystallization, early diagenesis, and radioactive decay of uranium. This makes the dating less precise than the analysis might indicate and makes it necessary to apply corrections to the estimated ages.

In light of these observations, correction models, categorized as 'open' system models were developed which take account of the influence of such disturbances on tropical corals. (Scholz et al. 2004, Szabo et al. 1994)

The focus of this chapter is to explore the dating of tropical corals and other carbonate climate archives using the U/Th method including open-system models, but excluding microstructure and biomineralization. We will examine the methodology in detail including the relevance of the open system and also some applications to geology and paleoclimatology.

This principle of dating by radioactive disequilibrium in the decay chain of uranium applies to many sources of climate records, such as deep-water corals, mollusk shells, and secondary carbonate precipitations on land (stalagmites and travertine). In each case, the formation process of these minerals favors the incorporation of uranium over its radioactive daughters. However, each mineral deposit has its own characteristics with regard to the incorporation of uranium and thorium and open system behavior Cheng et al. (2000a), Mallick and Frank (2002). Here, we will focus on one particular archive: tropical corals. At the end of the chapter, the possibility of dating other geological samples through uranium series disequilibrium will be analyzed briefly.

Methodology of ^{230}Th/^{238}U Dating

Principle of ^{230}Th/^{238}U Dating

This method is based on the radioactive decay chain of ^{238}U (Fig. 6.1).

The rates of decay in this radioactive chain are highly variable, ranging from a few billion years to hours. The radioactivity of ^{234}Th and ^{234}Pa decreases very rapidly with half-lives (also called periods) of 24.1 days and 6.7 h respectively. Therefore, these isotopes are insignificant relative to the periods of ^{234}U and ^{230}Th, key isotopes in ^{230}Th/^{238}U dating. During the formation of the skeleton, only the U isotopes are incorporated, hence ^{230}Th isotope is absent (^{230}Th concentration is zero at $t = 0$). It is therefore only by decay of ^{234}U that ^{230}Th accumulates over time in the coral structure.

To date a sample, it is essential to accurately measure the activities of ^{238}U, ^{234}U and ^{230}Th. Dating is performed by using the radioactive decay equations to define the activity ratios of ^{234}U/^{238}U and ^{230}Th/^{238}U so that the time elapsed since the formation of the coral skeleton can be calculated (Eqs. 6.1 and 2). The decay equation cannot be solved analytically but ages are estimated by iteration.

$$\left(\frac{^{230}Th}{^{238}U}\right) = 1 - e^{-\lambda_{230}t} + \frac{\delta^{234}U_m}{1000} \times \frac{\lambda_{230}}{\lambda_{230} - \lambda_{234}} \times \left(1 - e^{-(\lambda_{230}-\lambda_{234})t}\right) \quad (6.1)$$

$$\left(\frac{^{234}U}{^{238}U}\right) = \left(\frac{^{234}U}{^{238}U}\right)_{initial} \times \left(1 - e^{-\lambda_{234}t}\right) \quad (6.2)$$

with λ_{230} and λ_{234}, the constants of decay of ^{230}Th and ^{234}U respectively (Table 6.1), and t the time elapsed since the system closed.

In Eq. (6.1), the activity ratio (^{234}U/^{238}U) is expressed as ‰ compared to the radioactive equilibrium:

$$\delta^{234}U = \left[\frac{\left(\frac{^{234}U}{^{238}U}\right)}{\left(\frac{^{234}U}{^{238}U}\right)} - 1\right] \times 1000 \quad (6.3)$$

Table 6.1 Decay of the daughters of ^{238}U by Cheng et al. (2000b) with $T_{1/2}$ the half-life and λ the decay constant

	$T_{1/2}$ (years)	λ
^{238}U	4,468,314,000	1.55125×10^{-10}
^{234}U	245,250	2.82629×10^{-06}
^{230}Th	75,690	9.15771×10^{-06}

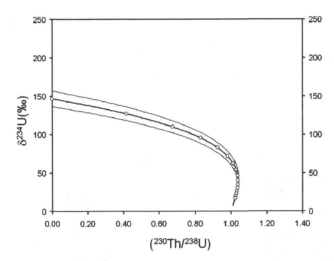

Fig. 6.2 Temporal evolution of the activity ratios (^{230}Th/^{238}U) and δ^{234}U in a closed system. The points (○) show the position of the coral every 50 thousand years

For the method to be applicable, two basic conditions need to hold:

1. During the formation of the aragonitic skeleton, only uranium is incorporated, and therefore ^{230}Th is absent.
2. The system remains closed to any exchange of uranium and thorium with the sedimentary environment after the aragonite is formed.

If a coral satisfies these essential conditions, the activity ratios of (^{234}U/^{238}U) and (^{230}Th/^{238}U) evolve over time (Fig. 6.2—assuming an activity ratio of (^{234}U/^{238}U) in seawater equal to 1.1468 (δ^{234}U = 146.8‰) at the moment of precipitation of the aragonite).

The accuracy of U/Th dating depends mainly on the accuracy with which the isotope ratios are measured, but the quality of the sample also plays a critical role because it determines whether the fundamental conditions for dating are met.

All of the radionuclides are alpha emitters. Historically, alpha spectrometry, i.e. the direct measurement of the alpha emission of each radionuclide, was the preferred method. However, in the late 1980s, the use of mass spectrometry, first with thermal-ionization, and nowadays with double-focusing inductively coupled plasma source mass spectrometry, became feasible leading to very high precision measurements of ^{238}U and its long-life daughters (^{234}U, ^{230}Th). The mass spectrometric methods have the advantage of directly measuring the number of atoms of an isotope, instead of measuring its radioactivity. This results in a very significant gain in accuracy because instead of measuring a few alpha emissions per minute, tens to millions of ions per second are detected. However, a very rigorous preliminary chemical treatment is essential as the presence of any other elements during ionization will decrease the quality of the measurement or even lead to isobaric interferences. From the number of atoms of each radionuclide measured by mass spectrometry, its activity A_i can be calculated according to the law of decay $A_i = \lambda \times N_i$; where N_i is the number of atoms measured by mass spectrometry and λ is the decay constant of the radionuclide. To calculate the activities of the daughters of ^{238}U from isotopic measurements, we use the decay rates shown in Table 6.1. These decay times have recently been updated Cheng et al. (2013), but we have opted to use the older values (Table 6.1) here because the re-evaluated values by Cheng et al. (2013) have still not been independently confirmed and are identical to the previous ones within a range of uncertainty.

In the following paragraphs, we will discuss the analytical aspect. However, we will firstly specify the sample selection criteria, before explaining the chemical process and finally the physical measurement (Fig. 6.3).

Selecting a Coral for ^{230}Th/^{238}U Dating

In order to get good accuracy on the age of the corals, it is crucial to select the best-preserved samples possible. The purpose of the analysis is to obtain ratios between ^{238}U, ^{234}U and ^{230}Th, produced exclusively by radioactive decay, and thus independent of any secondary addition or any depletion of uranium or thorium over the thousands of years the coral has spent as fossil in the reef environment. For this purpose, the selected coral fragments are physically and chemically cleaned of any metal oxide encrustations, traces of clay or organic residue identified beforehand by binocular microscope. This cleaning is carried out over several iterations of rinsing in an ultrasonic bath with ultrapure water, followed by diluted acid baths and oxidizing solutions. Finally, traces of potential bioerosion, caused for example by foraging mollusks, are sought out, and the contaminated parts eliminated.

To test the quality of the selection and cleaning processes, analysis by X-ray diffraction is conducted. This technique

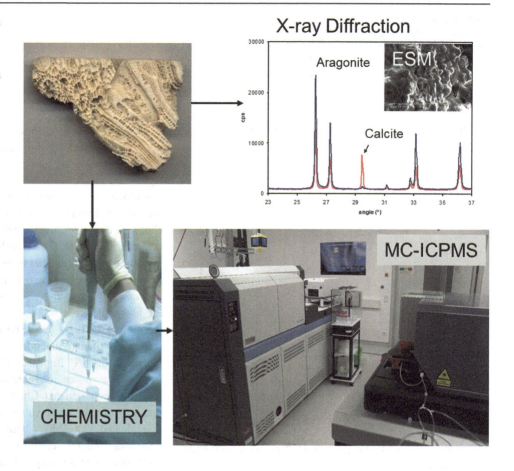

Fig. 6.3 Schematics of the U/Th disequilibrium methodology for fossil tropical coral. Subsamples of coral are taken from the skeleton in order to check that it is composed entirely of aragonite (analysis by X-ray diffraction). The scanning electron microscopy (ESM) identifies secondary aragonite fibers and signs of coral dissolution. Uranium and thorium are then extracted, chemically purified from the carbonate, and their isotopes are measured by inductively coupled plasma source mass spectrometry (MC-ICPMS)

identifies the abundance of various carbonate minerals in the sample, such as aragonite and calcite, the latter being either low or high in magnesium. Only samples identified with more than 98% of aragonite qualify for U/Th dating. More than 2% of calcite would indicate recrystallization and therefore poor preservation of the skeleton. In a less systematic approach, scanning electron microscopy may be carried out to determine the presence of micro traces of dissolution or precipitation of secondary aragonite fibers.

This selection process of the sample prior to dating is onerous, but often necessary, to ensure ages with the best accuracy and 'precision' possible. However, the samples are of macroscopic size, varying from a few tens to a few hundred milligrams, and a piece of coral is rarely perfectly preserved. The results of microanalysis on 1–5% aliquots of the sample, mean that the state of preservation was only tested on part of the sample later used for U/Th dating.

Chemical Procedure

The samples are placed in a strong acid solution (nitric acid or hydrochloric acid), and undergo a chemical treatment which involves several steps. Mass spectrometry measurements are of isotopic ratios, such as $^{234}U/^{238}U$ and $^{230}Th/^{232}Th$. The concentration or activity of the nuclides or the $^{230}Th/^{238}U$ ratio needed to calculate the age of the sample cannot be directly estimated. Consequently, a tracer, known as a 'spike', which contains isotopes of uranium and thorium that do not exist in the natural environment and have a well-established concentration, is added to the solution. These artificial isotopes allow the calculation of the concentration of natural nuclides. Generally, any U/Th dating by mass spectrometry depends on spikes containing ^{233}U, ^{236}U and ^{229}Th. Thus, measurements of the isotopic ratios $^{234}U/^{238}U$, $^{236}U/^{238}U$, $^{233}U/^{236}U$, $^{230}Th/^{229}Th$ and $^{232}Th/^{229}Th$ are needed to determine the concentration of ^{238}U, ^{230}Th and ^{232}Th in the sample, its isotopic ratios $^{234}U/^{238}U$, $^{230}Th/^{232}Th$ and $^{230}Th/^{238}U$, and the corresponding activity ratios.

Once chemical equilibrium between the spike and the sample in solution is reached, uranium and thorium are separated from the major, minor and trace elements by a column chemistry using an ion exchange resin. Several types of resins are used to purify the uranium and thorium from the sample. During the early days of Th/U dating most laboratories used successive series of anionic resin columns DOWEX 1X8. Nowadays, there are also separations for uranium and thorium using specifically designed resins such as the UTEVA resin, which allows for faster and highly effective purification. The sample dissolved in 3 N nitric acid is deposited on a column of UTEVA resin (0.5 ml), loaded in HNO_3 3 N. The column is rinsed several times in

3 N nitric acid (10 column volumes), then uranium and thorium are extracted with hydrochloric acid at different molarities (9–1 N) Douville et al (2010). This technique of chemical separation can be used to purify the uranium and thorium of any other component of the coral with an approximated yield of 100%. The entire chemical procedure can be done in approximately two days for about fifteen samples. For physical measurements, a similar time is required using the most effective tools, such as the MC-ICPMS described below. More recently it has become feasible to extract and purify uranium and thorium using automated extraction systems Wefing et al. (2017), but currently these tools remain the exception in the chemical preparation for U/Th dating. Lastly, the use of laser ablation systems has allowed for the direct extraction of material from samples, which are carried with a gas stream to the source of a mass spectrometer to detect the abundance of uranium and thorium isotopes. However, those techniques allow for rapid-age screening, but remain insufficiently precise for high precision age determination (Spooner et al. 2016).

Physical Measurement by Mass Spectrometry

After chemical purification, the fractions of uranium and thorium are deposited on a pure rhenium filament if analysis is performed on a mass spectrometer with a thermal source. For inductively coupled plasma mass spectrometry, the final solutions are diluted so that each sample has a similar concentration. The uranium and thorium atoms are ionized in the source of a mass spectrometer. Formerly thermal ionization of a solid sample from a heated filament was performed in thermal ionization mass spectrometry (TIMS) at temperatures of up to 1650 to 1800 °C. Most recent generations of mass spectrometers (MC-ICPMS) have substantially improved the ionization yield and hence the analytical accuracy due to their detection of the ions by simultaneous multi-collection and due to ionization temperatures in an Ar plasma of up to 8000 °C. Ions are accelerated by a high voltage and subsequently energy-filtered in an electrostatic filter and deflected by a magnet according to their mass to charge ratio. The ions are ultimately collected in a multi-collection system composed of several faraday cups and electron multipliers. With the advent of novel ultra-high resistances, faraday cups become available for low ion intensities otherwise typically measured on electron multipliers. The MC-ICPMS technology now makes it possible to determine isotopic ratios of uranium with an accuracy of less than 1‰, and of thorium with an accuracy of ± 1‰. This progress has resulted in a U/Th dating with a much higher analytical precision and accuracy than that obtained by TIMS.

Measurement routines using these complex instruments vary and require rigorous data processing. Data processing must, among other things, correct for the effects of mass fractionation related to the measurement, perform comparisons between standards and samples, and consider the impact of background noise from the instruments and from the chemical process (chemical blanks) on the measurements. Uranium standards, HU1 or NBL 112, with known isotopic ratios, are generally used to determine the reproducibility of physical measurements.

Overall, cutting edge mass spectrometry can achieve a dating with a higher precision than 100 years on corals of around 100,000 years. This possibility remains, however, theoretical due to the fact that the error on the final age not only depends on the quality of the physical measurement, but also, and above all, on the sample quality.

Once the isotopic ratios are measured, Eqs. 6.1 and 6.2 shown above are used to estimate the age of the coral and its initial $^{234}U/^{238}U$ ratio.

Limitations of the Method

Theoretically, the accuracy of U/Th dating is determined by the precision of the measurements by mass spectrometry of the isotopic ratios $^{230}Th/^{238}U$ and $^{234}U/^{238}U$. For this, it is assumed that the coral being studied was perfectly preserved and that the basic conditions mentioned above were respected. Therefore, the assumptions are that only uranium was incorporated into the aragonite skeleton at the time of its formation, that the system remained closed to any exchange of uranium and thorium with the sedimentary environment after the formation of aragonite, and finally, that the coral skeleton contained no ^{230}Th at the time of its formation. In

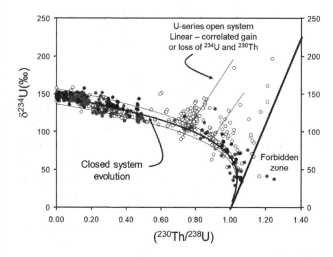

Fig. 6.4 Compilation of $\delta^{234}U$ activity ratios based on $^{230}Th/^{238}U$ ratios of fossil corals (black symbol: deep-water solitary corals; gray symbol: deep-water, reef-building corals; open symbol: tropical corals). The solid black line represents the variation in a closed system compared to the $\delta^{234}U$ ratio of present-day seawater. The dots on the right of the diagram in the 'forbidden' zone cannot be explained by the variation in a closed system. The straight dashed line indicates the distribution of isotopic ratios for co-genetic samples

addition, this dating method is based on the assumption that the activity ratio ^{234}U/^{238}U of sea water (equal to 1.1468) was constant over the last 500,000 years. For practical reasons, this ratio is commonly expressed as the relative deviation (in ‰) from radioactive equilibrium and is denoted δ^{234}U (Eq. 6.3).

Thus, today's δ^{234}U of seawater is 146.8 ± 0.1‰ (Andersen et al. 2010). Given the conditions mentioned above, using Eq. 6.2, we would expect to find an initial ratio of ^{234}U/^{238}U for the coral similar to that of seawater. However, it was observed that many corals more than 80,000 years old had a wide range of values for this ratio, and often values exceed the current value for seawater (Fig. 6.4). For example, a sample 125,000 years old, divided into several small pieces, can present age differences between its various fragments of more than 10,000 years, even though the measurement accuracy is ± 1000 years for each subsample. Variability of age within the same sample can be ten times larger than the measurement precision. In addition, the sub-samples of the same specimen can also show significant variability in their ^{234}U/^{238}U ratio over time. This highlights the crucial role of the ^{234}U/^{238}U ratio (or δ^{234}U) in the U/Th dating method for tropical corals.

Henderson and Slowey (2000) and Gallup et al. (1994) were the first to realize that the increase of ^{234}U and of ^{230}Th in ancient marine carbonates are often correlated (Fig. 6.4). This proved to be a very important observation because diagenesis, such as the dissolution of the skeleton or precipitation of secondary aragonitic fibers, cannot explain this correlation. For example, the dissolution of the coral skeleton during diagenesis will return uranium to pore waters, but the thorium will be quickly re-precipitated because of its inability to remain in solution. Consequently, the ^{230}Th/^{238}U ratio of the coral will increase and the age calculated will be overstated. This process does not involve a significant change in the isotopes of uranium, so it is expected that sub-samples of the same coral will have variable ^{230}Th/^{238}U ratios but a constant δ^{234}U. This equilibrium is reversed with the precipitation of secondary aragonitic fibers, as the fibers contain uranium but no thorium, leading to a reduction in the ^{230}Th/^{238}U ratio of a coral and an underestimate of age. Since these secondary fibers are younger than the skeleton and they supply uranium taken from seawater, the δ^{234}U of coral increases slightly but cannot exceed the δ^{234}U of seawater, i.e. 146.8‰ in total. It is clear that the disturbance in the dating system cannot be entirely attributed to these processes. In particular, this does not explain the observed depletion of ^{234}U and ^{230}Th.

With these points in mind, Henderson and Slowey (2000) and Gallup et al. (1994) proposed that the nuclear recoil effect resulting from the radioactive decay of ^{238}U and ^{234}U is responsible for the disturbance and developed a method to take this into account.

The Nuclear Recoil Effect and the 'Open' Dating System

During the radioactive decay of uranium isotopes, an α particle is ejected from the nucleus. The balance of kinetic energy requires that the nucleus produced (^{234}Th or ^{230}Th) recoils and therefore moves backward. In both cases, a thorium isotope is produced. The ^{234}Th in ^{234}U decays quickly ($T_{1/2}$ = 24.1 days) by emitting an electron. This process is not responsible for the movement of the nucleus, because the recoil energy is too low. Therefore, the backwards movement of the nucleus affects ^{234}U (which occupies the position of ^{234}Th) and ^{230}Th. The mobile nucleus can remain inside the crystal lattice but can also be ejected and pass through the pore fluids, or even into another neighboring crystal lattice (coral or sediment). This process, which occurs in both carbonates and sediments, or in organic matter present in a coral reef, is time-dependent: the more time has elapsed, the more the nuclei will have moved.

Obviously, this action is on a very small scale as displacement of the nucleus occurs over less than 20 nm. Therefore, it cannot be measured directly, but observed variations in concentration of the isotopes ^{234}U (^{234}Th) and ^{230}Th likely reflect the overall redistribution of radionuclides over the entire time elapsed since coral formation.

Thompson et al. (2003) and Villemant and Feuillet (2003) were the first to incorporate this process into the equation for dating (Eq. 6.1). Here, we restrict ourselves to the theoretical approach of Thompson and his colleagues, because, on the purely mathematical level, the two models of radionuclide redistribution are the same. The idea is simply to add a term to the laws of decay (Eqs. 6.1 and 6.2), that takes the recoil effect for ^{234}U and ^{230}Th into account. Equations 4 and 5 are essentially the same as Eqs. 6.1 and 6.2:

$$\left(\frac{^{230}Th}{^{238}U}\right) = \left(\frac{^{230}Th}{^{238}U}\right)_{initial} \times e^{(-\lambda_{230}t)} + f_{230}f_{234}\left(1 - e^{(-\lambda_{230}t)}\right) + f_{230}\frac{\lambda_{230}}{\lambda_{230} - \lambda_{234}}\left(e^{(-\lambda_{230}t)} - e^{(-\lambda_{234}t)}\right)(f_{234} - R_0)$$
(6.4)

$$\left(\frac{^{234}U}{^{238}U}\right) - f_{234} = \left(\left(\frac{^{234}U}{^{238}U}\right)_{initial} - f_{234}\right)e^{(-\lambda_{234}t)} \quad (6.5)$$

R_0 in Eq. 6.4 corresponds to the initial value of ^{234}U/^{238}U in seawater fixed at 1.148 ± 0.010 (current value ± 10‰ variability). f_{234} and f_{230} represent the proportions, expressed as activities, lost ($f < 1$) or gained ($f > 1$) following redistributions brought about by the nuclear recoil effect.

The redistribution factor f_{234} is estimated iteratively from the difference between the ^{234}U/^{238}U ratio resulting from the temporal evolution in a closed system and the corresponding evolution in an 'open' system.

The link between f_{230} and f_{234} (Eq. 6.6) is the difference of kinetic energy injected into the crystal lattice following the decay of ^{238}U and ^{234}U. The process of an α particle emission is the same with only the kinetic energy being different.

$$f_{234} = 1.157 \cdot (1 - f_{230}) \text{ (Villemant and Feuillet 2003)} \tag{6.6}$$

The definition of the redistribution factor f is purely mathematical, so gains and depletions in an open system for uranium are expressed as $f > 1$ and $f < 1$. Up to this point, the two calculations of Villemant and Feuillet (2003), and of Thompson et al. (2003) are identical. Assuming that the measured ratios can be used to determine f as defined above, it is possible to estimate the age of the coral in an open system (Eq. 6.4). This approach assumes that only the uranium and thorium in the coral are at the origin of the recoil process. However, this assumption is not always correct. In fact, the process of nuclear recoil, and thus the possible ejection of nuclei over time, does not allow a gain of either ^{234}U or ^{230}Th; the coral can only lose radionuclides. Therefore, a value of $f > 1$ is not expected. But the reality is quite different, because most of the corals show an increase ($f > 1$), and only very few corals show a depletion of radionuclides ($f < 1$).

Thompson and his colleagues considered this obstacle to be theoretical and therefore introduced further complexity in their approach. The gain of radionuclides can be explained, either by direct exchange of nuclei between corals very close to each other, or by fluids circulating in the reef, in other words, from an external source. Consequently, f is always < 1 and the gain in ^{234}U and ^{230}Th is the sum of the depletions over time and of the gain due to the retention of radionuclides ejected and/or transported in the reef. To account for this phenomenon, Thompson and his colleagues established the following equations derived for a simple exchange model:

Instead of estimating f directly from the activity ratios measured (Eq. 6.6), the values of f are fixed, but the excess of ^{234}U and redistribution slope m are estimated iteratively taking an external source into account (other corals or carbonates). Ultimately, we end up with equations that have the same form as those derived from Villemant and Feuillet's model (2003), because the source considered in this model is crucial and must have a composition similar to that of the corals. However, here it is possible to vary the parameters in the model and to find the slope m with the best fits for a set of samples of the same age.

These equations are used to correct for the nuclear recoil phenomenon and to place the values of the activity ratios measured, $(^{230}Th/^{238}U)$ and $(^{234}U/^{238}U)$, on a graph of variations in these ratios within a closed system to calculate the age of the corals.

For example, Fig. 6.5 shows the activity ratios $(^{230}Th/^{238}U)$ and $(^{234}U/^{238}U)$ measured in several corals of a coral reef on the Amedee Island, off the coast of New Caledonia (Frank et al. 2006). Ages calculated using Eqs. 6.1 and 6.2 and the measured activity ratios show a wide dispersion, from 123,600 to 146,000 years. These samples also show a high variability in the initial $\delta^{234}U$, ranging from 119.2 to 211‰. A linear correlation between the ratios $(^{230}Th/^{238}U)$ and $(^{234}U/^{238}U)$ is obvious. This part of the reef very likely developed during the last interglacial period (isotopic stage 5), corresponding to the last sea level maximum and dating back to about 125,000 years. The results therefore demonstrate that this is an 'open system'. Two of the samples were in deficit and twelve had an excess of ^{230}Th and ^{234}U. None of the measured values thus reflect an evolution within a closed system. However, all samples were selected in a rigorous way and have 99% aragonite with minor traces of dissolution and secondary aragonite precipitation. Therefore, early diagenesis of carbonate cannot explain these observations. By applying the models of

$$\left(\frac{^{230}Th}{^{238}U}\right) \text{measured} = 1 - e^{-\lambda_{230}t} + \frac{\lambda_{230}}{\lambda_{230} - \lambda_{234}}\left(\left(\frac{^{234}U}{^{238}U}\right)\text{initial} - 1\right)\left(e^{-\lambda_{234}t} - e^{-\lambda_{230}t}\right)$$

$$+ \frac{1}{m}\left\{\left(\frac{^{234}U}{^{238}U}\right)\text{measured} - \left(\left[\left(\frac{^{234}U}{^{238}U}\right)\text{initial} - 1\right]e^{-\lambda_{234}t} + 1\right)\right\} \tag{6.7}$$

$$m = \frac{(1 - f_{234})(1 - e^{-\lambda_{234}t})}{\left((1 - f_{234}f_{230})\left(1 - \frac{\lambda_{230}}{\lambda_{230}-\lambda_{234}}e^{-\lambda_{234}t} + \frac{\lambda_{234}}{\lambda_{230}-\lambda_{234}}e^{-\lambda_{230}t}\right) + (1 - f_{230})\frac{\lambda_{230}}{\lambda_{230}-\lambda_{234}}\left(\frac{^{234}U}{^{238}U}\right)\text{initial}\left(e^{-\lambda_{234}t} - e^{-\lambda_{230}t}\right)\right)} \tag{6.8}$$

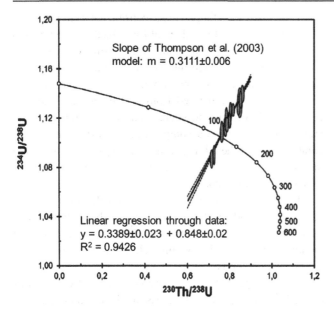

Fig. 6.5 Activity ratios measured in corals from the Amedee Island in New Caledonia. These corals came from a drilling site located between two zones of alteration, and show growth during the last stage of the interglacial (MIS 5.5), 125,000 years ago

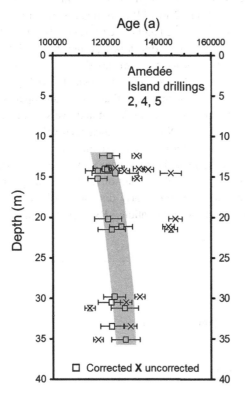

Fig. 6.6 'Raw' results from dating of the coral shown in Fig. 6.5 (× points) and adjusted by an open system model for the redistribution of U, subsequent to the recoil effect (□ points)

Thompson et al. and Villemant and Feuillet, ages are calculated which are consistent with the evolution of the reef during the last interglacial period.

Figure 6.6 shows the ages of the different samples taken from the reef according to depth. After correction using an open system model, it is obvious that the points are now organized in a linear trend, with older samples being located deeper in the core than the corals sampled above. The consistency in the linear relationship is such that it allows the aggregation rate of the reef to be assessed. We can therefore assume here that the process of nuclear recoil is the only mechanism disrupting the dating system. In this example, we applied the Thompson et al. model, with a value f_{234} of 0.975 ± 0.005 and an initial $\delta^{234}U$ of $148 \pm 2‰$, values determined by Thompson and his colleagues for corals from Barbados.

The Open System: Empirical Model

When dating a single coral, one option is to cut it into several sub-samples and to take them as independent samples to which corrections for an open system could be applied. If the corrected values show a linear correlation between the $^{230}Th/^{238}U$ and $^{234}U/^{238}U$ ratios indicating an excess or depletion of ^{234}U and ^{230}Th, it is possible to calculate the point where the regression line intersects the line of evolution in a closed system and to deduce the age of the coral.

In this way, a more accurate determination of age might be arrived at without having to apply a complicated theory. This idea was used by Scholz et al. (2004) to correct U/Th ages of corals from the Mediterranean. Scholz and his colleagues started with a result showing a variability of activity ratios ($^{230}Th/^{238}U$) and ($^{234}U/^{238}U$) between sub-samples of several corals that could not be explained by the Thompson model. The open system model, subsequent to the process of nuclear recoil, is logically limited to samples which have not undergone any physicochemical alteration. Once the coral skeleton is modified (dissolution, recrystallization, or other), uranium and thorium can be exchanged with the ambient environment. Hence, although still present, the recoil effect is no longer the only and predominant mechanism.

It is therefore obvious that theoretical models have strong limitations and coral selection becomes critical. Moreover, the open system models do not take small-scale variations in uranium concentration within the coral into account, even though these will also impact on the isotope redistribution process (Robinson et al. 2006).

Thus, it is essential to adapt the interpretation of results according to the quality of the selected samples to obtain the most accurate ages possible. It must be kept in mind that these ages, corrected by models, are approximations. In fact, models are created to bring the ages as close as possible to an unknown reality, and the researcher cannot know if the model chosen takes account of the real phenomena sustained by the sample during its geological history.

In conclusion, the analytical equipment available to us at present provides high precision and potentially accurate measures of U/Th ages of marine carbonates, with precisions on the permil magnitude for ages ranging from current day to several hundreds of thousands of years old. The limits on dating are not technical but primarily imposed by the quality of the samples. For recent samples, up to 15,000 years in age, the U/Th dating system can be considered to be a closed system, since the physicochemical alteration of corals and the nuclear recoil effect have not yet had a significant impact altering the U-series age.

For older corals, however, the impact of physicochemical alteration and especially that of the nuclear recoil displacement result in a system partially open to exchange of ^{234}U and ^{230}Th, requiring subsequent correction to obtain reasonable and accurate ages. The use of correction models reduces precision in dating, as not only the analytical error must be taken into account, but also the parameters of the correction model used, such as the initial δ^{234}U and the exchange factors f_{234} and f_{230}. Selection of the best-preserved specimens possibly ensures the highest quality of dating results that we can expect. If, however, the sample shows signs of significant alteration, it is a prerequisite to measure several subsamples to better identify the extent to which the system may have behaved open for uranium.

Estimating the Change in Sea Level from Tropical Corals

Changes in sea level come from the growth and melting of polar ice caps when the climate of the Earth varies. Knowing the exact variations in sea level during climate cycles can help to reconstruct a precise chronology of climate changes and to connect them with other parameters such as temperature recorded in polar ice cores, the temperature and salinity of the surface waters of the ocean, or even climate variations recorded in continental archives. A chronological framework established in this way helps to better constrain the phase relationships between the various components of the climate system and to determine the teleconnections between the hemispheres. In addition, knowledge of sea level during climate cycles is essential to study the variations in the continental surfaces (expansion or reduction of the continental margins) and their erosion. We have seen that U/Th dating of corals is a powerful tool to determine the evolution of coral reefs over time. But how can the ages of coral reef terraces contribute to a reconstruction of changes in sea level?

Linking the age (its time of formation) and location of a coral taken in a particular part of the world to sea level is complex. In most cases, tropical corals are located in volcanic areas that can be very tectonically active (subsidence or uplift). The sampled coral needs to have grown in situ and not moved by erosion. Also, corals can live at different depths depending on the species. On the continental margins, the processes of eustasy (change in volume of ocean water) and isostasy (variation in altitude of the continent) are also important. With changes in the volume of ice caps during the climate cycles, the continent may well rise or fall depending on the weight it supports. Given all these variables, the reconstruction of sea level from tropical corals requires the selection of corals from a habitat whose characteristics are well known. Therefore, the tectonic processes, which may have had an impact on their original position compared to where they were collected, need to be identified. The parameters required can be defined as follows:

- chronological age t;
- the sampling height above present-day sea level h (+ or −);
- the rate of uplift or subsidence $\Delta h/\Delta t$;
- the average depth of the habitat of the species under study d;
- the variability in the ecological habitat Δd.

The sea level over time $m(t)$ can then be determined by the following equation:

$$m(t) = h + d + \Delta h/\Delta t \times t.$$

The associated uncertainty is:

$$\Delta m(t) = \delta h + \Delta d + t \times \delta(\Delta h/\Delta t) + \Delta h/\Delta t \times \delta t.$$

The reconstruction of sea level is highly dependent on time and the $\Delta h/\Delta t$ values. Moreover, it assumes that the rate of uplift or subsidence has remained constant over the time interval of the reconstruction.

The localization h and depth of habitat d can be well constrained and are not, or only very little, influenced by time. To minimize the impact of subsidence or uplift rates, the ideal is to collect corals in the most stable places possible, in terms of tectonics, for the sea level (m) to be almost completely a function of t. However, most studies are located in places with strong uplift rates because these coral terraces emerged and become easily accessible to sampling. This sampling approach means that there is uncertainty about the uplift rate and therefore about the reconstructed sea level. Figure 6.7 shows reconstructions of sea levels obtained from U/Th dating of tropical corals, as they have been published for various regions ranging from the Pacific (Tahiti, the Marquesas, Vanuatu, Australia, New Caledonia) to the Atlantic (Barbados) (Bard et al. 1990; Cabioch et al. 2003; Frank et al. 2006; Gallup et al. 1994; Hanebuth et al. 2000; Thompson and Goldstein 2005). The graph also

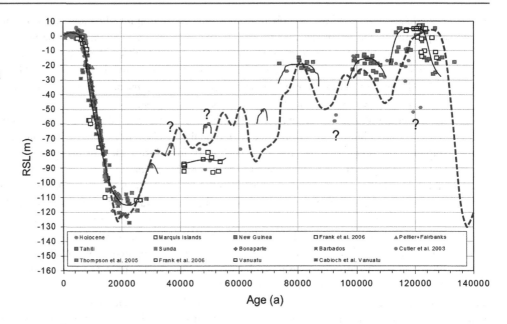

Fig. 6.7 Reconstruction of relative sea level (RSL) from U/Th dating of tropical corals (dots) and from oxygen isotopes in benthic foraminifera (black and gray lines)

shows a reconstruction of the variations in sea level and the associated uncertainty (black and gray lines) obtained from isotopic measurements of oxygen in marine sediments (Waelbroeck et al. 2002). This reconstruction is based on the assumption that the storage at the poles of a large quantity of freshwater depleted of ^{18}O causes an increase in the oceanic ^{18}O content. Therefore, the $^{18}O/^{16}O$ ratio recorded in 'deep water' foraminifera (less influenced by water temperature changes) can provide information on the size of ice sheets, and, by consequence the sea level during the lifetime of the foraminifera. With this approach, it is possible to reconstruct variations in sea level almost continuously. The various reconstructions show very similar variations in sea level, with a low level during cold climate periods and a high sea level during interglacial periods.

Over the past 140,000 years, the sea level has varied from +5 m (relative sea level, RSL) during the last interglacial maximum to about -120 m during the last and penultimate glacial maxima. A drop in sea level of this magnitude would correspond to a volume of ice of about 47 million km^3! By comparison, this is equivalent to an ice coverage over the entire North American continent with an average thickness of 2000 m. Knowing that an increase in sea level of one meter would have devastating impacts on much of the world population living today in coastal areas, it should be highlighted that natural changes in climate during the last transition (termination I) caused increases in sea level of 1.2 m every hundred years. It should further be noted that rapid changes in sea level are also evidenced during other climate transitions between glacial and interglacial periods. During the last interglacial, about 125,000 years ago, sea level was probably higher than the current level by about 5–9 m (Dutton and Lambeck 2012). Hence, in the context of the global warming observed since the early twentieth century, a rise in sea level of over one meter per century is quite conceivable if the ice sheets (Greenland, Antarctica) were to be destabilized.

The challenge for geochemists using U/Th dating of tropical corals is not to predict future changes in sea level. Rather, they hope to find new coral reefs to reconstruct past changes in sea level, and understand the pace and causes of these changes during climate transitions in more detail. Using corals has the great advantage that they can be dated more precisely than marine sediments. Hence, episodes of rapid climate change (Dansgaard-Oeschger cycle, Heinrich events, described in Volume 2) are the subject of special attention.

Other Geological Samples Datable by the U/Th Method

Other marine carbonates, such as deep-sea corals or shellfish shells can be dated by the U/Th method. In particular, deep-sea corals behave very similarly to their tropical counterparts, with a slightly higher concentration of uranium. However, there are two major differences that are very important for U/Th dating. First, the aragonite skeleton is often more robust in deep-water corals, especially for the species *Lophelia pertusa* and *Desmophyllum dianthus*. Secondly, the physicochemical alteration is less important because the skeleton remains underwater, from its formation to its removal. As a result, the openness of the uranium system due to the processes of nuclear recoil and physico-chemical alteration seems less important for these species. However, in deep waters, an increase in ^{230}Th resulting from decay of dissolved uranium is observed. Deep-water corals therefore incorporate small amounts of ^{230}Th during the formation of their skeleton, and a significant correction is required to achieve precision and accuracy in equivalent ages (Cheng et al. 2000a; Frank et al. 2004).

Shells of mollusks have another distinctive feature. The organism precludes incorporation of uranium during the formation of its aragonite shell through a still unknown biological process. A modern shell contains very little uranium (in the range of a few tens of ng/g), whereas an aragonite coral skeleton contains a few micrograms/g. This would not be problematic for U/Th dating if the shell remained in a closed system during its preservation in sediment. Unfortunately, it was observed that, after the death of the organism, the shell takes up uranium from its environment. The uranium concentration may increase by up to 10 mg/g, which is 100 times higher than at the time of its formation. As it is impossible to identify the source of the excess uranium and how it has accumulated over time, shells of mollusks are considered as open systems for uranium. Thus, one of the two basic conditions of dating mentioned above is violated (closed system), implying that based on our current knowledge, U/Th dating is not a suitable dating method for shells and mollusks (Kaufman et al. 1996).

For continental secondary carbonates, such as stalagmites, travertine or tuff, dating by U/Th follows the same 'analytical and theoretical' principles presuming the same two assumptions and using the same basic Eqs. (6.2) and (6.3) to determine the age. In groundwater a disequilibrium between ^{238}U, ^{234}U and ^{230}Th is created due to the fact that uranium is easily dissolved by the weathering of rocks, soils and sediments, while the less soluble thorium is essentially absent. In fact, the precipitation of calcite in the form of stalagmites, travertine or calcareous tuff leads to the co-precipitation of uranium in the carbonate without its radioactive daughter, ^{230}Th. Therefore, as in corals a radioactive disequilibrium is created during the formation of the carbonate. The concentration of uranium in these secondary carbonates is related to the uranium content of groundwater and the type of mineral formed. It can vary from a few ng/g to hundreds of μg/g. However, the application of the dating method is more complex than for marine organisms. The isotopic composition of uranium in groundwater can be highly variable depending on the weathering processes that come into play in an aquifer, leading to activity ratio values for (^{234}U/^{238}U) that vary between 0.8 and more than 5, or in delta notation between −200‰ and more than +5000‰ (Kaufman et al. 1996).

Another issue is that continental carbonates are often contaminated with clay or even organic particles. These may exhibit high uranium and thorium concentrations, without being in radioactive disequilibrium. Moreover, in porous carbonates, such as tuffs or travertines, several generations of crystals can be found in a single layer of carbonate, Mallick and Frank (2002). This means that U/Th dating of continental carbonates is far more difficult than for corals. The isotopy of uranium cannot be used to test if the system is open or closed, and the presence of an initial supply of thorium by groundwater or contaminants needs to be investigated. To estimate the importance of contaminants, the isotope—^{232}Th is used, Szabo et al. (1994). It is the most abundant isotope of natural thorium. It is at the origin of a decay chain called after ^{232}Th. As clay contains lots of thorium (>5 μg/g), the appearance of ^{232}Th in a carbonate sample is a sign of the presence of such contaminants (Ludwig and Titterington 1994).

Therefore, as for corals, sampling is a crucial stage of U/Th dating. It is essential to select a sample from a layer of carbonate containing a minimum of ^{232}Th, indicator of contaminants, and the sample must be representative of the original carbonate, i.e. the first generation deposited. Within these constraints, the U/Th dating of stalagmites, travertine and tuff can be accurate, and nowadays allows for precise determination of time frames for climate (Wang, et al. 2001), and archaeological (McDermott et al. 1993), studies.

Conclusions

Dating by U/Th methodologies has become a very powerful tool, widely used to obtain a fine chronology of the growth of coral reefs and many other carbonates. Although technical developments have been rapid in recent years, with precise isotopic measurements of radionuclides from the decay of uranium, the quality of the sample itself and the movement of radionuclides induced by their own radioactivity cause problems for dating. Today, we no longer refer to a dating system that is strictly closed, since the decay itself is partly the cause of exchanges of uranium and thorium with the coral environment. Correction models, known as 'open system age models' are emerging to incorporate this important issue. However, prior to being studied, the sampling process and characterization of the sample are crucial as dating samples significantly altered by diagenesis is doomed to failure.

A precise geochronological framework established under optimum analytical conditions and based on samples with as good a level of preservation as possible, provides exceptional opportunities to determine geological parameters such as the uplift or subsidence of a coral reef or variations in sea level. The U/Th dating method has now become a key element to place significant changes in major components of the climate system within a precise temporal context. This permits a direct comparison with astronomical forcings for the Quaternary.

References

Andersen, M. B., et al. (2010). Precise determination of the open ocean 234U/238U composition. *Geochemistry, Geophysics, Geosystems, 11*(12), n/a–n/a.

Bard, E., Hamelin, B., & Fairbanks, R. G. (1990). U-Th ages obtained by mass spectrometry in corals from Barbados: Sea level during the past 130,000 years. *Nature, 346,* 456–458.

Cabioch, G., Banks-Cutler, K. A., Beck, J. W., Burr, G. S., Corrège, T., Edwards, R. L. & Taylor, F. W. (2003). Continuous reef growth during the last 23 Kyr Cal BP in a Tectonically active zone (Vanuatu, South West Pacific). *Quaternary Science Reviews, 22,* 1771–1786.

Chen, T. Y., et al. (2016). Ocean mixing and ice-sheet control of seawater U-234/U-238 during the last deglaciation. *Science, 354*(6312), 626–629.

Cheng, H., Adkins, J. F., Edwards, R. L. & Boyle, E. A. (2000a). U-Th dating of deep-Sea Corals. *Geochimica et Cosmochimica Acta, 64,* 2401–2416.

Cheng, H., Edwards, R. L., Hoff, J., Gallup, C. D., Richards, D. A., & Asmeron, Y. (2000b). The half-lives of Uranium-234 and Thorium-230. *Chemical Geology, 169,* 17–33.

Cheng, H., et al. (2013). Improvements in 230Th dating, 230Th and 234U half-life values, and U-Th isotopic measurements by multi-collector inductively coupled plasma mass spectrometry. *Earth and Planetary Science Letters, 371–372,* 82–91.

Douville, E., Salle, E., Frank, N., Eisele, M., Pons-Branchu, E., & Ayrault, S. (2010). Rapid and precise ^{230}Th/U dating of Ancient Carbonates using inductively coupled plasma-quadrupole mass spectrometry. *Chemical Geology, 272,* 1–11.

Dutton, A., & Lambeck, K. (2012). Ice volume and sea level during the last interglacial. *Science, 337,* 216–219. https://doi.org/10.1126/science.1205749.

Esat, T. M. & Yokoyama, Y. (2006). Variability in the Uranium isotopic composition of the oceans over glacial-interglacial timescales. *Geochimica Cosmochimica Acta, 70,* 4140–4150.

Frank, N., Paterne, M., Ayliffe, L. K., Blamart, D., van Weering, T., & Henriet, J. P. (2004). Eastern North Atlantic Deep-Sea Corals: Tracing upper intermediate water D^{14}C during the Holocene. *Earth and Planetary Science Letters, 219,* 297–309.

Frank, N., Turpin, L., Cabioch, G., Blamart, D., Tressens-Fedou, M., Colin, C., et al. (2006). Open system U-series ages of corals from a subsiding reef in New Caledonia: Implications for sea level changes, and subsidence rate. *Earth and Planetary Science Letters, 249,* 274–289.

Gallup, C. D., Edwards, L., & Johnson, R. G. (1994). The timing of high sea levels over the past 200,000 years. *Science, 263,* 796–800.

Hanebuth, T., Stattegger, K., Grootes, P. (2000). Rapid flooding of the sunda shelf: A late glacial sea level record. *Science, 288,* 1033–1035.

Henderson, G. M. (2002). Seawater (234U/238U) during the last 800 thousand years. *Earth and Planetary Science Letters, 199,* 97–110.

Henderson, G. M., & Slowey, N. C. (2000). Evidence from U-Th dating against northern hemisphere forcing of the penultimate deglaciation. *Nature, 404,* 61–68.

Ivanovich, M. & Harmon, R. S. (1992). *Uranium-series disequilibrium: Applications to earth, marine, and environmental sciences* (2nd ed., p. 910), Oxford: Clarendon Press.

Kaufman, A., Ghaleb, B., Wehmiller, J. F. & Hillaire-Marcel, C. (1996). Uranium concentration and isotope ratio profiles within Mercenaria Shells: Geochronological implications. *Geochimica et Cosmochimica Acta, 60*(19), 3735–3746.

Ludwig, K. R. & Titterington, D. M. (1994). Calculation of ^{230}Th/U Isochrons, ages, and errors. *Geochimica et Cosmochimica Acta, 58*(22), 5031–5042.

Mallick, R., & Frank, N. (2002). A new technique for precise uranium-series dating of travertine micro-samples. *Geochimica et Cosmochimica Acta, 66*(24), 4261–4272.

McDermott, F., Grün, R., Stringer, C. B., & Hawkesworth, C. J. (1993). Mass-spectrometric U-series dates for Israeli Neanderthal/early modern hominid sites. *Nature, 363,* 252–255.

Reimer, P. J., et al. (2013). Intcal13 and Marine13 radiocarbon age calibration curves 0–50,000 Years Cal Bp. *Radiocarbon, 55*(4), 1869–1887.

Robinson, L., Adkins, J., Fernandez, D. P., Burnett, D. S., Wang, S. L., Gagnon, A. C., et al. (2006). Primary U distribution in scleractinian corals and its implications for U series dating. *Geochemistry Geophysics Geosystem, 7,* 1–20.

Scholz, D., Mangini, A., & Felis, T. (2004). U-series dating of diagenetically altered fossil reef corals. *Earth and Planetary Science Letters, 218,* 163–178.

Spooner, P. T., Chen, T., Robinson, L. F., Coath, C. D. (2016). Rapid uranium-series age screening of carbonates by laser ablation mass spectrometry. *Quaternary Geochronology, 31,* 28–39.

Szabo, B. J., Ludwig, K. R., Muhs, D. R., & Simmons, K. R. (1994). Thorium-230 ages of corals and duration of the last interglacial sea-level high stand on Oahu, Hawaii. *Science, 266,* 93–96.

Thompson, W. G., & Goldstein, S. J. (2005). Open-system coral ages reveal persistent suborbital sea-level cycles. *Science, 308,* 401–404.

Thompson, W. G., Spiegelman, M. W., Goldstein, S. L., & Speed, R. C. (2003). An open-system model for U-series age determinations of fossil corals. *Earth and Planetary Science Letters, 210,* 365–381.

Thompson, W. G., & Goldstein, S. L. (2006). A radiometric calibration of the SPECMAP timescale. *Quaternary Science Reviews, 25*(23–24), 3207–3215.

Villemant, B., & Feuillet, N. (2003). Dating open systems by the ^{238}U-^{234}U-^{230}Th method: application to quaternary reef terraces. *Earth and Planetary Science Letters, 210,* 105–118.

Waelbroeck, C., Labeyrie, L., Michel, E., Duplessy, J. C., McManus, J. F., Lambeck, K., et al. (2002). Sea-level and deep water temperature changes derived from benthic foraminifera isotopic records. *Quaternary Science Reviews, 21,* 295–305.

Wang, Y. J., Cheng, H., Edwards, R. L., An, Z. S., Wu, J. Y. Shen, C.-C. & Dorale, J. A. (2001). A high-resolution absolute dated late pleistocene monsoon record from hulu cave, China. *Science, 294,* 2345–2348.

Wefing, A. M., et al. (2017). High precision U-series dating of scleractinian cold-water corals using an automated chromatographic U and Th extraction. *Chemical Geology, 475,* 140–148.

Magnetostratigraphy: From a Million to a Thousand Years

Carlo Laj, James E. T. Channell, and Catherine Kissel

Since the publication in 1600 of the book *De Magnete* by William Gilbert, and the measurements by magnetic observatories progressively obtained from various parts of the globe, we know that the Earth's magnetic field is comparable to one that would be created by a bar-magnet placed at the center of the Earth and inclined by some 11° with respect to the axis of rotation (Fig. 7.1).

For each point on the Earth's surface, the intensity and direction of the Earth's magnetic field are defined in terms of two components: the declination, which is the angle on the horizontal plane between the magnetic north and the geographic North, and the inclination, which is the angle between the magnetic field vector and the horizontal plane. By convention, the declination is zero when the field vector points to the North (it is 180° if the field vector points to the South), and the inclination is positive when the field vector points downwards (which is the case today in the northern hemisphere).

Measurements by observatories, which began in 1576 in London and in 1617 in Paris, soon showed that declination and inclination were not stable throughout history: since measurements began, the inclination in Paris has changed by about 10° and the declination by about 30°.

The intensity has decreased by about 5% per century. This phenomenon is called the 'secular variation' of the geomagnetic field. Changes in declination are plotted on marine charts, essential tools for navigation by compass, before the development of GPS.

Secular variation was therefore identified as the first manifestation of the instability of the geomagnetic field. This was already surprising, but the biggest surprise was yet to come, when paleomagnetism methods were able to decipher the history of the geomagnetic field over prehistoric periods.

In 1906, Bernard Brunhes was the first to measure a direction of magnetization in rocks which was more or less opposite to that of the present geomagnetic field. Brunhes measured this magnetization both in a Miocene lava flow and in clays that had been baked when covered by this lava flow, which he called "natural brick". In doing so, Brunhes used for the first time a test in the field, now called the "baked contact test", which is based on the fact that, when a lava flow settles on a sedimentary layer, it re-magnetizes it, either partially or wholly, by heating it up.

If the direction of the magnetic field changed between the settling of the sediment and the arrival of the lava, the initial magnetic direction of the sediment will be replaced by the magnetic direction of the overlying lava. (To quote Bernard Brunhes "If, in the banks of natural clay, we have a well-defined magnetic direction and which differs from the direction of the current terrestrial field, it is reasonable to assume that the magnetic direction is that of the Earth's field at the time when the lava flow transformed the clays into "natural bricks"). Brunhes's conclusion that "at the time of the Miocene, around Saint-Flour, the North Pole was pointing upwards: it is the Earth's South Pole that was closest to Central France" is the first suggestion that the polarity of the magnetic field of the Earth could have reversed in the geological past.

Twenty years later, the Japanese scientist Motonori Matuyama was the first to attribute reverse magnetization of volcanic rocks in Japan and China to reversals of the Earth's geomagnetic field and to differentiate Pleistocene lava from Pliocene lava on the basis of the polarity of their magnetization. Matuyama was thus the first to use magnetic stratigraphy as a way to order sequences of rocks in time.

The modern era of magnetostratigraphy began in the 1950s in Iceland with the work of Hospers (1953). Hospers'

C. Laj (✉)
Department of Geosciences, Ecole Normale Supérieure, PSL Research University, 24 rue Lhomond, 75231 Paris Cédex, France
e-mail: carlo.laj@ens.fr

J. E. T. Channell
Department of Geological Sciences, University of Florida, 241 Williamson Hall, P.O. Box 112120 Gainesville, FL 32611, USA

C. Kissel
Laboratoire des Sciences du Climat et de l'Environnement, LSCE/IPSL, CEA-CNRS-UVSQ, Université Paris-Saclay, 91190 Gif-sur-Yvette, France

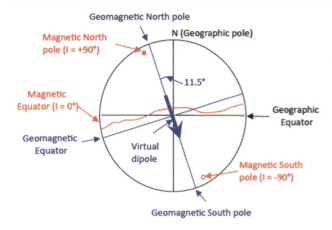

Fig. 7.1 Diagram of the geocentric, inclined dipole

work, followed by Wensink (1966), led to a subdivision of Icelandic Pliocene-Pleistocene lava in three polarity zones: normal-inverse-normal, from most recent to oldest. Afterwards, the joint use of paleomagnetic measures and potassium/argon dating (K-Ar, Chap. 5) by Cox et al. (1963) and McDougall and Tarling (1963a, b) marked the beginning of the development of the modern scale of magnetic polarities (Geomagnetic Polarity Time Scale, GPTS).

The first examples of magnetostratigraphic measures obtained from sedimentary sequences were those by Creer et al. (1954) and by Irving and Runcorn (1957), which demonstrated alternating positive and negative polarity in thirteen sites of Torridonian sandstone in Scotland, and in Devonian and Triassic rocks. In parallel, Khramov (1960) published magnetostratigraphic results from Pliocene-Pleistocene sediments in Turkmenistan, and proceeded to develop chromolithographic interpretations based on the assumption that the durations of the different polarity periods were identical. We now know that this assumption is unfounded. Other pioneering studies of magnetostratigraphy were carried out on red Triassic sandstone from the Chugwater Formation, on the European Triassic Bundsanstein and on the Moenkopi Formation from the Lower Triassic (see Opdyke and Channell 1996). All these studies were conducted on sandstone and silt mainly of continental origin, largely devoid of fauna, and therefore, the correlations based on the identification of polarity intervals were not supported by biostratigraphic correlations.

The first studies of marine sediments from the Pliocene-Pleistocene collected by coring in high southern latitudes (Opdyke et al. 1966) mark the beginning of modern magnetostratigraphy. These studies, combining magnetostratigraphy and biostratigraphy, improved and expanded the Geomagnetic Polarity Time-Scale (GPTS), which was obtained by the paleomagnetic study of basaltic outcrops and marine magnetic anomalies (MMA) (e.g. Heirtzler et al. 1968).

Over the past two decades, significant technical developments have enabled the dating of recent volcanic formations and sedimentary formations. In parallel, the development of corers like the Hydraulic Piston Corer from the Ocean Drilling Program (ODP) or the CALYPSO corer on board the research vessel *Marion Dufresne* of the French Polar Institute (IPEV) allowed the sampling of very long marine sedimentary sequences, with very high sedimentation rates. Thus, it was demonstrated that certain very brief changes in the magnetic polarity of the sediment, initially thought to be due to sampling artifacts, or to re-magnetization phenomena or to various sedimentary processes, were, in reality, reflecting coherent reversals in the polarity of the geomagnetic field, which are observed simultaneously at different locations on the surface of the globe. These "geomagnetic excursions", which occurred during periods previously considered "stable", such as the Brunhes or Matuyama periods, provide an exceptional tool for correlations over long distances. Provided that these excursions are well dated (on volcanic rocks with records of intermediary or reversed paleomagnetic directions), it is possible to establish a scale of geomagnetic instabilities (GITS = Geomagnetic Instabilities Time Scale), allowing a better temporal resolution than the scale of magnetic polarities.

Finally, over the last decade, a new method of magnetic correlation, based on the variations in intensity of the Earth's magnetic field, has been proposed. The records of geomagnetic paleointensity obtained from marine sediments contain a global signal, which is a global-scale correlation tool at an exceptional time resolution. Although this new method is more restrictive in its use (it requires a high level of homogeneity of the magnetic mineralogy of the sediments), it potentially allows cross-comparison with the stratigraphy of ice cores. As variations in geomagnetic paleointensity control the production of cosmogenic isotopes, like ^{10}Be and ^{36}Cl in the upper atmosphere, their flux, measured in ice cores from Greenland and Antarctica, is inversely correlated with these variations. This makes it possible to envisage a correlation between ice and marine sediments.

Below, we examine in turn the three major developments in magnetostratigraphy, first the polarity scale, then the scale of geomagnetic instabilities, and finally the development of the method of correlation by paleointensity. We adopt a "historical" approach, retracing the different stages and the main challenges encountered, and we will also attempt to highlight not only the advantages, but also the limitations of each method. We will illustrate, through examples, the unique role that magnetostratigraphy can play, especially in understanding the mechanisms of climate variability, by allowing the evaluation and possible quantification of temporal phase differences between various regions and between different records (ocean, cryosphere, land).

Establishing the Scale of Magnetic Polarities

First Coupled Measurements: Magnetization of Volcanic Rocks—K/Ar Dating; the McDougall and Tarling Scale and the Mankinen and Dalrymple Scale for the Plio-Pleistocene

In the 1960s, the first geomagnetic polarity time scales (GPTS) were developed using studies coupling magnetic polarity with radiometric dating (K/Ar, Chap. 5). These include the first studies conducted by Cox et al. (1963) in California and by McDougall and Tarling (1963a, b) on the Hawaiian Islands. Identification of the different zones of polarity of the Plio-Pleistocene was based solely on their radiometric age, as none of the volcanic sequences studied (apart from the Icelandic sequences studied by Watkins and Walker) had the continuity and duration to form a long, more or less continuous, stratigraphic sequence. The scale of polarities was therefore constructed by combining results from locations often very far from each other. So, there is no 'stratotype' (typical locality), as is the case for sedimentary outcrops.

The studies that marked this first phase of development of the scale were conducted from Jaramillo, the Cobb Mountain, in the Olduvai Gorge, Reunion, Mammoth, Cochiti and Nunivak (see Opdyke and Channell 1996). All these studies led to the gradual establishment of the magnetostratigraphic scale for the Plio-Pleistocene, shown in Fig. 7.2.

It was at this time that the custom of naming the long periods of polarity (polarity chrons) after the great geophysicists (Brunhes, Matuyama, Gilbert, Gauss) was introduced, while the shorter periods (polarity subchrons) were named after the place where they were discovered (Jaramillo, Mammoth, etc.).

The magnetic polarities scale by Mankinen and Dalrymple (1979) (Fig. 7.3), based on studies coupling magnetic polarity and K/Ar dating of volcanic rocks, is the last stage of this development phase of the magnetostratigraphic scale and remained the standard in the discipline covering the last five million years, for over ten years.

Further back than 5 million years, however, analytical uncertainties in radiometric dating become of the order of the duration of polarity periods, making their identification unreliable. Other methods are required to extend this scale to more ancient times.

Magnetic Stratigraphy in Pliocene-Pleistocene Sedimentary Series

It was the studies of magnetism of deep ocean sediments that first enabled this extension (Opdyke et al. 1974). The magnetic particles contained in the sediments were

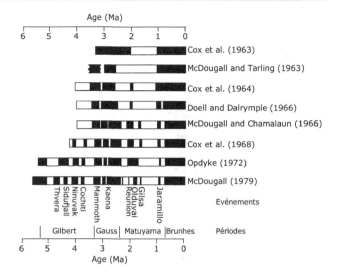

Fig. 7.2 The first scales of geomagnetic polarity for the Plio-Pleistocene period. After McDougall (1979). In black: normal periods, in white: inverse periods

deposited with a preferred orientation along the lines of the geomagnetic field prevailing at the time of deposition. In many cases, this orientation was preserved over time: just as for lava, sediments have a "magnetic" memory. Although the precise mechanisms by which sediments acquire magnetism are still poorly understood and this may limit their use in some applications, sediments, on the other hand, have the enormous advantage of registering the direction of the geomagnetic field almost continuously unlike lava, where the continuity of the signal is conditioned by the sporadic nature of volcanic eruptions.

Figure 7.4 shows an example of magnetic polarity recorded in the core RC12-65 collected by Opdyke et al. (1974), on the research vessel *R. Conrad* of the Lamont Doherty Earth Observatory, Columbia University, USA. While some details may differ, the sequence of polarity periods from the Mankinen and Dalrymple scale can be seen in the record (from the Brunhes to the Gilbert period).

However, the sediment core goes back further in time; the lowest level is from around 10 million years ago, based on paleontological estimates.

Magnetic Anomalies at Sea and the Heirtzler Scale of Magnetic Polarities

The study of magnetic anomalies at sea constituted the major step towards establishing the sequence of magnetic polarities, from the middle of the Mesozoic era through to recent times.

The hypothesis by Vine and Mathews (1963) of the spreading of the seafloor directly explains the formation of these magnetic anomalies on both sides of oceanic ridges (Fig. 7.5).

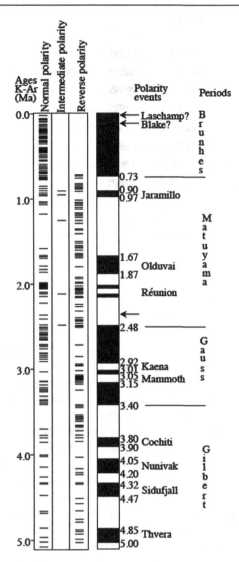

Fig. 7.3 Magnetostratigraphic scale established for the last five million years. After Mankinen and Dalrymple (1979)

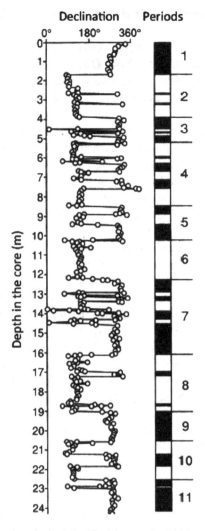

Fig. 7.4 Results obtained by Opdyke et al. (1974) on a marine sediment core showing changes in declination of natural magnetization illustrating the changes in magnetic polarity

During expansion, material from the upper mantle reaches the surface at the axis of the ocean ridges, solidifies, and is then pushed laterally in a symmetrical manner on either side of the ridge. The oceanic crust, composed of igneous rocks that contain magnetic minerals, acquires a thermoremanent magnetization during cooling and therefore registers the Earth's magnetic field at the time of cooling. Alternating normal and reverse polarities are recorded "continuously" in rocks at the ocean bottom, which gives the characteristic "pajama stripe" structure of the magnetic anomalies distribution at sea. This structure can be detected by trailing a magnetometer behind a research vessel.

This is exactly the type of work that was done by the Lamont Doherty Geological Observatory team led by Heirtzler, Pitman and Le Pichon. Their studies showed that magnetic anomalies, linear and symmetrical on both sides of the ridges, exist over large areas of the North and South Pacific and the Atlantic and South Indian oceans. In addition, assuming that the anomalies originate from a sequence of blocks magnetized in normal and reverse directions, they showed that these anomalies correspond to the same scale of magnetic polarities, the only difference being the rate of expansion of the various ocean basins. Heirtzler et al. (1968) also calculated the speed of expansion of the South Atlantic by using the date (known from another source) of 3.35 Ma for the Gauss/Gilbert boundary. Then, using various geophysical data and showing remarkable intuition, Heirtzler et al. (1968) considered that the rate of expansion of the South Atlantic remained constant for the last 80 million years. With this assumption, they proposed a scale of magnetic polarities going back to 80 Ma. With its new approach and results, this scale was a gigantic step forward. It extended the magnetostratigraphic scale from 5 Ma to 80 Ma and

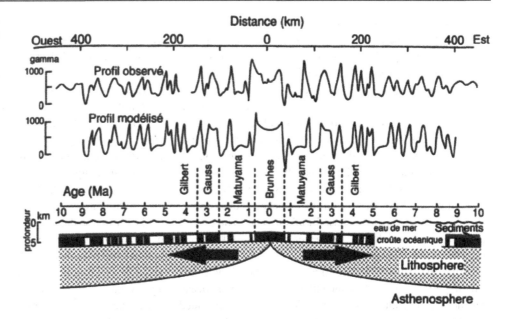

Fig. 7.5 Measurement of magnetic anomalies at sea on both sides of the mid-ocean ridge and comparison with the modeled profile proposed by Heirtzler et al. (1968)

proved to be very precise: in the thirty years since, studies have shown it to be out by only 5 Ma at 70 Ma!

In the scale of Heirtzler et al. (1968), the ages of the different polarity intervals were estimated solely assuming the rate of expansion of the seafloor. One of the objectives of the international program for ocean drilling (Deep Sea Drilling Project (DSDP), followed by ODP) was to link the scale of polarities inferred from the magnetic anomalies at sea to biostratigraphic ages. Sediments accumulate on the ocean crust as soon as this latter is created at the ridges. Therefore, the age of the crust is given by the sediment just above it.

Over the past forty years, hundreds of cores have been collected in the different oceans have allowed the Heirtzler scale to be extended to older ages, and to link it with biostratigraphic datings.

In parallel, the magnetostratigraphic scale has been improved by studies of emerged marine sediments, which allow specific intervals to be analyzed with a better temporal resolution than was achieved by studies at sea. These "segments" were then inserted in the magnetostratigraphic scale, thanks to the identification of specific characteristics.

The study by Lowrie and Alvarez, published in 1981, on the sediments of the Apennines in Umbria (Italy) is probably the most important of these. Not only has it led to a better understanding of the bio-magnetostratigraphic scale of the last 100 million years, but it also led to the discovery of a thin level of iridium-rich clay, from the Cretaceous-Tertiary boundary, which initiated the hypothesis of an extraterrestrial impact to explain the massive biological crisis that characterizes this boundary.

The Cande and Kent Polarity Scale

In a review of the Heirtzler scale, Cande and Kent (1992) also considered the magnetic profile of the South Atlantic as a starting point, in which they replaced some intervals with equivalent, more detailed intervals obtained from ridges in the Pacific and Indian oceans with a higher speed of expansion. However, they considered the rate of expansion of the South Atlantic to be continuous, but not constant as Heirtzler et al. (1968) had thought in their first approximation, and they adjusted the magnetic profile to match nine calibration points, whose ages had been estimated elsewhere (eight were high-precision $^{40}Ar/^{39}Ar$ ages, and the most recent age (2.60 Ma) was one proposed by Shackleton et al. (1990), at the Gauss/Matuyama boundary by astronomical calibration, see below).

In a more complete version of this work, Cande and Kent (1995) used astronomical calibration obtained by Shackleton et al. (1990), and Hilgen (1991a, b) for all the Pliocene/Pleistocene polarity reversals. They modified the Cretaceous-Tertiary boundary from 66 to 65 Ma. For a detailed description of this new scale (called CK95), which is the current benchmark for the Late Cretaceous and Cenozoic, we refer the reader to the original publication (Cande and Kent 1995).

The magnetostratigraphic scale available today is very reliable as far as the Late Jurassic. Traditionally, it is divided into two parts, the most recent corresponding to the Upper Cretaceous and Cenozoic eras, and the older corresponding to the Lower Cretaceous and Upper Jurassic eras. Polarity sequences for these two major periods are called C and M sequences, respectively. On the ocean floor, they are separated by the normal, calm Cretaceous period, between 118 and 84 Ma, during which no geomagnetic reversal occurred and therefore no magnetic anomaly is observed.

Some important features of the geomagnetic field appear evident when looking at the scale of magnetic polarities. On the one hand, the total time for normal and reverse polarity intervals is essentially the same, with no tendency for the field to remain in one or the other of the polarities. On the other hand, during the Cenozoic, the rate of reversal increased: there was a reversal every million years or so at the beginning of this period, but this reached four reversals per million years during the last 5 million years. The current normal period which has lasted for about 780,000 years therefore seems abnormally long.

Before the Upper Jurassic, the entire ocean floor corresponding to the current tectonic phase was absorbed in ocean trenches. As a result, the polarity scale can only be extended in time through the study of continental rocks and is much less continuous and precise than for the Cenozoic and Mesozoic eras. The most obvious and probably most documented characteristic is the long period of reverse polarity during the Permo-Carboniferous which lasted about 70 million years, and which is called the Kiaman interval. This interval was preceded and followed by periods where the field reversed frequently.

Astronomical Calibration of the Polarity Scale

In a famous article published in 1976, Jim Hays, John Imbrie and Nicholas Shackleton were the first to show that some indicators (proxies) of paleoclimate, such as isotopic records of oxygen, evolved over time depending on the orbital cycles of the obliquity, eccentricity and precession. These cycles, initially calculated by Milutin Milankovitch, were established much more precisely by André Berger, and gave paleoclimatologists precise solutions for the last three million years. Hays et al. have "adjusted" their initial age model to "match" the obliquity cycle in the record to that given by astronomical calculation. In doing so, they established the first 'cyclostratigraphy'. This method was widely used in the 1980s to constrain the ages of isotopic records of oxygen during the Brunhes period with a precision of a few thousand years.

Subsequently, the development of a hydraulic corer in the DSDP program resulted in the acquisition of previously inaccessible, very deep sediments, allowing the study of continuous sequences in even older sediments. Applied to the Matuyama period, astronomical calibration based on obliquity cycles did not initially show significant age differences between the limits of "astronomical" polarity and those of the Mankinen and Dalrymple scale. But when an astro-timeline, based on precession, was obtained at Site ODP 677, it became clear that the polarity intervals of the Mankinen and Dalrymple scale were not accurately dated.

This study paved the way for a complete revision of the ages of the polarity intervals, especially during the Gauss and Gilbert periods (Hilgen 1991a, b). The realization of the "youth" of the ages obtained by the K/Ar method for the Plio-Pleistocene era compared to those obtained by cyclostratigraphy has led to the extensive use of the $^{40}Ar/^{39}Ar$ method (Chap. 5), which had recently been developed to test the validity of cyclostratigraphic ages of this period. The age of the Brunhes/Matuyama reversal (initially set at 0.73 Ma by K/Ar) has been re-evaluated to 0.78 Ma due to a large number of independent $^{40}Ar/^{39}Ar$ measurements. The latter value is consistent with the cyclostratigraphic age. Good agreement is also observed for the Plio-Pleistocene era, where the new $^{40}Ar/^{39}Ar$ estimations coincide with the ages given by Shackleton et al. (1990), and Hilgen (1991a, b).

Although these new estimations, in general, confirmed the astronomical ages, the ages of geomagnetic reversals given by Cande and Kent (1992), based on magnetic anomalies at sea, did not seem to be in agreement with astronomical dating. The authors had in fact adopted durations for the Plio-Pleistocene polarity intervals that have proven inaccurate. As mentioned above, it is precisely the astronomical ages for the Plio-Pleistocene that Cande and Kent (1995) adopted in their new version of the scale. The consistency they then obtained between $^{40}Ar/^{39}Ar$ estimates, magnetic anomalies at sea and astronomical calculations over the entire time period covered by the new scale can be considered as a validation of this method. With this method, the age of a geomagnetic reversal can be estimated within the duration of one precession cycle, in other words, less than two or three times the length of a polarity reversal. This very high resolution led Renne et al. (1994) to calibrate the age of one of the standards of the $^{40}Ar/^{39}Ar$ method (Fish Canyon Tuff, FCTs) with the astronomical method, which reduced the uncertainty to 0.6% for the calibrated ages compared with the standard.

Principle and Practice of Magnetostratigraphy

The magnetic polarities scale shows that polarity reversals are largely random in time, which means that sequences of four or five successive reversals do not repeat themselves identically in time; therefore, they constitute a kind of "fingerprint" of specific geological periods. Magnetostratigraphy is based on this characteristic: if a sequence characteristic of the magnetostratigraphic scale can be identified within a particular series, then a specific age can be assigned to this section.

Although simple in principle, matching of a particular sequence to a segment of the magnetostratigraphic scale is far from easy in practice. The ideal condition for a record of magnetic polarities to be perfectly continuous would be for the "rain" of sedimentary particles to accumulate continuously over time. But this is rarely the case. In general, the rate of accumulation of a sedimentary sequence is variable, with potential differences of an order of magnitude, depending, for instance, on the changing climatic/environmental conditions over time. Hiatuses can also occur. All these factors change the appearance of the sequence of polarities, making its identification in GPTS complicated. However, in marine sequences, close to ideal conditions can be found for periods of about a few million years.

It is certainly a major advantage for magnetostratigraphic studies to have thick sections (10^2–10^3 m). Indeed, the thicker a section is, the more likely it is that it will contain several polarity zones, and therefore can be more easily correlated with a defined segment of the reference magnetostratigraphic scale. Our colleague, Robert Butler, made an amusing analogy between identification of a particular sequence in the magnetic GPTS and fingerprinting in a police investigation. Usually, a full fingerprint can identify a person, half a fingerprint leaves identification open to discussion but a quarter fingerprint is unlikely to be accepted as irrefutable proof in a court of law.

In practice, even in favorable cases, the researcher is faced with a series of zones of normal and reverse polarity, which is often difficult to unambiguously correlate with the GPTS scale. The assumption made at the outset is that the rate of sedimentation of the sampled section is more or less constant. So, different ways of correlating the sequence to GPTS are tested. In most cases, however, it is necessary to have an independent marker in time (radiometric dating or biostratigraphic data already independently correlated to the GPTS) so that magnetostratigraphy can correlate the full section being studied to the GPTS, and thus, to specify its age, temporal thickness and rate of accumulation.

Cross-correlation has been used by some authors to assess the agreement between the sequence being studied and a particular segment of the GPTS, by calculating the correlation coefficient corresponding to the various, visually evaluated, solutions. A coefficient of maximum value indicates the most likely correlation, while a value close to zero allows rejection of that hypothesis. In practice, this method is only usable when the studied section contains a large number of polarity intervals.

Regardless of the method used to associate the studied sequence with a particular segment of the GPTS, the quality of a magnetostratigraphic study is primarily based on unambiguous identification of the magnetic polarity at each level of the studied sequence. This can sometimes be achieved if there is agreement between determination of polarity in two parallel and close sections. In general, however, this requires a complete paleomagnetic study (determination of the magnetic mineralogy, field test, progressive demagnetizations either by heating or by applying alternating fields) to establish the stability of the magnetization and its acquisition at the time of deposition. For this purpose, Opdyke and Channell (1996) established ten criteria to assess the quality of a study. The authors themselves recognize that it is very difficult to simultaneously comply with all ten for one section, but at least five should be present in a modern magnetostratigraphic study.

A High-Quality Magnetostratigraphic Study: The Siwalik Sequences in Pakistan

The deposits in the foothills of the Siwalik Basin in northern India and Pakistan are among the most studied fluvial sediments. Indeed, in them, many fossils of mammals, including primates, have been discovered. Their precise chronological study and the correlation of different outcropping sections is of crucial importance for understanding the evolution and migration of these hominids and their land use.

From a paleomagnetic point of view, the main problem with this study is that the deposits of Siwalik are "red beds" and magnetization is carried by hematite. The magnetization of this type of sediment is very complex, and it has been shown in some cases that magnetization was acquired as a result of chemical reactions in the sediment, considerably after deposition. It is therefore essential for a magnetostratigraphic study of Siwalik to establish that magnetization was acquired during deposition or immediately afterwards so as to prevent invalidating correlation with the magnetic polarity scale (GPTS).

The study of the magnetic mineralogy showed that the components of magnetization of these red beds fall under two different categories of hematite: a red pigmentation phase and a specular hematite phase. The red pigmentation phase acquired its magnetization well after deposition (at least one polarity interval later). However, the magnetization carried by specularite was acquired during deposition or immediately afterwards. These two components can be separated by thermal demagnetization, the magnetization of the specularite being isolated between 525 and 600 °C. This component can then be used to establish the magnetic polarity of the sampled sections (Fig. 7.6) which is then correlated with the scale of Mankinen and Dalrymple (1979) (Fig. 7.7).

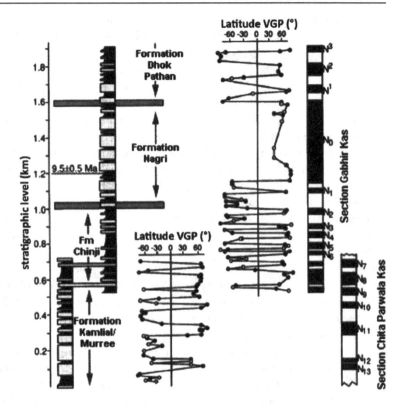

Fig. 7.6 Magnetostratigraphic results obtained by Johnson et al. (1985) on two sections from Chita Parwala and Gabhir Kas. Some levels of hard sandstone (represented in gray) clearly connect the two cuts. The magnetic data is shown as the virtual geomagnetic pole (VGP) latitude, which is located either in the southern hemisphere (reverse polarity) or in the northern hemisphere (normal polarity). The filled-in circles denote the most reliable estimates, the open circles represent the estimates of lower, but acceptable, "quality". The column on the right shows the sequence of polarities obtained along both cuts and their correlation

Geomagnetic Excursions and the Scale of Magnetic Instabilities (GITS)

Discovery of Geomagnetic Excursions

Throughout the process of establishing the scale of magnetic polarities, the presence of very short excursions of the field polarity was observed in lavas and sedimentary sequences. These excursions were often characterized by intermediate magnetization directions between direct and reverse polarities, and did not correlate to polarity intervals recognizable in the polarity scale well established on other sites. Therefore, these excursions were initially regarded with some skepticism by paleomagnetic researchers and were sometimes attributed to episodes of extreme amplitude of the secular variation or even to artifacts due to sampling, re-magnetization or sedimentological processes. The short duration of these excursions also meant that their registration in volcanic or sedimentary sequences was often random, and until very recently, no excursion was observed in more than one site.

A change of attitude in the community occurred about fifteen years ago, and can be largely attributed to a combination of two main factors: firstly, considerable progress in the dating methods of both volcanic rocks and sedimentary sequences allowed the unambiguous correlation of some of these excursions recorded in these two media, and, on the other hand, the discovery of rapid climate changes recorded in ice cores encouraged paleoceanographers to collect marine sedimentary sequences with a very high sedimentation rate, and with the longest possible series (IMAGES programs, ODP, IODP). This was made possible by the development of new corers as previously mentioned. These high rates of accumulation significantly increased the probability of recording rapid changes in magnetization, characteristic of an excursion.

Despite this progress, detailed records of excursions remain largely limited to the last two million years. The best documented excursions occur as double changes in polarity which define a very short polarity interval in the opposite direction to the initial and final polarities. The total duration of an excursion is very short: the most recent (and most accurate) estimates converge around a duration of around 1500–2000 years (Laj et al. 2000, 2004, 2014; Laj and Channell 2007). In addition, directional changes related to the best documented excursions like the Laschamp (LE) excursion or the Iceland Basin (IBE) excursion, seem to have occurred more or less simultaneously in the various places where they were recorded. Therefore, geomagnetic excursions, as well as being of fundamental interest as a probe into the phenomena of magnetic instability in the Earth's core, also provide very precise points of correlation with a global significance.

It is by now well established that excursions are accompanied by a decrease in the intensity of the geomagnetic

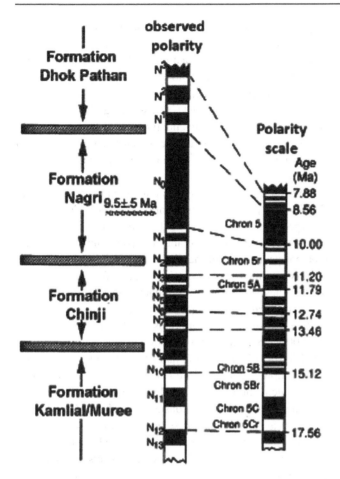

Fig. 7.7 Magnetic polarities observed in the Siwalik cut correlated with the Mankinen and Dalrymple (1979) polarities scale, the most complete and reliable at the time

field. The field acts as a screen for cosmic radiation (consisting essentially of charged particles), and this reduction causes a significant increase in the amount of cosmonuclides (^{14}C, ^{10}Be, ^{26}Al...) formed in the upper atmosphere through the impact of cosmic radiation on the various particles air. As ^{10}Be can be measured in polar ice cores, this opens up new opportunities for ice-sediment correlation.

A Scale of Geomagnetic Instabilities?

The use of geomagnetic excursions as precise temporal tie points could be particularly useful in at least two important areas of Earth sciences. This is especially useful to understand the operating mechanisms of the terrestrial dynamo which is the origin of the field itself. Recent models, theoretical and numerical, of the Earth's dynamo (Glatzmaier and Roberts 1995) indeed, give accurate assessments of time constants, frequencies and geometries of the transitional field during excursions. It is of prime importance to characterize the role of the solid inner core and the lower mantle in the mechanism of excursions and reversals. It is also important in paleoclimatology studies, where an independent chronology of climatic/environmental phenomena needs to be of higher resolution that the scale of polarities to allow evaluation of the synchronicity and phase shifts, either early or late, of climate events in different parts of the globe (examples are described below).

Currently, the major obstacle to the widespread use of this method is probably the difficulty of integrating sedimentary data and volcanic data into a single unified scale of geomagnetic instabilities. The brevity of the excursions is both an advantage (giving very precise temporal tie points), and paradoxically, it also represents an obstacle to the creation of this scale. In fact, a specific excursion is not systematically recorded in all sequences, including those with a medium to high sedimentation rate. Studies to develop a high-resolution chronology of geomagnetic instabilities, especially of excursions, are often based on accurate dating, using the K/Ar and $^{40}Ar/^{39}Ar$ methods, of as large a number as possible of lava flows recording either an abnormal direction, or a geomagnetic field with a very low intensity, or both characteristics together. It is important to keep in mind that $^{40}Ar/^{39}Ar$ datings are obtained by reference to standards whose ages were defined based on astro-chronological calibrations. So, several ages have been proposed for the same standard (e.g. Fish Canyon Sanidine, commonly used for dating in the Quaternary). It sometimes appears that none of them provides good agreement between the $^{40}Ar/^{39}Ar$ ages and the glaciological or astronomical scales. Further work is therefore necessary to "reconcile" these various approaches.

Today, at least seven geomagnetic excursions from the Brunhes period have been inventoried in detail: the excursions of Mono Lake (34 ka), Laschamp (41 ka), Blake (120 ka), Iceland Basin (188 ka) Pringle Falls (211 ka), Big Lost (560–580 ka) and Stage 17 (670 ka) (Laj and Channell 2007). Other excursions from the same period are being studied. Studies are also underway for earlier periods, such as the Matuyama period during which at least eleven excursions seem to have occurred. All of these excursions act as specific temporal tie points, that greatly increase the temporal resolution of the magnetostratigraphic scale.

Magnetostratigraphy Based on Variations in the Intensity of the Geomagnetic Field

Introduction

Over recent years, the stratigraphy of climate records has undergone a major change, particularly due to the discovery of rapid and precise markers, both lithostratigraphic (Heinrich events in marine sediments in the North Atlantic) and climatic (Dansgaard-Oeschger events in Greenland ice cores).

Sedimentary records from the North Atlantic can be correlated with a very good approximation to records obtained in ice cores, by using anchor points common to both types of records, such as levels of volcanic ash or melting events (Bond et al. 1993). However, the correlation of ice cores with sedimentary records or sedimentary records between themselves in very diverse parts of the world, is a far more complex problem: a direct correlation, based on the recognition of climate signals may not take into account the possibility of a time difference between the occurrence of a particular climatic event in two distant regions or by to different media. And yet an accurate calculation of these phase shifts (leads and lags) is of central concern to paleoclimatologists, since it is essential to understand the mechanisms of global climate change, and to the highlight of chains of causality and the spatio-temporal spread of a climate event.

At the scale of a basin, the magnetic susceptibility measured continuously in sediment cores, directly on-board research ocean vessels, usually allows the correlation of the different cores with a resolution of about one centimeter, which is sufficient for most studies. This physical measure allows splicing of records obtained from different cores, taken from the same site or in neighboring sites, to join together the different sections of cores obtained by the hydraulic piston corer, aboard the *Joides Resolution* within the ODP international program. Magnetic susceptibility reflects the ability of the sediment to acquire a magnetization induced by a weak magnetic field. Measurement of magnetic susceptibility is commonly used because it has the advantage of being non-destructive, as the induced magnetization measured disappears as soon as the imposed field is switched off. It varies depending on the magnetic content of the sediment, and particularly on the concentration of magnetic particles, their nature and size. As changes in these physical and chemical parameters are generally highly dependent on the paleo-environmental context, these are the same within the same basin or water mass, and so the magnetic susceptibility can be used as a local (or regional) correlation tool. Over long distances, oxygen isotopic ratios are the most commonly used stratigraphic tool. However, these ratios also include regional climate components, precisely those that researchers are trying to fix in time relative to each other. Correlating these ratios between distant locations could therefore mask the phase shifts which may really exist and that climatic studies attempt to quantify. Independent time constraints of climate variations with a global value are required for this exercise.

Initially proposed a decade ago, a new method, based on relative variations in the intensity of the geomagnetic field in the past recorded in sediments, is now being recognized as capable of revealing phase shifts over thousands of years at most, in climate records obtained from sites geographically very distant from each other. In this method, the relative changes in the paleointensity curve obtained from a specific sedimentary core is compared with a reference curve depicting the relative variations of the geomagnetic dipole field. The latter is, in fact, the only component of the field which varies synchronously across the globe. After synchronization of two magnetic profiles with the reference curve, the phase shifts and/or synchronicity of the two paleoclimate records from these two distant locations can be evaluated.

This method needs to be applied with some caution. Firstly, everywhere in the world, the local field is the superposition of the dipole field on multipolar components, which are much more variable in space and have shorter time constants. To plot the dipole field variations curve, the authors compiled records obtained from different parts of the world. These compilations eliminate the non-dipolar components which are averaged out in space. Compilations of relative paleointensity variations in the field for the last 800,000 years (Sint 800) (Guyodo and Valet 1999) and the last 1.5 Ma (PISO-1500) (Channell et al. 2009) show synchronization of characteristic events globally and the widespread attenuation of the non-dipolar field. However, only characteristics with time constants of the order of 10^4–10^5 years are apparent on these two compilations, as some of the individual records have been obtained from sediment with accumulation rates of a few centimeters per thousand years, insufficient to record short-term characteristics and rapid changes.

These rapid characteristics are, however, clear in other compilations characterized by a higher sediment accumulation rate. Initially proposed for the North Atlantic Ocean, the NAPIS-75 compilation (North Atlantic Paleointensity Stack for the last 75,000 years) (Laj et al. 2000), then SAPIS (South Atlantic Paleointensity Stack) (Stoner et al. 2002) and finally GLOPIS-75 (Global Paleointensity Stack) (Laj et al. 2004) have shown that rapid components of the variations in the intensity of the geomagnetic dipole field can be recognized globally in sedimentary records (Fig. 7.8). The outer limit of temporal resolution appears to be of the order of 400 years, which is the time constants of the dipolar field. GLOPIS-75 is currently the reference for the last 75,000 years because it is precisely placed on the accurate ice age model (Cf. Section "A Correlation Between Sediment and Polar Ice"). This new age scale for Greenland ice was developed by counting annual levels (Greenland Ice Core Chronology or GICC05) (Andersen et al. 2006; Svensson et al. 2006). For earlier periods, up to 1.5 Ma, PISO-1500 is used.

Although, in principle, the method is simple, its implementation is far from it. Firstly, a linear relationship between the intensity of the geomagnetic field existing at the time of deposition and the magnetization of the sediment only exists if a single magnetic mineral carries the magnetization and the

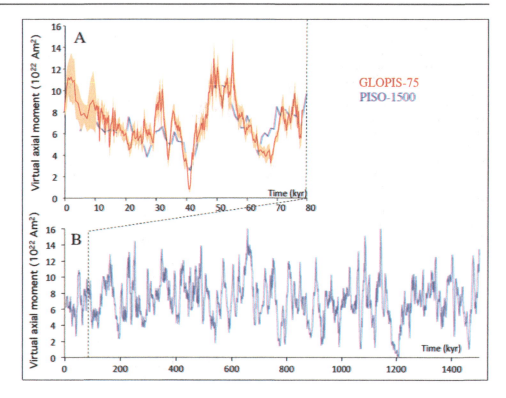

Fig. 7.8 Changes in the intensity of the geomagnetic dipole field. **a** Over the last 75 thousand years: global curve GLOPIS-75 (Laj et al. 2004) here placed on the most recent GICC05 ice age model (Laj et al. 2014) and compared to PISO-1500 on the same period of time; **b** Over the last 1.5 million years (PISO-1500) (Channell et al. 2009)

size of magnetic grains is constant along the sedimentary column, in other words, if the magnetic mineralogy of the sediment is uniform. All of these characteristics must be verified by a magneto-mineralogical study of the sediment (this verification can lead to the rejection of a significant proportion of cores, depending on the basin being studied), in order to obtain a reliable record of the relative variations of the field. This is obtained by dividing, at each stratigraphic level, the intensity of the measured natural remanent magnetization by a standardization parameter, which will take into account the variations in the concentration of magnetic grains. Of the three magnetic parameters related linearly to concentration: susceptibility; isothermal remanent magnetization (IRM) and anhysteretic remanent magnetization (ARM),[1] it is the latter that is most often used, since it depends primarily on the same magnetic grains as the ones carrying natural magnetization.

Three recent examples of long-distance, multi-archive correlations are described below showing the combined use of well-dated geomagnetic excursions and paleointensity records in paleoclimatology studies. The first one concerns a sediment-ice correlation. Of the other two, one is based on the propagation time of the North Atlantic deep-water mass from north to south, and the other on the anti-correlation between the intensities of the North Atlantic and Circum-Antarctic deep currents.

A Correlation Between Sediment and Polar Ice

The Laschamp geomagnetic excursion was the first to be discovered and is certainly the most studied excursion of the Brunhes period. Discovered in 1967 by Bonhommet and Babkine (1967), in lavas of the Puy de Laschamp, then in the Olby flow in the French Massif Central, it was originally dated between 20 and 8 ka. A series of studies conducted by different research groups demonstrated the difficulty of dating lavas as recent as these. Recently the combined use of K/Ar and ^{40}Ar/^{39}Ar methods (Chap. 5) on basalt samples collected at Laschamp and at Olby led to the establishment of a critical requisite for reliable dating of these recent samples: the absence of excess argon in the initial composition of the lava, or of any other potential disruption of the K/Ar system pre- or post eruption (Guillou et al. 2004).

Considering the uncertainty of 2.4% on the decay constant of ^{40}K, Guillou et al. (2004) proposed a date of 40.4 ± 2.0 ka, which was a considerable improvement, in terms of accuracy, over previous radiometric dating. Shown in Fig. 7.9 is a series of "snapshot data" obtained from lava flow from the Massif Central, the Canary Islands and New Zealand superimposed on the GLOPIS-75 curve. The

---

[1]Isothermal remanent magnetization (IRM) is the magnetization acquired by a sample at a given temperature (most often room temperature), after application of a constant magnetic field and subsequent cancellation. Anhysteretic remanent magnetization (ARM) is obtained at room temperature through the combined action of a stationary field at a similar level to the Earth's geomagnetic field and a strong alternative field in the same direction. The acquired magnetization is measured after cancellation of the two fields.

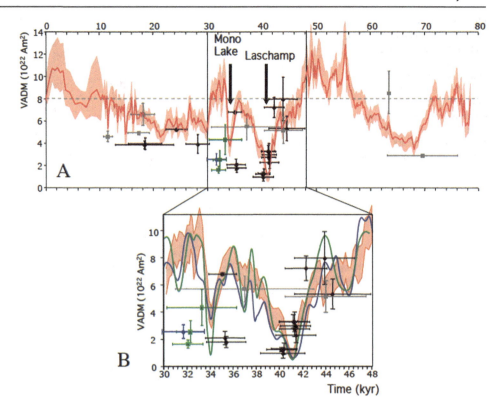

Fig. 7.9 **a** GLOPIS-75 curve (red) (Laj et al. 2004) together with Earth's magnetic field intensity data obtained from dated lava flows from the Massif central (black), the Canary Islands (green) and New Zealand (blue) (the grey dots are intensities reported on the age given by the geological map) (Laj et al. 2014). **b** Zoom on the 20–48 ka interval of the above diagram **a** with, in addition, the Earth's magnetic field intensity reconstructed from the ^{10}Be data (blue) and ^{36}Cl data (green) obtained from the Greenland ice core

absolute Earth's magnetic field intensity values obtained from the lava flows are precisely dated using the coupled K/Ar and ^{40}Ar/^{39}Ar methods. This figure also shows that the coupled volcanic intensity/dating is consistent with the one obtained by GLOPIS-75, placed on the GICC05 age model (Laj et al. 2014).

It also shows that geomagnetic variations observed in the sediments used to build NAPIS-75 and then GOPIS-75 are similar to those of the geomagnetic field, recalculated based on the concentrations of cosmogenic isotopes registered in Greenland ice (Wagner et al. 2000).

This results from the fact that the Earth's magnetic field shields our planet from cosmic and solar radiations and therefore modulates the production of cosmogenic isotopes (^{10}Be, ^{36}Cl, ^{14}C) in the upper atmosphere. Assuming a constant solar activity, a reduction (increase) in the intensity of the geomagnetic field leads to an increase (decrease) in this production. The concentration of these radionuclides, especially ^{36}Cl, recorded in Greenland ice, compared to the reference curve GLOPIS-75, has enabled a precise correlation, independent of climate, to be established between the ice and sedimentary records.

Transferred to the GICC05 age model, the maximum peak flow of ^{10}Be corresponding to the intensity minima of the Laschamp excursion, is dated at 41.25 ka, in perfect agreement with the radiometric dating (41.2 ± 1.6 ka; Laj et al. 2014). In addition, the width of this peak can be accurately estimated and thus gives a measure of the duration of the excursion as 1.5 ka, again in perfect agreement with sedimentary data.

If the Laschamp excursion (and also the more recent Mono Lake excursion) constitutes a critical tie point for all the stratigraphies (in ice and sediment), it is also observed that, even outside these instability periods, the intensity profile and that of ^{10}Be and ^{36}Cl are extremely similar in time and amplitude for the 20–50 ka interval.

These studies show that it is possible to accurately transfer the ice age model to sedimentary sequences, independently of climate variations. The magneto-stratigraphic tool based on variations in the intensity of the geomagnetic field is therefore a very powerful tool, within the limits of its application.

Paleo-Oceanographic Implications of High-Resolution Magnetic-Assisted Stratigraphy

The last glacial period was characterized by very rapid climate changes which have been observed in the upper and middle northern latitudes. Heinrich (H) and Dansgaard-Oeschger (DO) rapid climatic events, are associated with iceberg discharges into the ocean and large amplitude oscillations of air temperature over Greenland (Dansgaard

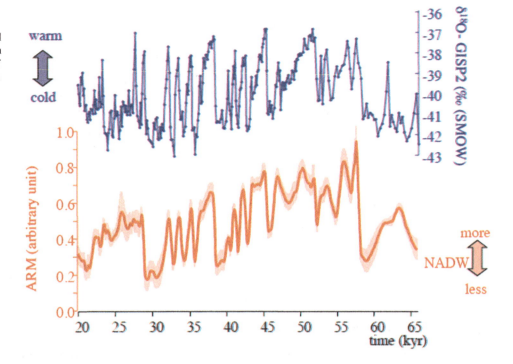

Fig. 7.10 Changes in the intensity of the North Atlantic Deep Water during the last glacial period (below) reconstructed from variations in the concentration of magnetic particles along the path of this deep-water mass. Top: curve showing air temperature changes over Greenland during the same period. From Kissel et al. (2008)

et al. 1993; Bond et al. 1993). The role of ocean circulation in these rapid climate changes remained to be defined. Consequently, the evolution of the deep North Atlantic water mass was investigated, in particular by the study of the magnetic properties of a number of sediment cores collected along the path of the Nort Atlantic Deep Water (NADW) (Kissel et al. 2008).

By this approach, changes in the strength of the North Atlantic deep current (NADW) in the North Atlantic have been shown to be completely synchronous with changes in atmospheric temperatures. This shows that the dynamics of deep ocean can change at a very high speed. This is shown in Fig. 7.10, which illustrates the amount of small magnetic grains, mobilized by the overflow water at the Faroe-Iceland and Iceland-Greenland sills and transported by the deep current, before being gradually deposited at each sampling site. Its oscillations therefore reflect the relative changes in the intensity of the convection in northern seas and of the deep current, as well as its ability to transport the magnetic particles from their basaltic source in the north (Iceland–Faroe) towards the south.

In the South Atlantic, during the same period, analysis of the carbon isotopic ratio of the benthic foraminifera shells (living on top of the sediment) and of the neodymium isotopic ratio showed that the NADW also varied in intensity, and was replaced by Antarctic deep water (AABW) as soon as it weakened (Charles et al. 1996; Piotrowski et al. 2005). These two water masses have indeed a different isotopic signature. These variations were supposed to match with the variations in air temperature over Greenland. The NADW activity in the Deep Cape basin was therefore correlated by the authors with warm events over Greenland.

However, this long-range correlation of climate parameters is entirely based on the assumption of synchronization between hemispheres. A new correlation between these paleoceanographic records from the north and south of the Atlantic was proposed, using global variations in the intensity of the Earth's magnetic field as a long-distance, climate-independent correlation tool, measured for each of the studied sedimentary sequences (Fig. 7.11). This showed, for each warm event over Greenland, that when the NADW becomes stronger in the north, it takes about 860 ± 220 years to spread to the depths in the south, "chasing out" and replacing the AABW (Kissel et al. 2008) (Fig. 7.11). This has been the only experimental attempt to quantify this time lag.

Compared with data obtained in the north from the Greenland ice sheet, those obtained from ice cores in central Antarctica appear to show a slightly different story, with less abrupt variations than in the north.

More important, after synchronization with the Greenland cores, using variations in the abundance of atmospheric methane trapped in the ice (Blunier and Brook 2001), gradual warming, called type A events, began around 1500 years before the main warm events (interstadials) in the northern hemisphere. A bipolar seesaw mechanism has been proposed, according to which the southern hemisphere warms up when the northern hemisphere cools (Broecker 1998). More recently, the EPICA community obtained a new record of the changes in air temperature from an ice core from the

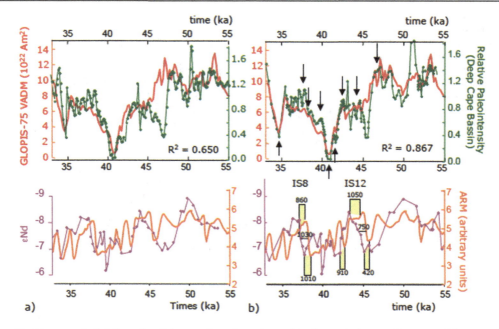

Fig. 7.11 a Top: Earth's magnetic field intensity of the core from the Deep Cape Basin (green) reported on its own age model and compared to GLOPIS-75 (red). Bottom: The age model of the core from the Deep Cape basin is based on the correlation between tracers of variations in relative intensity of deep waters of the North Atlantic based on the εNd in purple in the south, and variations in the quantity of magnetic grains (orange) in the north; **b** top: adjustment of the magnetic field intensity from the Cape Basin (green) onto the reference curve GLOPIS-75 (red). The R^2 is the correlation coefficient. Bottom: The lag between the two paleoceanographic curves after this adjustment is of the order of 880 years on average (underlined in yellow). From Kissel et al. (2008)

Antarctic Dronning Maud Land site (EDML). This core, because of its location, recorded the climate of the South Atlantic, with a resolution comparable to that of the Greenland ice cores. After synchronization, once again based on methane, it appears that all the warm events perfectly reflect the Dansgaard-Oeschger events in Greenland: the seesaw mechanism therefore also works on short time scales.

A study of marine cores from the South Indian ocean showed how the interhemispheric seesaw mechanism also concerns the Antarctic Circum Current (Mazaud et al. 2007). The record of relative paleointensity of core MD94-103 taken from the Kerguelen Plateau was precisely correlated with the reference curve GLOPIS-75, between 30 and 45 ka, thanks, in particular, to the identification of the minima of the Laschamp and Mono Lake excursions (Fig. 7.12). This correlation allowed all the paleoclimatic and paleoenvironmental records obtained from that core and other nearby cores to be transferred to the time scale defined for the Greenland ice cores (GISP2 at the time of the publication).

Similarly, to what is observed for cores located along the path of the NADW, the concentration of fine magnetic grains in the sediment shows, at these sites, located east of the volcanic islands of Kerguelen, the ability of the current to remobilize particles coming from erosion of basaltic series rich in magnetite and to transport them downstream, i.e. towards the east. This is the deep Antarctic Circumpolar Current (ACC). As in the north, the maxima of magnetic

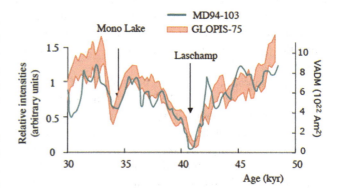

Fig. 7.12 Normalized relative intensity obtained from core MD94-103 (turquoise) compared and plotted on the same time scale as the reference curve GLOPIS-75 (Laj et al. 2004). Modified from Mazaud et al. (2007)

concentration observed during the study period and accurately fixed in time, correspond to increases in intensity of the ACC. These maxima are found to be in phase with periods of warming over Antarctica, events of type A, and in antiphase with those of Greenland. Similarly, these periods of intense activity of the ACC are in opposition to the periods of intense activity of the NADW (Fig. 7.13) demonstrating that the inter-hemispheric ocean seesaw mechanism, suggested by orbital scale models, was also functioning at the millennial scale at least during the last glacial period (Mazaud et al. 2007).

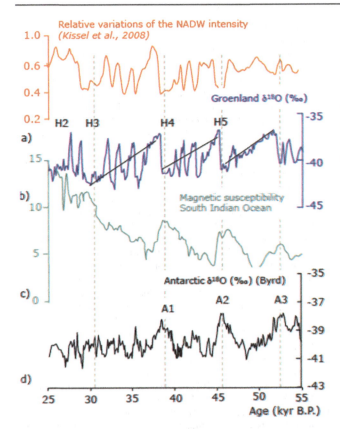

Fig. 7.13 a Past variations in the intensity of the NADW between 30 and 50 ka based on the magnetic concentration in North Atlantic sediments; **b** oxygen isotope record obtained from the GISP2 ice core in Greenland (H is for Heinrich events); **c** variations in the concentration of magnetic minerals observed east of the Kerguelen Plateau (compilation of three separate records); **d** oxygen Isotopic curve from Byrd ice (Antarctica) showing the A1 to A2 events. In this figure, all records are placed on the GISP2 age scale (following the correlation by Blunier and Brooks (2001) for Greenland and Antarctica and that shown in Fig. 7.12 for North Atlantic and Indian Ocean). From Mazaud et al. (2007)

Conclusions

During the years 1960–1980, the main objective of paleomagnetic studies of marine sediments and of the magnetization of the oceanic crust was to establish the sequence of magnetic polarities. The resultant magnetostratigraphic scale, whose basic features were well established by the late 1970s, constitutes a huge step forward in an impressive number of stratigraphic studies as well as tectonic studies focused in specific geographic areas or more globally at the scale of plate tectonics.

In its original form, magnetic stratigraphy, based on identifying polarity reversals of the geomagnetic field, typically had a million years as its unit of time.

The discovery, followed by the accurate documentation of geomagnetic excursions, was an important first development, by providing tie points of well-defined age and very short duration.

The scale of geomagnetic polarity reversals resolves the time intervals of the order of several hundreds of thousands of years and the short instabilities like excursions allow a resolution of a few hundreds of centuries to be achieved. Furthermore, the significant decrease in the intensity of the total geomagnetic field, documented in sediments during excursions, permitted a first connection with other natural archives such as ice cores from Greenland or Antarctica. The reduction in the field led to an increase in the production and therefore flux of cosmogenic isotopes such as ^{10}Be, ^{36}Cl and ^{14}C, arriving at the surface after their formation in the upper atmosphere, whose concentration is accurately measured in ice.

Finally, in recent years, identification of paleointensity profiles in sedimentary sequences provides an extremely accurate means of correlation. This opened up a new phase, still in development, of high-resolution magnetic stratigraphy. Since the understanding of mechanisms of climate change through the analysis of phase shifts between hemispheres and/or between low and high latitudes is of critical importance, it is clearly essential to dispose of a correlation tool independent of climate with a resolution of about a few hundred years.

By facilitating the unification of time scales for different types of records, both continental and oceanic, high resolution magnetic stratigraphy, based on changes in the relative intensity of the field contributes to a better understanding of the chronology and dynamics of mechanisms responsible for climate and geomagnetic variations at the scale of the recent Quaternary, between 0 and 500 ka.

References

Andersen, K. K., et al. (2006). The greenland ice core chronology 2005, 15–42 ka. Part 1: Constructing the time scale. *Quaternary Science Reviews, 25*, 3246–3257.

Blunier, T., & Brook, E. J. (2001). Timing of millennial-scale climate change in antarctica and greenland during the last glacial period. *Science, 291*, 109–112.

Bond, G., et al. (1993). Correlation between climate records from North Atlantic sediments and Greenland ice. *Nature, 365*, 143–147.

Bonhommet, N., & Babkine, J. (1967). Sur la présence d'aimantation inverse dans la Chaîne des Puys. *Comptes rendus des séances de l'Académie des sciences, série B, 264*, 92–94.

Broecker, W. S. (1998). Paleocean circulation during the last deglaciation: A bipolar see-saw? *Paleoceanography, 13*, 119–121.

Brunhes, B. (1906). Recherches sur la direction de l'aimantation des roches volcaniques. *Journal de physique, V*, 705–724.

Cande, S. C., & Kent, D. V. (1992). A new geomagnetic polarity timescale for the late cretaceous and cenozoic. *Journal of Geophysical Research, 97*, 13917–13951.

Cande, S. C., & Kent, D. V. (1995). Revised calibration of the geomagnetic polarity timescale for the late cretaceous and cenozoic. *Journal of Geophysical Research, 100*, 6093–6095.

Channell, J. E. T., et al. (2009). Stacking paleointensity and oxygen isotope data for the last 1.5 Myr (PISO-1500). *Earth and Planetary Science Letters, 283,* 14–23.

Charles, C. D., et al. (1996). Climate connections between the hemisphere revealed by deep sea sediment core/ice core correlations. *Earth and Planetary Science Letters, 142,* 19–27.

Cox, A., et al. (1963). Geomagnetic polarity epochs and pleistocene geochronometry. *Nature, 198,* 1049–1051.

Creer, K. M., et al. (1954). The direction of the geomagnetic field in remote epochs in Great Britain. *Journal of Geomagnetism and Geoelectricity, 6,* 164–168.

Dansgaard, W., et al. (1993). Evidence for general instability of past climate from a 250-kyr ice-core record. *Nature, 364,* 218–220.

Glatzmaier, G. A., & Roberts, P. H. (1995). A three-dimensional self-consistent computer simulation of a geomagnetic field reversal. *Nature, 377,* 203–209.

Guillou, H., et al. (2004). On the age of the laschamp geomagnetic event. *Earth and Planetary Science Letters, 227,* 331–343.

Guyodo, Y., & Valet, J.-P. (1999). Global changes in intensity of the Earth's magnetic field during the past 800 Kyr. *Nature, 399,* 249–252.

Hays, J. D., et al. (1976). Variations in the Earth's orbit: Pacemaker of the ice ages. *Science, 194,* 1121–1132.

Heirtzler, J. R., et al. (1968). Marine magnetic anomalies, geomagnetic field reversal and motions of the ocean floor and continents. *Journal of Geophysical Research, 73,* 2119–2136.

Hilgen, F. J. (1991a). Extension of the astronomically calibrated (polarity) time scale to the miocene/pliocene boundary. *Earth and Planetary Science Letters, 107,* 349–368.

Hilgen, F. J. (1991b). Astronomical calibration of Gauss to Matuyama sapropels in the Mediterranean and implications for the geomagnetic polarity time scale. *Earth and Planetary Science Letters, 104,* 226–244.

Hospers, J. (1953). Reversals of the main geomagnetic Field I, II, and III. *Proceedings of the Koninklijke Nederlandse Akademie van Wetenschappen B, 56,* 467–491.

Irving, E., & Runcorn, S. K. (1957). Analysis of the palaeomagnetism of the torridonian sandstone series of North-West Scotland. *Philosophical Transactions of Royal Society, London, A250,* 83–99.

Johnson, N. M., et al. (1985). Paleomagnetic chronology, fluvial processes, and tectonic implications of the Siwalik deposits near Chinji Village, Pakistan. *The Journal of Geology, 93,* 27–40.

Khramov, A. N. (1960). *Palaeomagnetism and stratigraphic correlation. Gostoptechjzdat* (218 p.), Leningrad. Geophys. Dept., A.N.U., Canberra.

Kissel, C., et al. (2008). Millennial-scale propagation of atlantic deep waters to the glacial southern ocean. *Paleocanography, 23,* PA2102. https://doi.org/10.1029/2008pa001624.

Laj, C., & Channell, J. E. T. (2007). Geomagnetic excursions. In M. Kono (Ed.), *Treatise on geophysics* (Vol. 5, pp. 373–416).

Laj, C., et al. (2000). North atlantic paleointensity stack since 75 ka (NAPIS-75) and the duration of the laschamp event. *Philosophical Transactions of the Royal Society of London. Series A: Mathematical, Physical and Engineering Sciences, 358,* 1009–1025.

Laj, C., et al. (2004). High-resolution global paleointensity stack since 75 kyrs (GLOPIS-75) calibrated to absolute values. *Timescales of the Geomagnetic Field (American Geophysical Union, Washington, C, 2004) Geophysical Monograph, 145,* 255–265.

Laj, C., et al. (2014). Dynamics of the earth magnetic field in the 10–75 kyr period comprising the Laschamp and Mono Lake excursions: New results from the French Chaîne des Puys in a global perspective. *Earth and Planetary Science Letters, 387,* 184–197.

Lowrie, W., & Alvarez, W. (1981). One hundred million years of geomagnetic polarity history. *Geology, 9,* 392–397.

Mankinen, E. A., & Dalrymple, G. B. (1979). Revised geomagnetic polarity time scale for the interval 0–5 m.y.b.p. *Journal of Geophysical Research, 84,* 615–626.

Mazaud, A., et al. (2007). Variations of the ACC-CDW during MIS3 traced by magnetic grain deposition in Midlatitude South Indian Ocean Cores: Connections with the Northern Hemisphere and with Central Antarctica. *Geochemistry, Geophysics, Geosystems, 8,* Q05012. https://doi.org/10.1029/2006GC001532.

McDougall, I. (1979). The present status of the geomagnetic polarity time scale. In M. W. McElhinny (Ed.), *The earth: its origin, structure and evolution* (pp. 543–566). London: Academic Press.

McDougall, I., & Tarling, D. H. (1963a). Dating of reversals of the Earth's magnetic field. *Nature, 198,* 1012–1013.

McDougall, I., & Tarling, D. H. (1963b). Dating of polarity zones in the Hawaiian Islands. *Nature, 200,* 54–56.

Opdyke, N. D., & Channell, J. E. T. (1996). *Magnetic stratigraphy* (346 p.). San Diego, CA: Academic Press.

Opdyke, N. D., et al. (1966). Paleomagnetic Study of Antarctic Deep-Sea Cores. *Science, 154,* 349–357.

Opdyke, N. D., et al. (1974). The extension of the magnetic time scale in sediments of the Central Pacific Ocean. *Earth and Planetary Science Letters, 22,* 300–306.

Piotrowski, A. M., et al. (2005). Temporal relationships of carbon cycling and ocean circulation at glacial boundaries. *Science, 307,* 1933–1938.

Renne, P. R., et al. (1994). Intercalibration of astronomical and radioisotopic time. *Geology, 22,* 783–786.

Shackleton, N. J., et al. (1990). An alternative astronomical calibration of the lower pleistocene timescale based on ODP Site 677. *Earth and Environmental Science Transactions of the Royal Society of Edinburgh, 81,* 251–261.

Stoner, J., et al. (2002). South Atlantic and North Atlantic geomagnetic paleointensity stacks (0–80 ka): Implications for inter-hemispheric correlation. *Quaternary Science Reviews, 21,* 1141–1151.

Svensson, A., et al. (2006). The Greenland ice core chronology 2005, 15–42 ka. Part 2: Comparison to other records. *Quaternary Science Reviews, 25,* 3258–3267.

Vine, F. J., & Mathews, D. H. (1963). Magnetic anomalies over oceanic ridges. *Nature, 199,* 947–949.

Wagner, G., et al. (2000). Chlorine-36 evidence for the mono lake event in the summit GRIP ice core. *Earth and Planetary Science Letters, 181,* 1–6.

Wensink, H. (1966). Paleomagnetic stratigraphy of younger basalts and intercalated Plio-Pleistocene tillites in Iceland. *Geologische Rundschau, 54,* 364–384.

Dendrochronology

Frédéric Guibal and Joël Guiot

Dendrochronology differs from other absolute dating methods in that the age assignment is not based on a simple, automatic count of annual deposits i.e. the rings, but on a set of intercomparisons of a large number of chronologies so as to ensure the annual status of each tree-ring (also known as growth rings), after eliminating the potential pitfall of anomalies in the anatomy of rings which may result, some years, in the absence of a ring or the formation of double rings (also known as false rings).

To understand the principle of this method, some fundamental aspects related to the formation of growth rings in trees in temperate regions should be recalled. Because of the marked climatic seasonal contrast of temperate regions an annual status can be assigned to each ring (with the exception of some accidents in growth).

A Bit of Botany and Ecology

The annual growth of woody plants is composed of an axial component which leads to the lengthening of branches (primary growth) and a radial component which leads to the formation of a ring (secondary growth). The radial growth of the trunk, branches and roots results from a layer of actively-dividing cells, the cambium, immediately beneath the bark. This gives rise to vascular tissues: wood, on the inside, responsible for, among other functions, the upward flow of the sap, and phloem, on the outside, responsible for the downward movement of the elaborated sap (Fig. 8.1). Year after year, the previously formed tissue is pushed inwards for wood and outwards for phloem.

In areas with a temperate climate, fluctuations in the physical aspects of the atmosphere (temperature, humidity, sunshine) mean that vegetation has a period of activity and a period of rest within the same calendar year. Cambial activity is discontinuous in time: in deciduous oaks on the plains in western France, cambial activity lasts from April to September; for larches which grow in the internal Alps above 1500 m, cambial activity extends from mid-June to mid-August.

A ring is made up of two parts: the earlywood which develops at the beginning of the growing season, and the final latewood which develops later in the growing season. These differ in terms of the cells that compose them, their dimensions, their disposition and the thickness of their walls. Variations in the thickness of the cell walls have consequences for the density of the wood, in the form of intra-annual and inter-annual variations more or less linked to changes in climate conditions. These conditions act according to the principle of limiting factors. Growth cannot proceed faster than is allowed by the most limiting factor. This limiting effect may be continuous, variable or sporadic depending on the case. The action of climatic factors is attenuated or amplified by other factors, both abiotic (soil, topography) and biotic (age, competition, pest attacks, phenology).

Crossdating

Aristotle, Buffon and Leonardo mentioned the existence of annual tree rings. Leonardo da Vinci, in particular, observed a relationship between ring widths and the weather conditions of the year. However, it is the American astronomer Andrew E. Douglass (1867–1962) who, in laying out the methodological foundations, is considered the father of

F. Guibal (✉)
Institut Méditerranéen de Biodiversité et d'Ecologie marine et continentale, UMR7 263 CNRS/Aix-Marseille Université/IRD/Univ Avignon, Europôle de l'Arbois, BP 80, 13545 Aix-en-Provence Cedex 04, France
e-mail: Frederic.guibal@imbe.fr

J. Guiot
European Centre for Research and Teaching in Environmental Geosciences CEREGE, Aix-Marseille University, CNRS, IRD, INRAE, Collège de France, BP 80, 13545 Aix-en-Provence Cedex 04, France

© Springer Nature Switzerland AG 2021
G. Ramstein et al. (eds.), *Paleoclimatology*, Frontiers in Earth Sciences,
https://doi.org/10.1007/978-3-030-24982-3_8

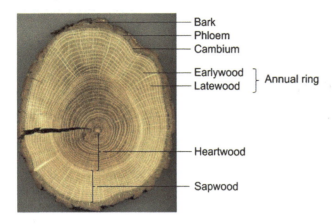

Fig. 8.1 Cross-section of deciduous oak

dendrochronology (*dendron*: tree, *chronos*: time, *logos*: study). Dendrochronology remained fairly discreet until it gained in notoriety about 1929, when Douglass succeeded for the first time in dating beams from ruins of Indian buildings in the US state of New Mexico, qualifying it as a dating method with an annual resolution (Robinson 1976). Despite this, it was not until the arrival of the computer in the 1960s, that dendrochronology truly took off, with a proliferation of research laboratories.

In the regions where the first dendrochronological studies were conducted (semi-arid regions of the southwestern USA and cold regions), the existence of a single limiting climate factor (rainfall in semi-arid regions, summer temperatures in cold regions) was instrumental in creating a series of rings whose width varied from one year to the next. However, in regions with a temperate climate, where growth depends on several factors, one factor can compensate for another, and the annual ring width series are less variable, making dendrochronological studies more complicated. In fact, if weather conditions are adequate, year after year, to meet the ecological requirements of the tree, the rings form a temporal series of constant width, and do not provide any chronological information, since it is extreme variations that serve as landmarks.

Dating with dendrochronology is based on a fundamental stage, called crossdating, which is relative dating assuring the proper placement in time of each ring. Crossdating is established by intercomparison of different pieces of wood on which sequences exhibiting similar ring patterns are identified. For this, similar sequences of coinciding narrow and wide rings, that is, separated by the same number of rings, need to be identified. This is possible only with two conditions. Firstly, the limiting factors for radial growth must vary in intensity from one year to another, with an unrepeatable series over time, so that the succession of ring widths are also variable in such a way as to be irreproducible. Secondly, the limiting factors must act in a similar way on trees with the same environmental requirements and over a wide enough geographical area to cause ring widths to vary the same way in many trees. This principle is important because ring widths can be crossdated only if one environmental factor becomes critically limiting.

Crossdating, or synchronization, is essential to check the accuracy of the ring count and the presence of any growth abnormalities. Abnormalities may appear as double rings (false rings) in the same calendar year or as missing rings. Indeed, some years, after a cold winter, possibly followed by a late spring or preceded by severe defoliation in the previous year, the ring may be partially or totally absent. In other years, as a result of the early onset of a summer drought (as in the Mediterranean region), the cambium may develop latewood elements and, then, thanks to improved weather conditions through the summer, may start producing earlywood elements again before producing latewood at the end of the normal growing season. In this case, the "first" ring is identified as supernumerary or false. Crossdating remains largely subjective and various methods have been developed to describe the observed similarities more objectively (McCarthy 2004).

In living trees of the same species, with a confirmed contemporaneity between the samples, synchronization is established by identifying sequences of similar rings over several series under a microscope (in cores or cross sections of trunk), firstly from the same tree, and then, between series from different trees. The operator compares the series from the bark inwards, records the rings and counts the sequences of narrow rings or ones with a distinguishing feature (color or width of the final wood, presence of any traumatic scars or ducts etc.). This stage allows the identification of each ring in terms of its vintage, after any anatomical abnormalities such as missing rings or double rings have been detected. After synchronization of the series has been established, ring width series are measured (1/100–1/1 000 mm).

On samples of unknown date, i.e. samples of wood from trees felled at an unknown date, the measurement of ring width series allows the establishment of digitized series from which graphs are drawn to compare the temporal variations in ring widths. The graphs are then compared in pairs and sequences of similar ring patterns are sought; maximum similarity between two curves is obtained when contemporary years are superimposed: maximum values, minimum values and the number of times two series show the same upward or downward trend in relation to the preceding year.

This analysis allows us to date the year of formation of each ring and identifies the felling date of the tree. The felling date of the tree is ensured when the outermost ring is still present; for species where the anatomical difference between sapwood (functional wood nearest the bark) and heartwood persists over time (oak, ash, elm, larch, etc.), the felling date of the tree is estimated based on the date assigned to the last ring.

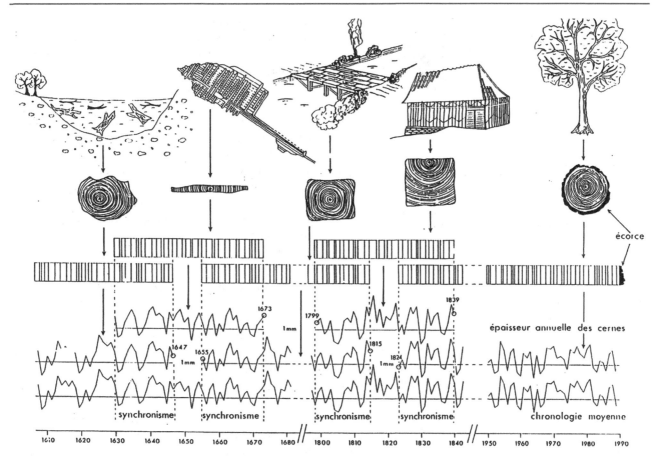

Fig. 8.2 Diagram of the theoretical construction of a master chronology

The dating of samples from trees which died at an unknown date requires synchronization between the analyzed series and a chronology of previously dated ring widths, called a master chronology. This master chronology must be composed of several series of rings from the same tree species as the one to be dated and from trees exposed to the same climatic factors, i.e. geographically close. Building a master chronology involves assembling several mean chronologies, homogeneous in their ecological and geographical origin, partially synchronous, based on the presence of ring patterns common to these timelines, and one of which is constructed from living trees for which the year of formation of the outermost ring is known. This permits each ring in the master to be assigned the year of its formation (Fig. 8.2). The representativity of a master chronology is related to its sample depth, i.e. the number of series included in the calculation of the mean ring width value. Even in climatically homogeneous regions, the geographic area covered inevitably leads to the inclusion of trees from forest stands subject to a variety of local climates, due to differences in altitude, exposure, continental character or even having grown in different site conditions (bedrock, exposure, phyto-ecological communities, degree of clearing of the site etc.) or having experienced more or less different stresses in the form of local disturbances. This summation results in a master chronology that is the average of annual wood layers over time of a given species in a region exposed to the same macro-climate, over a more or less extensive area.

Temporal and Spatial Extension

For the Holocene period, the longest chronologies, covering several thousand years, come from North America (Ferguson 1969; Ferguson and Graybill 1983), the British Isles (Pilcher et al. 1984; Baillie and Brown 1988), Central Europe (Leuschner 1992; Krapiec 1998; Schaub et al. 2008; Kaiser et al. 2011), North-West Europe (Eronen et al. 2002; Grudd et al. 2002) and Siberia (Naurzbaev and Vaganov 1999; Rashit et al. 2002). In the southern hemisphere, several groups have built thousand-year chronologies in Argentina and Tasmania (Barbetti et al. 1995; Roig et al. 1996; Cook et al. 2000).

In the same way that the representativeness of a master chronology is related to the quality of the climate signal evidenced by a high frequency of pointer years, the

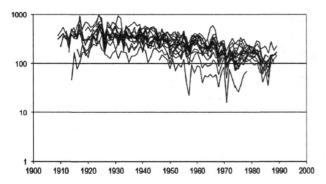

Fig. 8.3 Mean chronology (thick line) built up from multiple synchronous ring-width series of variable length (in line with the custom in dendrochronology, the y-axis is shown on a logarithmic scale to make the thinner rings more distinct)

chronological representativeness of the sample to be dated depends on the number of pointer years it contains. For this reason, trying to date a chronology composed of too few rings is usually an exercise doomed to failure. For a given site, a multiplicity of samples is essential to acquire a representative mean chronology for the site in which individual variances are minimized; in the field, this means sampling at least a dozen cases presumed to be contemporaneous, in order to achieve, whenever possible, a mean chronology of at least 80 years (Fig. 8.3). This methodological requirement explains the negative outcome of repeated attempts to date isolated pieces of wood, regardless of the context of their discovery, even if, in certain exceptional conditions (very long ring series, particularly well-documented period), statues and dugout canoes may have been dated by dendrochronology (Arnold 1996; Eckstein 2006)!

Synchronization between different tree species, called heteroconnexion, although discouraged because of differences in climate response and ecological requirements between species, is sometimes carried out between species with very similar ecological requirements. For example, comparisons are commonly made between oak and chestnut, oak and elm, larch and spruce.

Teleconnection or comparison of tree ring series over long distances, from areas subjected to different climate conditions, although theoretically just as frowned upon as the previous exercise is nevertheless, often done if the study is initiated in an area previously never investigated and for which there is no knowledge base.

In this way, the first master chronologies of oak representing the North East of France and Brittany were initiated. In a first example, master chronologies already in place for the South West of Germany were used to calibrate the first samples in Franche-Comté and Burgundy, which were analyzed by the Laboratory of Chrono-Ecology in Besançon, in the early 1980s. In a second example, master chronologies representative of the South West of England have allowed dating of sites located in the Loire valley, the Penthièvre and the Rennes basin by the City of London Polytechnic and Queen's University Belfast.

We should also mention that before starting to crossdate a piece of wood, it is often necessary to start with an approximate age of the piece, provided by ^{14}C dating which crossdating will then refine until accuracy to the year is achieved. It should also be noted that although dendrochronology is an absolute dating process accurate to a single year, this does not prevent occasional dating failures, especially when master chronologies for the species and/or region are lacking.

Contribution of ^{14}C to Calibration

Extremely valuable for its ability to date wood vestiges by establishing, under the conditions detailed above, the year of formation of each ring, even the year of death of the tree, dendrochronology has the undeniable advantage of contributing to the calibration of radiocarbon dates by converting ^{14}C age to the true calendar age.

In the 1950s, when the first radiocarbon datings were obtained on objects from past human societies, the match between the ^{14}C dates and the calendar dates was considered adequate. However, the archaeological material used was not very well dated or very old, and the ranges of uncertainty were so great that they masked potential minor deviations. According as the accuracy of ^{14}C dating improved and the body of datings grew, it quickly became obvious that the ^{14}C dates obtained were more recent than the dates obtained independently, in particular those obtained from remains from ancient Egypt. Given that any uncertainty inherent in relative dates obtained on such material could not be ruled out, and that in a living tree, only the outermost ring has a ^{14}C content in balance with that of the atmosphere, a program of ^{14}C dating of tree rings dated to the year by dendrochronology was initiated on long-living Methuselah pine (*Pinus aristata*) from the slopes of the White Mountains in California (Fig. 8.4). The results confirmed the disparity between ^{14}C dates and calendar dates (de Vries 1958). Irregular fluctuations were noted in the dates obtained, and led Suess (1965) to establish a calibration procedure for the ^{14}C dating of tree rings, in order to express the raw dates (conventional dates) in chronometrically calendar dates (calibrated dates). Measurements carried out on sequences of five or ten consecutive rings collected on the California pines have shown that the gap between the two calendars, ^{14}C years and actual years, remained low for the last 2500 years,

Fig. 8.4 View of a several thousand year old Methuselah pine (*Pinus aristata*) from the White Mountains (California)

of rings from American pines, oaks and European Scots pines. After publication in 1993, these curves, called radiocarbon calibration curves, established by thousands of measurements, constituted the basis on which corrections are now possible for the entire Holocene period, and currently, for the past 12,400 years (Stuiver et al. 1998). Extended into the past through the dating of tropical reef corals, varved sediments and speleothems (see Chap. 4), the calibration curves obtained from tree ring data are now being extended using series of dated tree rings series from Bølling Allerød and from Younger Dryas in Germany, the area around Zürich, Northwest Italy and watersheds of tributaries in the mid Durance valley (France).

The correction curve for ^{14}C dates in calendar years shows that the actual time seems compressed by about 15% before the sixth millennium BC, that there are plateaus along the curve (for example, around 500 BC. or during the ninth and tenth millennia BC.), and that multiple small fluctuations can, for some periods, affect the rectilinear shape of the curve. The practical consequences of these types of variations, showing that variations in concentration of atmospheric ^{14}C have been erratic, are very different: small fluctuations, after correction, can result in particularly inaccurate dates; but in some cases, very precise dates can be achieved for those periods particularly affected by fluctuations in the atmospheric content of ^{14}C.

but that from 500 BC, the gap increased sharply to almost 800 years at 5500 years BC.

A systematic study of this phenomenon was then carried out in the 1980s and 1990s by several laboratories on blocks

Examined more closely, the radiocarbon calibration curve with its very twisted appearance reflects a stochastic process which constitutes, at certain times, a particularly valuable

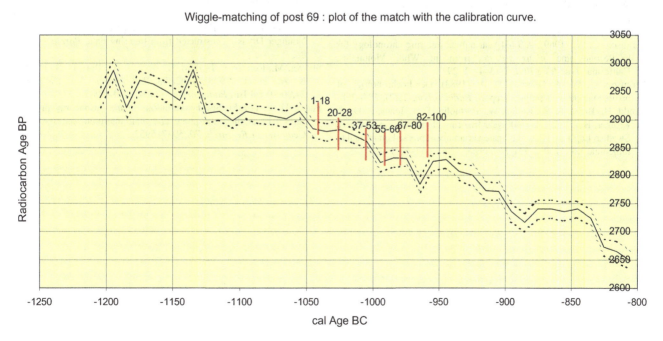

Fig. 8.5 Dating by wiggle-matching of the late Bronze age oak post n° 69 from the submerged coastal habitat of Montpenèdre, Hérault (Oberlin et al. 2004). The x-axis corresponds to the calendar years; the y-axis corresponds to radiocarbon years. Vertical lines = standard deviation of the measure

chronological marker. Indeed, the matching of these kinks (known as wiggles), based on a multiplicity of ^{14}C dates obtained from blocks of tree rings, separated by a known number of calendar years, allows, by the method of "wiggle-matching" (Pearson 1986), a greatly improved accuracy of radiocarbon dating, as illustrated by the example of the post n° 69 (Fig. 8.5) from the submerged coastal habitat of Montpenèdre (Hérault, France).

References

Arnold, B. (1996). *Pirogues monoxyles d'Europe centrale. Construction, typologie, évolution* (Vol. 2). Coll. Archéologie aujourd'hui, Archéologie Neuchâteloise.

Baillie, M. G. L., & Brown, D. M. (1988). An overview of oak chronologies. In E. A. Slates & J. O. Tate (Eds.), *Science and archaeology* (196, pp. 543–548), Glasgow 1987, *Brit. Arch. Rep. Brit.*

Barbetti, M., Bird, T., Dolezal, G., Taylor, G., Francey, R. J., Cook, E., et al. (1995). Radiocarbon variations from tasmanian conifers: Results from three early Holocene logs. *Radiocarbon, 37*(2), 361–369.

Cook, E. R., Buckley, B. M., D'Arrigo, R. D., & Peterson, M. J. (2000). Warm-season temperatures since 1600 BC reconstructed from Tasmanian tree rings and their relationship to large scale sea-surface temperature anomalies. *Climate Dynamics, 16,* 79–91.

de Vries, H. (1958). Variation of the concentration of radiocarbon with time and location on Earth. *Proceedings of the Koninklijke Nederlandse Akademie van Wetenschappen: Proceedings Series B, 61,* 267–281.

Eckstein, D. (2006). Human time in tree-rings. *Dendrochronologia, 24* (2–3), 53–60.

Eronen, M., Zetterberg, P., Briffa, K. R., Lindholm, M., Meriläinen, J., & Timonen, M. (2002). The supra-long scots pine tree-ring record for Northern Finnish Lapland. Chronology construction and initial inferences. *The Holocene, 12*(6), 673–680.

Ferguson, C. W. (1969). A 7404-year annual tree-ring chronology for bristlecone pine, *pinus aristata*, from the White Mountains, California. *Tree-Ring Bull, 29*(3–4), 1–29.

Ferguson, C. W., & Graybill, D. A. (1983). Dendrochronology of bristlecone pine: A progress report. *Radiocarbon, 25*(2), 287–288.

Grudd, H., Briffa, K. R., Karlén, W., Bartholin, T. S., Jones, P. D., & Kromer, B. (2002). A 7 400-year tree-ring chronology in Northern Swedish Lapland: Natural climate variability expressed on annual to millennial time scales. *The Holocene, 12*(6), 657–665.

Kaiser, K. F., Friedrich, M., Miramont, C., Kromer, B., Sgier, M., Schaub, M., et al. (2011). Challenging process to make the late glacial tree-ring chronologies from Europe Absolute—An Inventory. *Quaternary Science Reviews,* 13 p. https://doi.org/10.1016/j.quascirev.2010.07.009.

Krapiec, M. (1998). Oak dendrochronology of the neoholocene in Poland. *Folia Quaternaria, 69,* 5–133.

Leuschner, H.-H. (1992). Subfossil trees. In T. S. dans Bartholin, B. E. Berglund, D. Eckstein, & F. H. Schweingruber (Eds.), *Tree rings and environment. Proceedings of the International Dendrochronological Symposium, Ystad, South Sweden, 3–9 September 1990* (pp. 193–197). Lund: Lund University, Department of Quaternary Geology.

McCarthy, B. C. (2004). *Introduction to dendrochronology,* Ohio University, World Wide Web homepage. http://www.plantbio.ohiou.edu/epb/instruct/ecology/dendro.htm.

Naurzbaev, M. M., & Vaganov, E. A. (1999). 1957-year chronology for Eastern Taimir. *Siberian Journal of Ecology, 6,* 67–78.

Oberlin, C., Leroy, F., & Guibal, F. (2004). High precision ^{14}C dating of a bronze age tree-ring chronology from the pile-dwelling settlement of Montpenèdre, Hérault, Southern France. In *Proceedings of the IVth Int. Symp. Radiocarbon and Archaeology*, Oxford, 9–14/04/2002, *Oxford University School of Archaeology Monograph* (Vol. 62, pp. 193–200).

Pearson, G. W. (1986). Precise calendrical dating of known growth-period samples using a 'curve fitting' technique. *Radiocarbon, 28*(2A), 292–299.

Pilcher, J. R., Baillie, M. G. L., Schmid, B., & Becker, B. (1984). A 7,272-year tree-ring chronology for Western Europe. *Nature, 312,* 150–152.

Rashit, M., Hantemirov, M., & Shiyatov, S. G. (2002). A continuous multimillenial ring-width chronology in Yamal, Northwestern Siberia. *The Holocene, 12*(6), 717–726.

Robinson, W. J. (1976). Tree-ring dating and archaeology in the American South-West. *Tree-Ring Bull, 36,* 9–20.

Roig, F., Jr., Roig, C., Rabassa, J., & Boninsegna, J. (1996). Fuegan floating tree-ring chronology from subfossil *Nothofagus* Wood. *The Holocene, 6*(4), 469–476.

Schaub, M., Kaiser, K. F., Frank, D. C., Buentgen, U., Kromer, B., & Talamo, T. (2008). Environmental change during the Allerød and Younger Dryas reconstructed from tree-ring data. *Boreas, 37,* 74–86.

Stuiver, M., Reimer, P. J., Bard, E., Beck, J. W., Burr, G. S., Hughen, K. A., et al. (1998). IntCal98 Radiocarbon Age Calibration, 24,000-0 cal BP. *Radiocarbon, 40*(3), 1041–1083.

Suess, H. E. (1965). Secular variations in the cosmic ray produced carbon-14 in the atmosphere and their interpretation. *Journal of Geophysical Research, 70,* 5937–5952.

The Dating of Ice-Core Archives

Frédéric Parrenin

The wealth of testimony about past variations in our climate and environment found in deep ice cores in Antarctica and Greenland is acknowledged well beyond the limits of glaciological research. Uniquely, both local climate variations and global atmospheric composition can be reconstructed from a single archive: the ice. Effective use of the information provided by the glacial archives requires dating as precisely as possible of these various records. To do this, the specific characteristics of ice need to be considered.

The first characteristic results from the compaction of snow layers under their own weight. At the surface, the snow is not very dense (0.3–0.4 g/cm^3): air circulates freely in the first meters of this porous milieu, the firn, and then with more difficulty as the density increases and the porosity decreases. When the density is greater than about 0.83 g/cm^3 (below about 100 meters in the center of Antarctica), air is trapped in bubbles in the ice and insulated from the atmosphere. In the depths, under the effect of pressure, the bubbles become compressed and are then transformed into clathrates, i.e. the gas molecules become incorporated into the crystalline structure of the ice. This means that the air is younger than the ice that imprisons it. Therefore, to date ice core archives, which have some signals recorded in the ice and others recorded in the air bubbles, two distinct chronologies are required. The evaluation of the age difference between gas and ice is discussed in Section "Ice-Air Age Difference".

Moreover, ice does not lend itself to the use of radioactive methods. Carbon-14 dating can only be used in exceptional cases, for example, on plant debris or when sufficient amounts of carbon dioxide are extracted. Although the quantities of ice necessary for carbon-14 dating have decreased since the advent of accelerator mass spectrometry, the dates obtained are only averages over a few meters of ice. In addition, this method is not applicable beyond a few tens of thousands of years because the period of radioactive decay of carbon-14 is 5730 years.

Datings developed by glaciologists are then based on complementary methods such as counting annual layers (Section "The Counting of Annual Layers"), comparison with other dated records (Section "Identification of dated horizons") and with variations in insolation (Section "Orbital Tuning and Indicators of Local Insolation"), and glaciological modeling (modeling of the accumulation of snow and the flow of ice, Section "Flow Modeling"). After these methods have been presented, we will describe, in Section "The Inverse Method: A Collective Approach", a statistical technique, known as the 'inverse method', which consists of collecting these different sources of chronological information to achieve an optimum date and to assess its confidence interval.

Ice-Air Age Difference

Introduction

The firn is the porous upper area of the ice caps. It marks the transition from snow on the surface to the ice below. Depending on its location, its thickness can vary from roughly 50 m (Greenland) to 120 m (central Antarctica). Its density varies from the surface density (typically 0.4 g/cm^3) to the density at the *close off* depth, i.e. the depth at which the pores close (typically 0.83 g/cm^3). At this depth, air is trapped in isolated bubbles and no longer circulates.

The study of transport of air in the firn has led to the development of a simple model (Sowers et al. 1992) from which we can distinguish four zones in the firn (Fig. 9.1).

- *The convective zone* is located just below the surface. Convection in this zone is caused partly by the thermal gradient and partly by surface winds. The depth of this zone varies from one site to another, and may reach 20 m

F. Parrenin (✉)
Institut des Géosciences de l'Environnement,
St Martin d'Hères, France
e-mail: frederic.parrenin@univ-grenoble-alpes.fr

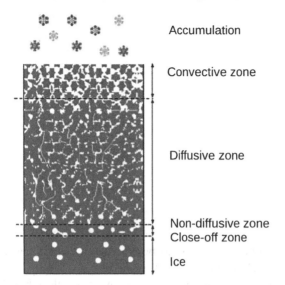

Fig. 9.1 Diagram showing the different parts of the firn. Adapted from Sowers (1992)

in sites with high winds and rugged relief (dunes) at the surface.
- The diffusive zone corresponds to an area where the air column is static (no convection). Movement happens at the molecular level (diffusion), and in this zone, elementary and isotopic fractionation is observed. For example, the heaviest molecules are located, in priority, in the cold deeper part, while the lighter molecules are more numerous in warmer, higher areas.
- The non-diffusive zone is the zone where the pores are almost closed and where molecular diffusion is negligible. From this point, the composition of the bubbles no longer changes, even though the total pressure can still evolve.
- The close-off zone is located at the base of the firn. The bubbles of air are closed off, and the air is trapped inside. This close off zone has a well-defined density of about 0.83 g/cm^3. This close-off depth will increase as the surface temperature decreases and will vary according to the rate of accumulation on the surface of the firn. As a result, the depth of the close-off zone will vary during glacial and interglacial periods.

The top of the non-diffusive zone is a depth of critical importance, above which the air contained in the pore space of the firn is still in contact with the surface atmosphere. Although the snow at this depth fell hundreds or even thousands of years earlier, the gas is still at age 'zero', i.e. at the age of the most recent snow. Consequently, at any depth of an ice core, the gas in the trapped air bubbles is younger than the ice surrounding it. The difference between the age of the ice and the age of the gas bubbles is denoted as Δage. In other words, the gas of the same age as the ice is found lower down, and this difference in depth is denoted Δdepth.

In reality, it is not possible to attribute an exact age to the gas at a given depth. As the air travels through the diffusive column, it mixes the gases from atmospheres of different periods, with a typical average time of a few decades. Moreover, as the close-off boundary is not attributed to a specific depth, but extends over several meters, gas trapped at the same depth may have become imprisoned at slightly different times. Therefore, the signal produced may be diffuse, all the more so if the accumulation of snow is low.

Modeling the Densification of the Firn

To evaluate the Δage, one must evaluate:

- *The density at the close-off* which is often calculated from the surface temperature, using observations carried out at different sites (Salamatin et al. 2009); it can also be deduced from the concentration of air in the ice;
- *The age of the gas at the close-off* which is often ignored in the case of Antarctic cores, as it is very tiny compared with the age of the ice; it can be assessed using a gas diffusion model for the firn in the case of Greenland cores, where the age of the ice at close-off is only a few hundred years (Schwander et al. 1997);
- *The density profile in the firn*, which is derived from a mechanical model; various mechanical models have been published (see, for example, Salamatin et al. 2009); they generally take into account the slippage of snow grains relative to each other, a dominant process at the surface, and the deformation of the grains which becomes dominant at greater depths.

As shown in Fig. 9.2, the calculated depth of the close-off increases when accumulation increases (vertical advection increases) or as the temperature decreases (densification happens more slowly). These models were validated using current data (especially density profiles) from sites with very varied average temperature and accumulation conditions, both in Antarctica and in Greenland (Fig. 9.2. See Salamatin et al. 2007). However, it is worth noting that no site included corresponds to the conditions of the last ice age in Antarctica, which had very cold temperatures and very low accumulation. Also, these validations only pertain to the present, with current orbital parameters and thus with very specific daily and seasonal insolation distributions.

Fig. 9.2 Depth of the close-off of bubbles based on the rate of accumulation and the surface temperature, assumed to be at a steady state, as calculated by the model by Arnaud et al. (2000). The conditions in different polar sites are indicated by crosses. Adapted from Landais et al. (2006)

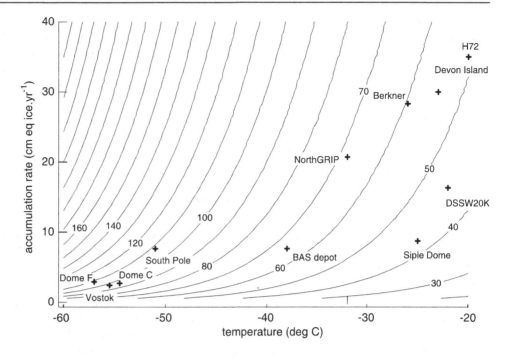

Application of the ^{15}N and ^{40}Ar Isotopes in the Bubbles

Nitrogen and argon have isotopic compositions (in ^{15}N and ^{40}Ar) in the atmosphere almost constant over the timescales studied in ice cores. But the isotopic composition of the air bubbles varies due to a process occurring in the firn.

Because of the mixing, no fractionation occurs in the convective zone. In the diffusive zone, two types of fractionation take place:

- *Gravitational fractionation*, under the influence of gravity, draws the heavy isotopes towards the bottom of the firn, according to the equation:

$$\delta_g = \left[\left(\frac{T}{T_0}\right)\right]\left[\exp\left(\frac{\Delta mgz}{RT}\right) - 1\right] \times 1000 \quad (9.1)$$

where Δm is the difference in mass between the two isotopes, g is the gravitational acceleration, z is the height of the diffusive column, R is the constant of perfect gases and T is the temperature, expressed in Kelvin. This fractionation will therefore depend primarily on the height of the diffusive column, and to a lesser extent, on the temperature of the firn. All things being equal, the gravitational fractionation is proportional to the difference in mass between the two isotopes under consideration. So, it is four times higher for argon (^{40}Ar and ^{36}Ar) than it is for nitrogen (^{15}N and ^{14}N).

- *Thermal fractionation* draws the heaviest types towards the cold extremity. Thermal fractionation in equilibrium may be written as:

$$\delta_g = \left[\left(\frac{T}{T_0}\right)^\alpha - 1\right] \times 1000 \quad (9.2)$$

where T and T_0 are the temperatures at either end of the diffusive column and α is the thermal diffusion coefficient, which depends in a complex way on the temperature.

Nitrogen-15 and argon-40 can thus be used in two different ways to constrain the age differences between ice and gas.

Firstly, abrupt changes in temperature can be identified in both the ice (where it is recorded in the variations in the isotopic composition of oxygen and of hydrogen in the H$_2$O molecule, see §11.3, Chap. 11), and in the air (in the isotopic composition anomaly due to thermal fractionation). Thus, an estimate of Δdepth may be deduced. This method was used to validate firn models in Greenland during major rapid changes in temperature, called Dansgaard-Oeschger events. In Antarctica, temperature variations are less abrupt, and so detection of the temperature anomaly remains ambiguous.

Secondly, assuming that the convective column is known and that no fractionation takes place during the pore closing process, we can calculate the thickness of the diffusive column. This technique also served to validate the firn models in Greenland. For sites on the Antarctic plateau, the situation is more complex, because nitrogen-15 and argon-40 suggest a decrease in the diffusive column during the glacial periods, although the firn models calculate an increase in the thickness of the firn (Landais et al. 2006). Three hypotheses may explain this discrepancy: (1) the height of the convective zone increased during glacial periods; (2) firn models do not apply to glacial conditions of the Antarctic plateau; (3) another fractionation process occurs at the closure of the pores. This issue remains undecided at present.

Synchronization of Two Ice Cores

As explained above, the ice/gas differential in cores with a low rate of accumulation from the Eastern Antarctic shelf and during the ice ages is still poorly constrained. An alternative way to obtain an estimate, for both ice and gas recordings, is synchronization with a core with a higher rate of accumulation, wherein the ice/air differential is better constrained.

Loulergue et al. (2007) and Parrenin et al. (2012) have applied this method to constrain the Δage of the EDC (EPICA Dome C) from the EDML (EPICA Dronning Maud Land) and the TALDICE (Talos Dome Ice Core) cores. Gas synchronization is based on the rapid variations in methane, and ice synchronization uses volcanic signatures. So, this study shows that the firn model, forced with temperature and accumulation scenarios as for dating ice (Parrenin et al. 2007b), overestimates the Δage at EDC by 500–1000 years, during the last glacial period. Consequently, the densification mechanism during glacial periods at EDC is poorly understood and the models need to be improved.

The Δage, during glacial periods, for Antarctic plateau sites with low accumulation is therefore an open question. Further studies are needed to clarify this issue.

The Counting of Annual Layers

On the polar ice caps and glaciers, many of the properties of snow differ depending on whether it accumulates in summer or winter. For example, in summer, dust is more abundant in the snow, because during this season, the winds are more conducive to dust transport towards the poles. The annual layers can therefore be identified, either visually, or by chemical analysis or by isotopic analysis. Counting annual layers is a simple method of dating, provided that the accumulation of snow is sufficient, so that the stratigraphy is not destroyed by winds mixing the layers near the surface. For this reason, the counting of layers is impossible in the central regions of the Antarctic plateau where the deep drilling of Vostok, Dome C and Dome F are located, but it is possible over Greenland and the coastal regions of Antarctica.

A large project for systematic counting called Greenland Ice Core Chronology 2005 (GICC05) has been undertaken by a Danish team at the Niels Bohr Institute in Copenhagen. It is based on the cores of DYE-3, GRIP and NorthGRIP, and currently extends over the last 60,000 years (Svensson et al. 2008). More recently, the WAIS (West Antarctic Ice Sheet) Divide ice core has been counted back to 31,000 years (Sigl et al. 2016).

Glaciologists use various records to identify annual layers. Where possible, the isotopic variations in the ice ($\delta^{18}O$ and δD), which are dependent on the temperature at the moment of the precipitation, provide the most reliable recording of the changing of the seasons. However, water molecules diffuse in the form of vapor through the firn, then more slowly through the ice. This diffusion smooths out the seasonal isotopic signal until it disappears at a certain depth, even more rapidly when accumulation is low and the temperature is high. Thus, the seasonal cycle of isotopes is hardly recognizable on the NorthGRIP core, which has a low accumulation; it is quite muted in the GRIP core. The longest sequence on which the seasonal cycle oxygen-18 was used was obtained from the Dye-3 core in Greenland: 67,000 isotopic analyses allowed the dating of the core year by year over the last 7900 years. Beyond that, the thickness of annual layers is insufficient and the isotopic diffusion through the ice makes counting inaccurate.

Other data taken from the content of impurities in Continuous Flow Analysis (CFA), from the Electrical Conductivity Measurement (ECM), from the insoluble dust content, and from Visual Stratigraphy (VS) complete the isotopic information when this is available (Fig. 9.3). The CFA allows the various soluble compounds, such as Na^+, Ca^{2+}, H_2O_2, NH_4^+, NO_3^- and SO_4^{2-} to be separated. The ECM is a non-destructive measurement, conducted continuously in the field, but it only provides information on an amalgamation of these different soluble compounds. The VS uses the fact that impurities diffuse the light in the ice. However, this recording generally shows several peaks in a year and is therefore not easy to interpret. In Greenland and during the Holocene, a typical year is characterized by Na^+ (dominated mainly by marine inputs) showing a peak in late winter. Spring has a high dust content, high Ca^{2+} and low H_2O_2. Summer is characterized by high concentrations of NH_4^+, NO_3^-, and sometimes SO_4^{2-}. This method, based on data from CFA, ECM and VS from GRIP and NorthGRIP, was the one principally used to establish GICC05 in the period between 7900 and 14,800 years b2k (this notation means 'years before 2000') (Rasmussen et al. 2006). In the older part (14,800–60,000 Years b2k), the method is the same, but only NorthGRIP data were used (Svensson et al. 2008).

In summary, none of the individual indicators is perfect, but combined, they permit an annual dating, as long as the thickness of the layers, which thin out as they sink into the ice cap, remain sufficient. For GICC05, counting was carried out by different people and on different cores, and the independently obtained results were compared so as to

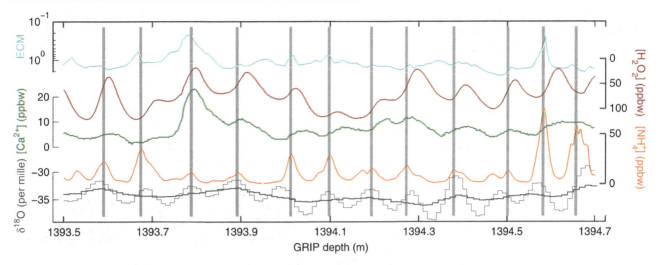

Fig. 9.3 Example of a 1.2 m section from the GRIP core of about 8.8 ka with the annual layers marked by the gray vertical bars. From top to bottom, the records used to identify annual layers are: ECM, H_2O_2, Ca^{2+}, NH_4^+ and $\delta^{18}O$. For this last indicator, the thick line represents the raw data and the fine line represents data after correction for the diffusion effect. Adapted from Rasmussen et al. (2006)

minimize subjective errors. Moreover, each uncertain layer was listed so as to obtain a confidence interval on the final chronology. The listed errors are few: under 2% as far back as the last deglaciation, and around 5% before for older ages (Fig. 9.4).

Identification of Dated Horizons

Even if ice cores cannot be dated directly by the conventional radiochronological methods, events that have been dated elsewhere can be identified in these cores: this is what we call the identification of dated horizons. In the following, we will detail the principal types of horizons used to date the polar cores.

Volcanic Horizons

Volcanic horizons can be identified in the cores, both in Antarctica and in Greenland. Some events are sufficiently intense or occur sufficiently close to the core to deposit volcanic dust (ash) visible to the naked eye. However, most events only deposit fine aerosols. These can then be identified with chemical analyses performed on the ice: sulfate, in particular, has several peaks corresponding to volcanic inputs that are easily identified because they far exceed the usual levels. These volcanic horizons also change the dielectric properties of the ice, which enables them to be identified by measuring its conductivity (this is a non-destructive measurement, generally carried out in the field, immediately after the cores have been brought to the surface).

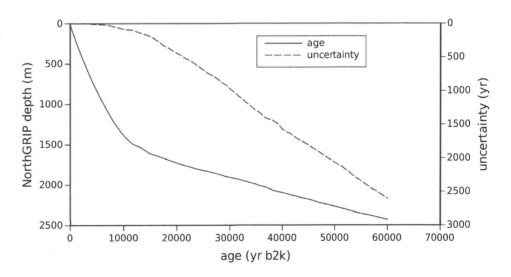

Fig. 9.4 Depth/age relationship on the NorthGRIP core according to the GICC05 scale (solid line) with associated uncertainty (dotted line). The notation 'b2k' means before 2000

For the last few thousand years, a large number of these volcanic events have been dated by different methods (Zielinski et al. 1994): either by counting the layers in sites with a sufficiently high accumulation or through historical writings, or by dating volcanic material near the volcano in question, especially by the carbon-14 method applied to biological debris (e.g., trees caught in the lava). Starting with an approximate glaciological dating, these events can generally be identified by their intensity in the cores. They can also allow two ice cores to be stratigraphically connected (see Fig. 9.5).

Before the Holocene, only a few layers of ash (Fig. 9.6) can be dated accurately. Chemical analysis of these ashes enables unambiguous identification of the relevant volcanic eruption. This signature can sometimes be connected to that of the volcanic material located close to a volcano. This volcanic material, in larger quantity, can be dated by conventional radiochronological methods.

In this way, a visible ash layer in the ice cores of Dome C and Dome Fuji could be dated by an American team using the argon/argon method (see Chap. 5) to date ash found near the volcano (Mt Berlin, Antarctica; Narcisi et al. 2006). For Greenland, notable volcanic horizons were identified that are referred as: Saksunarvatn, Vedde, Fugloyarbanki, '33 ka ^{14}C' and Z2. These horizons were dated by the carbon-14 method or by the argon/argon method (see Svensson et al. 2008 and included references for further detail).

Fig. 9.6 Ash layer present at about 239 m in the core drilled at Talos Dome (East Antarctica). Copyright F. Parrenin (frederic.parrenin@univ-grenoble-alpes.fr)

Dansgaard-Oeschger Events

Dansgaard-Oeschger events (D-O) were identified for the first time in Greenland ice cores and correspond to abrupt changes in temperature during the last glacial period (see §3.2, Chap. 4). Synchronous variations (within a few decades) were also observed in the atmospheric content of methane (Severinghaus et al. 1998) measured in air bubbles from both Antarctic and Greenland ice. These events can therefore be dated by the Greenland cores which are relatively precisely dated through counting of the annual layers (See Section "The Counting of Annual Layers"). This dating can then be transferred to the Antarctic cores thanks to the methane records.

The changes in climate associated with the Dansgaard-Oeschger events, although their maximum impact was probably at the level of the North Atlantic, are visible in many locations on the planet, in other climate

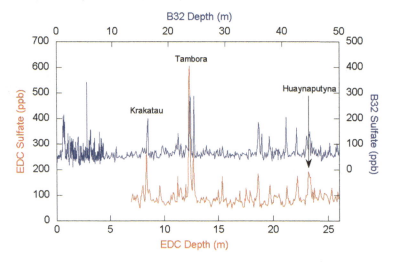

Fig. 9.5 Sulfate profiles for the first tens of meters in two ice cores from Eastern Antarctica: EDC (EPICA Dome C, bottom) and B32 (top), the latter located near the EDML site (EPICA Dronning Maud Land). Several well-known eruptions can be identified. Moreover, these volcanic profiles can be used to synchronize the ice cores between each other. Adapted from Severi et al. (2007)

archives. They are found in the isotopic composition of speleothems (stalactites and stalagmites found in caves) in Europe and Asia, in South America and the Indian Ocean. These speleothems can be dated with a precision of a few hundred years to one or two thousand years, using the uranium/thorium method, which provides a specific age for the marked transitions of the D-O events. Figures 9.7 and 9.8 summarize these recordings with their respective dating.

Variations in the Magnetic Field and in Solar Activity

Beryllium-10 and carbon-14 are both produced in the upper atmosphere by the flux of cosmic particles. This flux is modulated, partly by the magnetic field of the solar wind which deflects the charged particles, and partly by the terrestrial magnetic field. Unlike beryllium-10 whose deposition on the surface of the Earth is almost directly related to its production in the upper atmosphere, the composition of carbon-14 in the atmosphere is also influenced by the exchanges between the different carbon reservoirs on Earth. But the major changes in these two indicators (carbon-14 and beryllium-10) are simultaneous.

Beryllium-10 can be accurately measured in ice cores, both from Antarctica (see Raisbeck et al. 2007 and included references) and Greenland (see Beer et al. 2006 and included references). This allows the dating achieved by counting the annual layers in Greenland to be transferred to Antarctic cores for the Holocene (Ruth et al. 2007) and for the Laschamp anomaly in the geomagnetic field, which occurred about 41 ka ago (Raisbeck et al. 2007).

Carbon-14, meanwhile, is measured in tree rings, which are very accurately dated for the last 12.4 ka using dendrochronology (see Chap. 8). We can then import this dating to the ice cores when the variations in solar activity are significant enough so that beryllium-10 and carbon-14 can be synchronized. This method has been used to date the Holocene part of the Antarctic ice cores, where counting of the layers is not possible (Ruth et al. 2007).

Finally, significant anomalies in the geomagnetic field can be identified in other paleoenvironmental archives such as volcanic lava, which can be dated by the argon/argon or potassium/argon methods. The Laschamp anomaly is thus dated with a relatively good accuracy (Guillou et al. 2004), while the older Bruhnes-Matuyama transition is only very crudely dated (Raisbeck et al. 2006). Chap 7 provides more detail on magnetic stratigraphy.

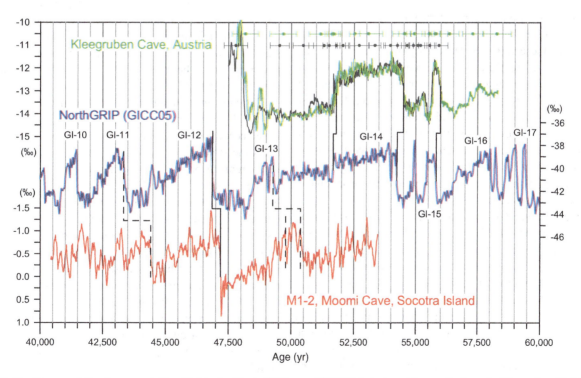

Fig. 9.7 The Dansgaard-Oeschger events identified in the North-GRIP core (GICC05 dating), and in records from the Kleegruben (Spötl et al. 2006) and Moomi (Burns et al. 2003) caves, the latter being dated by a uranium/thorium method. Adapted from Svensson et al. (2008)

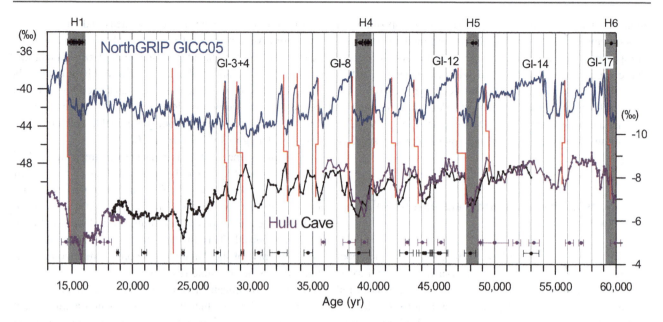

Fig. 9.8 The Dansgaard Oeschger events identified in the NorthGRIP core (GICC05 dating) and records from the Hulu cave (Wang et al. 2001). The points dated in the records from Hulu Cave by the uranium/thorium method are marked at the bottom of the figure (with their error bar). The shaded areas represent Heinrich events as identified in speleothems in Brazil (Wang et al. 2004). These points dated by the uranium/thorium method are also shown at the top of the figure. Adapted from Svensson et al. (2008)

Orbital Tuning and Indicators of Local Insolation

The evolution of orbital parameters of the Earth over the last million years is known with great precision (Laskar et al. 2004), and these variations leave a signature in most climate records. Therefore, the use of variations in insolation to date polar cores is a natural choice. Uncertainty in this dating is due, in part, to the assumption that there is a constant phase difference between orbital and climatic variations, and, secondly, to the evaluation of this phase difference. The advantage of this dating is that it has an almost constant uncertainty of a few thousand years along the length of the core, especially for the deeper parts, as long as we are able to recognize the orbital cycles.

Several parameters recorded in the ice cores show strong variations in the orbital frequencies. These were then used to align the ice cores with orbital cycles: these are the D/H ratio in the ice, an indicator of local temperature (for example, Parrenin et al. 2004) and the $^{18}O/^{16}O$ ($\delta^{18}O_{atm}$) ratio in the air bubbles (for example, Dreyfus et al. 2007). The variations in $\delta^{18}O_{atm}$ are a reflection of two aspects of the environment (Landais et al. 2010). The first one is that the variations in the $\delta^{18}O$ of the ocean, directly related to the volume of land ice, impact fully on the $\delta^{18}O$ of the atmospheric oxygen through the process of photosynthesis by seaweed. The second one is that a portion of the variations in $\delta^{18}O_{atm}$ is determined by the behavior of the terrestrial biosphere. This is the Dole effect which depends in a complex way on the reactions of photosynthesis and respiration. The $\delta^{18}O_{atm}$ signal shows variations mainly related to the precession, so it is very easy to use it to 'count' these cycles, at least for the periods during which they are sufficiently important variations in the insolation signal. However, there is no reason to assume that the phase difference between $\delta^{18}O_{atm}$ and insolation remained constant over time.

To avoid this limitation in the 'traditional' methods of orbital alignment, more direct indicators of local insolation have recently been proposed. Local insolation in summer alters the structure of the snow at the surface, and this signature remains present down to the close-off zone, regardless of the densification process. These structural parameters have an impact on the volume of the pores at close-off (and thus the air content), and also on the molecular fractionation processes between O_2 and N_2 when the pores close.

Bender (2002), was the first to suggest that the O_2/N_2 ratio in the air bubbles analyzed in the Vostok core was dependent on the local insolation at the summer solstice (Fig. 9.9). Recently, this link has been confirmed in the first core drilled at Dome Fuji (Kawamura et al. 2007) (Fig. 9.9) and recent measurements in the second core, with an improved analytical process, show an almost perfect correlation.

Raynaud et al. (2007) also studied in more detail the air content in the EDC core and suggested that it depended on local insolation averaged over a period centered on the

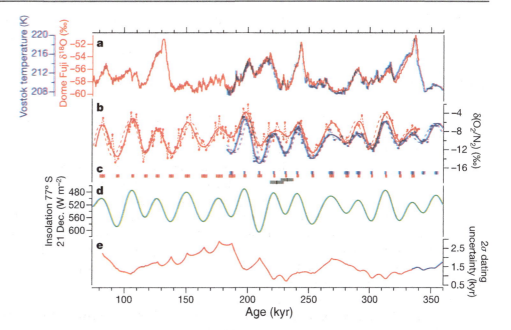

Fig. 9.9 a Isotopic composition of the ice and temperature at Vostok and Dome Fuji. b O_2/N_2 ratio measured in the air bubbles at Dome Fuji and Vostok. The dots are the raw data; the thick lines represent the filtered data. c Age markers deduced from the alignment with local insolation, with 2σ error bars. d Insolation at the summer solstice at 77° S used as an alignment target. e 2σ uncertainty of the O_2/N_2 dating. Adapted from Kawamura et al. (2007)

summer solstice. However, for this second indicator, it is important to correct for variations in the altitude at which the bubbles formed, as this will have an impact on atmospheric pressure and thus the air content of these bubbles.

Although the specific physical link between these indicators and insolation is still subject to debate and research, we can make two observations. Firstly, no signal with a 100 000 year periodicity is present in the O_2/N_2 record, so that the O_2/N_2 proxy does not seem to be dependent on climate. Secondly, although these indicators are measured in the gas bubbles, they are caused by modifications in the structure of the snow at the surface, and therefore provide a dating of the ice (and not of the bubbles!). This avoids uncertainty associated with Δage.

Flow Modeling

The ice has an enormous advantage over other archives in that it can be dated using physical models that take the variations in the rate of accumulation of snow and the flow of ice into account. The age in the ice core at an altitude of z can be written as:

$$\chi^{(z)} = \int \frac{d(z')}{T(z')a(z')} dz' \quad (9.3)$$

where χ is the age of ice, d is the relative density of ice (compared with pure ice), a is the initial accumulation of snow (expressed in cm equivalent to pure ice per year, denoted as cm-i.e./year) and T is the thinning function, i.e. the thickness of an annual layer relative to its initial thickness at the time it fell. d can be measured from the ice core.

The parameter a is generally calculated from indicators measured in the ice core, while the parameter T is obtained from a flow model. These two steps are detailed below.

Evaluation of Accumulation on the Surface

For the top few hundreds of meters, the thinning of the layers of snow and ice is minimal (T close to 1) and well-assessed by modelling. So, accumulation at the surface may be determined from well-dated horizons such as layers of volcanic ash (described in Section "Volcanic Horizons") using the formula (9.3). Below this depth, the isotopic composition of the ice (D/H or $^{18}O/^{16}O$) is generally used. As for the surface temperature, the field measurements in Antarctica and Greenland show a good correlation between isotopic composition and the surface accumulation of snow. In a review of measurements in Antarctica, Masson-Delmotte et al. (2008) derived a relationship as follows:

$$a = a_0 \exp(\beta(\delta D - \delta D_0)) \quad (9.4)$$

where a and δD are respectively, a reference accumulation and a reference isotopic composition, and where $\beta = 0.0152$. This relationship is derived from the saturated vapor pressure of the ice and can be calculated from a simple model of precipitations of an air mass. Note, however, that it does not take into account the phenomenon of re-deposition of the snow by the wind, which modifies the accumulation without altering the isotopic composition of the snow. On the other hand, when we extrapolate this relationship to temporal variations in accumulation, it is important to consider the temperature variations and isotopic composition at the source of the air masses which also alter the isotopic composition of the ice (Parrenin et al. 2007b).

Ice Flow Models

Ice has a solid exhibiting viscoplastic behavior where the relationship between stress and strain can be determined experimentally and theoretically. It is thus possible to simulate the trajectory followed by a particle of ice within the glacier over time in order to establish a chronology. Modeling of behavior of ice within an ice sheet requires not only a good knowledge of the viscoplastic properties of the material, but of the conditions at the boundaries of the cap. These boundary conditions are: (1) the temperature and surface accumulation over time; (2) basal conditions, such as the geothermal flow and the rate of friction on the bedrock; (3) the lateral conditions for the area under consideration, since local models are used for dating purposes. These lateral conditions generally result from global simulations of the polar cap over time (Ritz et al. 2001). In this way, the thinning function adapted to the ice core drilling site is obtained.

Below is a qualitative description of how this function varies. For a stationary dome, the thinning function can be written:

$$T = \frac{1 - \frac{m}{a}}{\omega + \frac{m}{a}} \quad (9.5)$$

with $\mu = m/a$ the ratio of basal fusion to surface accumulation and with ω the standard vertical profile of horizontal flow (see Parrenin et al. 2007b, for details). ω varies almost linearly from 0 at the ice base interface to 1 at the surface, because the deformation is concentrated at the base of glacier. For certain domes, the Raymond effect causes more deformation at the top of the ice and therefore a less linear ω profile.

In a non-stationary case, variations in the thickness of the ice (related to climatic variations) cause bumps in the thinning function (Fig. 9.10). Moreover, for ice core drilled along the flow line, like Vostok, the ice comes from upstream and more complex deformation effects exist. The parameter that most influences the thinning function is the thickness of ice at the place of origin of the ice: if this thickness is large compared to the thickness at the drill site, then the column of ice has become compressed overall, resulting in a strong thinning (that is, a low thinning function). And reciprocally.

The Limitations of Modeling

Unfortunately, dating using models becomes increasingly inaccurate as it approaches the base of the ice cap, for various reasons. Firstly, the mechanical properties of the ice are not perfectly understood. They depend not only on pressure and temperature conditions, but also on the size and orientation of the crystals that make up the ice. Secondly, the conditions at the base of the bedrock cannot be measured directly in situ. Finally, the lateral conditions throughout the past, the outcome of a large-scale model, may also be tainted by a significant error. These lateral conditions determine the position of the domes and dividing lines in the domain, and therefore the direction of the ice particles.

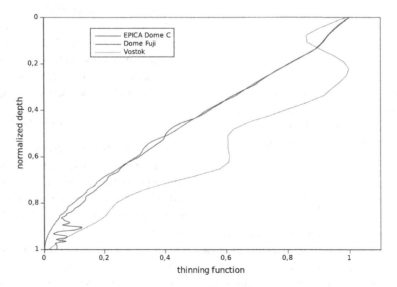

Fig. 9.10 Thinning functions for ice cores from Dome C (Parrenin et al. 2007b), Dome Fuji (Parrenin et al. 2007a) and Vostok (Parrenin et al. 2004)

The Inverse Method: A Collective Approach

All of absolute dating methods described in the preceding sections have advantages and disadvantages. The counting of layers and flow modeling are precise methods in terms of the duration of events (at least for relatively recent periods) because they are based on an assessment of the thickness of the annual layers. However, errors accumulate and inaccuracy in the absolute age increases rapidly with depth.

Orbital tuning is generally applicable over the entire length of a core, as long as the stratigraphy is maintained. In addition, accuracy does not diminish with depth, and so far, this is the most accurate method for dating the lower part of polar cores. Unfortunately, this method is not very accurate in terms of duration of events, and the accuracy in terms of absolute ages is limited by the assumption that there is a constant phase difference between the record being studied and the insolation (and obviously does not allow the variations in this phase to be reconstructed). Research into indicators of local insolation paves the way for a significant improvement in methods of orbital alignment, with a possibility of uncertainties below 1000 years. However, we still lack sufficient distance to assess the real accuracy of these methods. In addition, other methods will always be more accurate for very recent times.

Volcanic eruptions provide important dated horizons. This is particularly true for the last millennium, but beyond this, only a few have an absolute age that is sufficiently precise. Comparison with other dated records is particularly useful for the dating of Dansgaard-Oeschger events, which can be precisely localized using variations in the isotopic composition of ice (in Greenland cores only), or variations in the methane content in air bubbles. Most of these events have been dated accurately for the last ice age, especially from speleothems and the U/Th method (Chap. 6). Other studies are underway to improve this accuracy and to study previous ice ages. In any case, these dated horizons do not provide continuous dating and are mainly relevant to recent periods.

As these different sources of chronological information are complementary, it is clear that to obtain an optimum dating from the ice records, it is essential to combine them. This is what glaciologists have attempted to do with Antarctica ice cores. Initially, the poorly known parameters in flow models (such as melting and sliding at the base of the glacier) were adjusted by trial and error to obtain a good agreement with dated horizons. But this approach quickly becomes difficult when there are several free parameters and when the different error bars for the dated horizons must be taken into account.

In the early 2000s, an inverse method to formalize and systematize this optimization of dating in a probabilistic framework was developed (Parrenin et al. 2001). In the context of dated horizons with a certain error bar, it facilitates the determination of the probability density called '*a posteriori*' both for the uncertain parameters in the flow models and for the final dating. Concretely, this probability density not only provides an optimum dating, but also a confidence interval. This probabilistic method, based on the Metropolis-Hastings algorithm, was applied to the Vostok (Parrenin et al. 2004), Dome C (EDC3 dating, Parrenin et al 2007a, b) and Dome Fuji (Parrenin et al. 2007b) ice cores.

This inverse approach, however, has several limitations. Firstly, it only takes into account the errors related to the lack of understanding of poorly known parameters. In other words, the model is considered perfect once an optimal estimation of these parameters is achieved. This is the same as saying that the model is able to describe all the relevant flow mechanisms and is therefore, in agreement with all sets of markers whose age errors are correctly estimated. In reality, however, many physical phenomena influencing flow are not taken into account in the model, either because they are not properly understood, or because the level of complexity necessary to describe them is incompatible with inverse modeling (direct model is too costly in computing time with too many parameters to inverse). This limitation of the models appears clearly at the base of the EDC core where the model, even after optimization of its parameters, is unable to reproduce the age markers obtained by orbital alignment (Dreyfus et al. 2007). A second limitation of this inverse method is that it can be applied to only one core at a time, and the optimal dating obtained is different for each core, making it difficult to compare climate and environmental signals. In reality, these cores can be synchronized more accurately in the ice phase (for example, by volcanic horizons), as well as in the gas phase (for example, CH_4 and $\delta^{18}O_{atm}$).

As a result, a new method of optimization was developed (Lemieux-Dudon et al. 2010; Parrenin et al. 2015). This considers the information gleaned from modeling to be weakly constrained (the model is not deemed perfect) and applies to several cores simultaneously, taking into account the stratigraphic links between these cores, both in the ice and in the air. This method can thus provide an optimum dating common to the different cores from Antarctica and Greenland.

Conclusion

The dating of glacial archives is a complex problem which, in the absence of the radioactive methods, is based on several complementary techniques. For the Holocene and in high accumulation sites, dating by counting layers is accurate to about 1%. This dating was confirmed by comparison with dendrochronology using beryllium-10, and with volcanic horizons dated by carbon-14. The last millennium can also be dated to within a few years through identification of known volcanic eruptions. For the rest of the Holocene, synchronization with the dendrochronological scale using beryllium-10 gives us dating accuracy to within a few decades.

For the last glacial period, the accuracy of counted time scales decreases to about 5% up to 60 ka. This dating provides ages for the Dansgaard-Oeschger events which are confirmed by uranium/thorium dating of speleothems in Europe and Asia. It is also compatible with some volcanic horizons dated by carbon-14 and argon/argon. The main methods of dating for the last 60 ka therefore now seem to be in agreement to within a few hundred years rather than a few thousand years as was the case until recently.

Further back than the last glacial period, counting of layers is no longer possible, and the chronologies are mainly based on speleothems, volcanic horizons and orbital alignment. Orbital tuning has a precision of about 5 ka. Local insolation indicators (O_2/N_2 ratio, air concentrations) could lead to an improved accuracy of 1–2 ka, but this must be confirmed by independent methods.

References

Arnaud, L., Barnola, J.-M., & Duval, P. (2000). Physical modeling of the densification of snow/firn and ice in the upper part of polar ice sheets. In T. Hondoh (Ed.), *Physics of Ice CoreRecords* (pp. 285–305). Japan.: Hokkaido University Press, Sapporo.

Beer, J., Vonmoos, M., & Muscheler, R. (2006). Solar variability over the past several millennia. *Space Science Reviews, 125*(1), 67–79.

Bender, M. L. (2002). Orbital tuning chronology for the Vostok climate record supported by trapped gas composition. *Science Letters, 204,* 275–289.

Burns, S. J., Fleitmann, D., Matter, A., Kramers, J., & Al-Subbary, A. (2003). Indian Ocean climate and absolute chronology over dansgaard/oeschger events 9–13. *Science, 301,* 1365–1367.

Dreyfus, G. B., Parrenin, F., Lemieux-Dudon, B., Durand, G., Masson-Delmotte, V., Jouzel, J., et al. (2007). Anomalous flow below 2700 m in the EPICA Dome C ice core detected using $\delta 18O$ of atmospheric oxygen measurements. *Climate of the Past Discussions, 3*(2), 341–353. https://doi.org/10.5194/cp-3-341-2007.

Guillou, H., Singer, B. S., Laj, C., Kissel, C., Scaillet, S., & Jicha, B. R. (2004). On the age of the Laschamp geomagnetic excursion. *Earth and Planetary Science Letters, 227,* 331–343.

Kawamura, K., Parrenin, F., Uemura, R., Vimeux, F., Severinghaus, J. P., Matsumoto, K., et al. (2007). Northern hemisphere forcing of climatic cycles over the past 360,000 years implied by absolute dating of antarctic ice cores. *Nature, 448,* 912–917.

Landais, A., Barnola, J., Kawamura, K., Caillon, N., Delmotte, M., Ommen, T. V., et al. (2006). Firn-Air $\delta^{15}N$ in modern polar sites and glacial-interglacial ice: A model-data mismatch during glacial periods in antarctica? *Quaternary Science Reviews, 25*(1–2), 49–62.

Laskar, J., Robutel, P., Joutel, F., Gastineau, M., Correia, A. C. M., & Levrard, B. (2004). A Long-term numerical solution for the insolation quantities of the earth. *Astronomy & Astrophysics, 428,* 261–285.

Lemieux-Dudon, B., Blayo, Petit, J. R. E., Waelbroeck, C., Svensson, A., et al. (2010). Consistent dating for Antarctica and Greenland ice cores'. *Quaternary Science Reviews, 29*(1–2), 8–20.

Masson-Delmotte, V., Hou, S., Ekaykin, A., Jouzel, J., Aristarain, A., Bernardo, R. T., et al. (2008). A review of Antarctic surface snow isotopic composition: Observations, atmospheric circulation, and isotopic modeling. *Journal of Climate, 21*(13), 3359–3387.

Narcisi, B., Petit, J.-R. & Tiepolo, M. (2006). A volcanic marker (92 ka) for dating deep east Antarctic ice cores. *Quaternary Science Reviews, 25,* 2682–2687.

Parrenin, F., Barker, S., Blunier, T., Chappellaz, J., Jouzel, J., Landais, A., et al. (2012). On the gas-ice depth difference (Δdepth) along the EPICA Dome C ice core. *Climate of the Past, 8*(4), 1239–1255 https://doi.org/10.5194/cp-8-1239-2012.

Parrenin, F., Barnola, J.-M., Beer, J., Blunier, T., Castellano, E., Chappellaz, J., et al. (2007a). The EDC3 chronology for the EPICA dome C ice core. *Climate of the Past, 3,* 485–497.

Parrenin, F., Bazin, L., Capron, E., Landais, A., Lemieux-Dudon, B., & Masson-Delmotte, V. (2015). IceChrono1: A probabilistic model to compute a common and optimal chronology for several ice cores. *Geoscientific Model Development, 8*(5), 1473–1492. https://doi.org/10.5194/gmd-8-1473-2015.

Parrenin, F., Dreyfus, G., Durand, G., Fujita, S., Gagliardini, O., Gillet, F., et al. (2007b). Ice flow modelling at EPICA dome C and dome Fuji, East Antarctica. *Climate of the Past, 3,* 243–259.

Parrenin, F., Rémy, F., Ritz, C., Siegert, M., & Jouzel, J. (2004). New modelling of the Vostok ice flow line and implication for the glaciological chronology of the Vostok ice core. *Journal Geophysical Research, 109,* D20102.

Raisbeck, G. M., Yiou, F., Cattani, O., & Jouzel, J. (2006). ^{10}Be Evidence for the Matuyama-Brunhes geomagnetic reversal in the EPICA dome C ice core. *Nature, 444*(7115), 82–84.

Raisbeck, G. M., Yiou, F., Jouzel, J., & Stocker, T. F. (2007). Direct north-south synchronization of abrupt climate change record in ice cores using beryllium 10. *Climate of the Past, 3*(3), 541–547.

Rasmussen, S. O., Andersen, K. K., Svensson, A. M., Steffensen, J. P., Vinther, B. M., Clausen, H. B., et al. (2006). A new greenland ice core chronology for the last glacial termination. *Journal Geophysical Research, 111,* D06102.

Raynaud, D., Lipenkov, V., Lemieux-Dudon, B., Duval, P., Loutre, M.-F. & Lhomme, N. (2007). The local insolation signature of air content in Antarctic ice. A newstep toward an absolute dating of ice records. *Earth and Planetary Science Letters, 261*(3–4), 337–349.

Ritz, C., Rommelaere, V., & Dumas, C. (2001). Modeling the evolution of antarctic ice sheet over the last 420,000 years: Implications for altitude changes in the Vostok Region. *Journal Geophysical Research, 106*(D23), 31943–31964.

Ruth, U., Barnola, J.-M., Beer, J., Bigler, M., Blunier, T., et al. 'EDML1: A chronology for the EDML ice core, Antarctica, over the last 150 000 Years. *Climate of the Past, 3,* 475–484 (2007).

Salamatin, A. N., Lipenkov, V. Y., Barnola, J. M., Hori, A., Duval, P., & Hondoh, T. (2009). Snow-Firn Densification in Polar Ice Sheets. In T. Hondoh (Ed.), *Physics of Ice Core Records-2.* Sapporo: Hokkaido University Press.

Schwander, J., Sowers, T., Barnola, J.-M., Blunier, T., Fuchs, A., & Malaizé, B. (1997). Age scale of the air in the summit ice: Implication for the glacial-interglacial temperature change. *Journal Geophysical Research, 102,* 19483–19493.

Severi, M., Castellano, E., Morganti, A., Udisti, R., Ruth, U., Fischer, H., et al. (2007). Synchronisation of the EDML1 and EDC3 timescales for the Last 52 Kyr by volcanic signature matching. *Climate of the Past, 3,* 367–374.

Severinghaus, J., Sowers, T., Brook, E., Alley, R., & Bender, M. (1998). Timing of abrupt climate change at the end of the Younger Dryas interval from thermally fractionated gases in polar ice. *Nature, 391,* 141–146.

Sigl, M., Fudge, T. J., Winstrup, M., Cole-Dai, J., Ferris, D., McConnell, J. R., et al. (2016). The WAIS Divide deep ice core WD2014 chronology—Part 2: Annual-layer counting (0–31 ka BP). *Climate of the Past, 12*(3), 769–786. https://doi.org/10.5194/cp-12-769-2016.

Sowers, T. A., Bender, M., Raynaud, D. & Korotkevich, Y. L. (1992). 'The $\delta^{15}N$ of O_2 in air trapped in Polar ice: A tracer of gas transport in the firn and a possible constraint on ice age-gas age differences. *Journal of Geophysical Research, 97*(15), 15683–15697.

Spötl, C., Mangini, A., & Richard, D. A. (2006). Chronology and Paleoenvironment of marine isotope stage 3 from two high-elevation speleothems, Austrian Alps. *Quaternary Science Reviews, 25*(9–10), 1127–1136.

Svensson, A., Andersen, K. K., Bigler, M., Clausen, H. B., Dahl-Jensen, D., Davies, S. M., et al. (2008). A 60 000 year greenland stratigraphic ice core chronology. *Climate of the Past, 4* (1), 47–57.

Wang, X., Auler, A. S., Edwards, L., Cheng, H., Cristalli, P. S., Smart, P. L., et al. (2004). Wet periods in Northeastern Brazil over the past 210 Kyr linked to distant climate anomalies. *Nature, 432,* 740–743.

Wang, Y. J., Cheng, H., Edwards, R. L., An, Z. S., Wu, J. Y., Shen, C. C., & Dorale, J. A. (2001). A high-resolution absolute-dated late pleistocene monsoon record from Hulu Cave, China. *Science, 294* (5550), 2345–2348.

Zielinski, G. A., Mayewski, P. A., Meeker, L. D., Whitlow, S., Twickler, M. S., Morrison, M., et al. (1994). Record of volcanism since 7000 B.C. from the GISP2 greenland ice core and implications for the volcano-climate system. *Science, 264*(5161), 948–952.

Parrenin, F., Jouzel, J., Waelbroeck, C., Ritz, C. and Barnola, J.-M (2001) Dating the Vostok ice core by an inverse method. *Journal of Geophysical Research: Atmospheres 106,* (D23), 31831–837851.

Reconstructing the Physics and Circulation of the Atmosphere

Valérie Masson-Delmotte and Joël Guiot

The variability and evolution of the physical parameters of atmospheric circulation are currently monitored in real time and on a global scale thanks to a dense network of weather stations (over 13,000 measurement sites on land and sea), and to satellite observations of the Earth. This 'instrumental' period, during which the physical parameters of the atmosphere were directly monitored, began in the mid seventeenth century following the invention and use of thermometers, barometers, rain gauges etc. However, standardization of measurement tools and their wide-scale use took a long time and was due to a continuous effort by the meteorological services. The outputs of ancient instruments from before 1950 must be homogenized to modern observation standards, and gaps exist in regional temperature information due to changes in the spatial monitoring network. The oldest meteorological series of data available are in Europe, where temperature series for the center of England start in 1659. Intensive work was carried out on the weather records of the Alpine region, providing access to accurate measurements, from 1780, of average monthly temperatures and cumulative monthly precipitations.

Work is underway to extend the use of these historical measurements to study monthly variability in temperature and precipitation, and to assess other parameters (pressure, sunshine etc.). Use of these old measurements involves working on documents of the time, computerizing the data, and statistical analysis of regional databases. With the exception of Europe, where instrumented measurements were conducted particularly early, weather information is generally only available from 1860, except for the most inaccessible areas, such as Antarctica, where systematic meteorological monitoring did not start until the International Geophysical Year 1957–1958. The 'instrumental period' is therefore very short compared to the time frame of the climate system and does not permit an understanding of the natural climate variability on a global scale for the period prior to when human activities affected the composition of the atmosphere.

In order to characterize natural climate evolution and to place the climate change of recent decades within a broader context, continental paleoclimatology has established methods of quantifying ancient climates by taking advantage of a large number of natural archives, in soils, lakes, vegetation, continental and polar ice. These archives have allowed qualitative or quantitative indices of the main parameters describing climate to be defined. These indices are often referred to as proxies. Below, we briefly review all the climate parameters reconstructed from these continental archives.

The atmospheric parameters most commonly determined from continental paleoclimate archives are surface air temperature (or surface lake water temperature) and parameters related to soil hydrology (e.g. precipitation, drought indices, etc.). In some cases, these parameters can be estimated over a season, when a resolution of less than a year can be detected in the archives (tree rings, ice cores in sites with a high level of snow accumulation) or when the archive is particularly sensitive to seasonal effects (temperature of the coldest month, temperature of the growing season for vegetation, etc.). Most continental proxies do not directly record the amount of precipitation, but reflect the local water balance (precipitation minus land-based evaporation, runoff into lakes, net accumulation of snow on the glaciers at the drilling sites). Quantification of these climate parameters from the records is often made difficult by the discontinuous nature of geological recording, for example, the process of sedimentation in lakes. Some records, such as tree rings, function as threshold systems and identify an atmospheric signal once the threshold is reached (low temperature, dry

---

V. Masson-Delmotte (✉)
Laboratoire des Sciences du Climat et de l'Environnement, LSCE/IPSL, CEA-CNRS-UVSQ, Université Paris-Saclay, 91190 Gif-sur-Yvette, France
e-mail: valerie.masson@lsce.ipsl.fr

J. Guiot
Centre for Research and Teaching in Environmental Geoscience CEREGE, Aix-Marseille University, CNRS, IRD, INRAE, Collège de France, BP 80 13545 Aix-en-Provence Cedex 04, France

season). Moreover, many proxies are not sensitive to a single atmospheric variable but to the combination of effects related to temperature and hydrology. The combined use of multiple markers within a single medium or multiple archives from the same site allows these effects to be separated out. Finally, comparison between proxies, paleoclimate reconstructions, modeling of climate and proxies all improve our understanding of how climate dynamics and proxies operate.

The dynamics of the atmosphere can also be estimated from the continental paleoclimate records. During the instrumental period, it is possible to determine how certain modes of atmospheric circulation such as the El Niño-Southern Oscillation (ENSO), the North Atlantic Oscillation (NAO), the Southern Annular Mode (SAM) or the Pacific-North American oscillation (PNA) modulate the spatial response of proxies, just as they affect the spatial distribution of temperature and rainfall. Provided that the spatial distribution of proxy records is sufficient in key areas, and depending on the stability of these tele-connections through time, this fingerprint can then be used to estimate past inter-annual variations in pressure indices characteristic of patterns of atmospheric circulation in recent centuries (PNA, NAO, ENSO, SAM).

Over large time scales, loess deposits and dunes reflect the prevailing wind direction. Similarly, concentrations of marine and continental aerosols, the size distribution of continental dust particles in polar ice, reflect changes in the aridity of the regions of origin as well as changes in the efficiency of transport of aerosols in the atmosphere. However, quantitative estimates of the intensity of surface winds remain a challenge. Past changes in other atmospheric parameters such as cloudiness are difficult to determine from proxy records in natural archives.

Continental paleoclimatology can also help to characterize the frequency and intensity of 'extreme' events. The intensity and amplitude of past droughts have been estimated using dendrochronological databases in North America and Europe. Sedimentary and geochemical markers from lake sediments are used to determine the intensity and occurrence of flooding by the great rivers. High-resolution analysis of lagoon sediments and the isotopic composition of tree rings or speleothems are currently being used in an attempt to characterize past variations in the activity (trajectories, intensity, frequency) of tropical cyclones and extratropical storms.

We have briefly presented the atmospheric variables that can be estimated from continental paleoclimate records. In the following chapters, we will explain the reconstruction methods used; the assumptions upon which they are based; their limitations and uncertainties; and finally, we will present several specific examples reflecting the diversity of continental archives (lakes, vegetation, ice) whose dating techniques were described in Part II. We will present some archives and some proxies particularly well suited to each of the interfaces under consideration. We have considered high latitudes and low latitudes separately. Several archives (lacustrine cores, speleothems) and proxies (pollen, diatoms) are present in both temperate and tropical regions, but their interpretation is specific to the particular geographical area.

In the high latitudes of both hemispheres, polar ice is of paramount importance in paleoclimatology because it records both climate forcings and some local and global climatic variations. Sedimentary archives (from lakes and bogs) with their pollen records have long been studied in paleoclimatology and cover almost all of the continents. They were initially understood to mostly reflect a local signal. However, the comparison with ice and marine records revealed the broad geographical spread of many events known to palynologists, such as the Younger Dryas cold episode which lasted for close to a millennium and was felt throughout the Atlantic area of the northern hemisphere. The loess covers a significant area of the continents; their sequences are excellent indicators of atmospheric circulation. On a shorter timescale, some archives provide information with annual or near annual resolution. Among these, tree rings provide a wealth of information on temperate regions. They are supplemented by multi-centennial archives from historical written documents, such as wine harvesting dates, in Europe, or cherry blossom dates, in Japan.

At low latitudes, sedimentological tracers collected in tropical lakes, complemented by biological proxies, such as the abundance of diatoms or pollen, contribute to a better understanding of the functioning of the major inter-tropical climate systems and give an insight into sometimes discontinuous records. Diatoms are good indicators of the characteristics of lake water and, complemented by adequate hydrological modeling, they allow an assessment of water resources in watersheds. At high altitudes, tropical glaciers are very sensitive to climate variations in the long and medium term.

Cave records, for instance from speleothems, offer well-dated, albeit discontinuous, records particularly sensitive to changes in the hydroclimate and vegetation cover above the caves, sometimes with very high temporal resolution.

Interpretation of Records, Limitations and Uncertainties

Paleoclimatology draws its information mainly from two types of approaches with their advantages and limitations. The first approach is the most basic and consists of using simple equations to interpret a climate signal from a univariate series. This is the preferred approach for geochemistry which often uses scaling to transform an isotopic signal into a temperature curve (in the case of polar and tropical glaciers). This approach is also possible with data from historical documents, which, when they do not provide direct climatic information, recount events related to a climate variable (floods, droughts, freezing) that may be standardized semi-quantitatively (Pfister 1980).

Another approach is required for analysis of the living world. The climate signal recorded by microorganisms is complex and is a reaction to a combination of several climatic variables; temperature, salinity, and nutrients in the ocean; temperature and precipitation on land. It is therefore not possible to decode this signal with a simple equation and a multivariate approach is needed to interpret the changes in sets of pollen and diatoms in a continental environment, and of diatoms and foraminifera in a marine setting (Chap. 21). On a smaller timescale, the thickness or density of a tree ring, also influenced by a complex environment, can seldom be interpreted with a simple calibration equation. Several series from the same region need to be used to get a clear climate signal.

Uncertainties on the Temporal Scale

Before discussing the uncertainties and limitations associated with the interpretation of series of proxies, we will review the uncertainties related to their temporal resolution. Figure 10.1 shows the temporal characteristics of climate forcings (internal and external) and the different types of supports of commonly studied proxies. Many forcings and components of the climate system have characteristic times of less than a year, while most proxies have longer characteristic times. In addition, with the uncertainty of dating (see Part II), it is clear that time is an important factor of error in the study of the interactions between climate and proxy.

Among proxies providing a seasonal resolution, tree rings are prominent, but this is at the expense of the robustness of the long term signal. Although it is possible in theory to reach 10,000 years, trees have a much shorter lifespan and the long extended data series are achieved by splicing many short series together. The behavior of trees in the low frequency range is not exclusively due to climate, and this can lead to significant disturbances in reconstructions. Glaciers, another paleoclimate indicator, often have a high resolution for recent periods, due to compaction of the ice, but this reduces progressively as we go back in time. The same goes for marine and lake cores. In some cases, the deposition of their sediments may have an annual resolution (varved sediments), but bioturbation (disturbance of the sediment by small aquatic animals) often prevents this level being achieved in practice.

Historical records are often very accurate, but they have strong differences related to changes in instruments or observers. This provides series with a high resolution but over short periods. This review of climate records shows that no single proxy is perfect, and that without multi-proxy comparison, errors of interpretation could easily occur.

Uncertainties Associated with Geochemical Indicators: The Specific Case of the Isotopic Composition of Precipitations

Over the past forty years, quantitative reconstructions of temperature changes have been obtained from the estimation of past changes in the isotopic composition of precipitation. These are measured in various continental archives which hold ancient precipitation directly (glaciers, ice caps, groundwaters) or indirectly. Indeed, past changes in the isotopic composition of precipitation can leave a fingerprint in the isotopic composition of molecules formed using this water such as the calcite fossil skeletons of lake microorganisms, the calcite of speleothems, or the cellulose of tree rings.

Different stable isotopic forms of the water molecule are present on Earth. Their abundance is expressed by reference to the international standard SMOW (Standard Mean Ocean Water), which has 0.038% of $H_2^{17}O$, 0.310% of HDO, 0.2005% of $H_2^{18}O$ and 99.762% of the principal form, $H_2^{16}O$. The different isotopic molecules are characterized by a different number of neutrons, and therefore, different masses; different vapor saturation pressures; as well as differences in symmetry. During each change of phase (condensation, evaporation), the water molecules undergo isotopic fractionation, which includes processes at equilibrium (exchanges between infinite reservoirs) and kinetic processes: during evaporation on the surface of the ocean or re-evaporation of rain drops during their precipitation, or during the formation of ice crystals in the clouds, the processes are faster than the diffusion time of water molecules and cause kinetic effects associated with the molecular diffusivity of the different isotopic forms.

Since 1958, the International Atomic Energy Agency coordinates a network of observations and a database of the isotopic compositions of precipitations in modern times (Fig. 10.2a). Since the 1960s, the measurements have revealed a close relationship between the isotopic composition of precipitations and air temperature (the 'isotopic thermometer', Fig. 10.2b). This relationship at a local level has been used intensively to quantify the changes in past temperatures.

The use of natural archives to estimate variations in past climates from the isotopic composition of oxygen or of deuterium nevertheless poses many problems which are sources of uncertainty for the quantification of climate reconstructions:

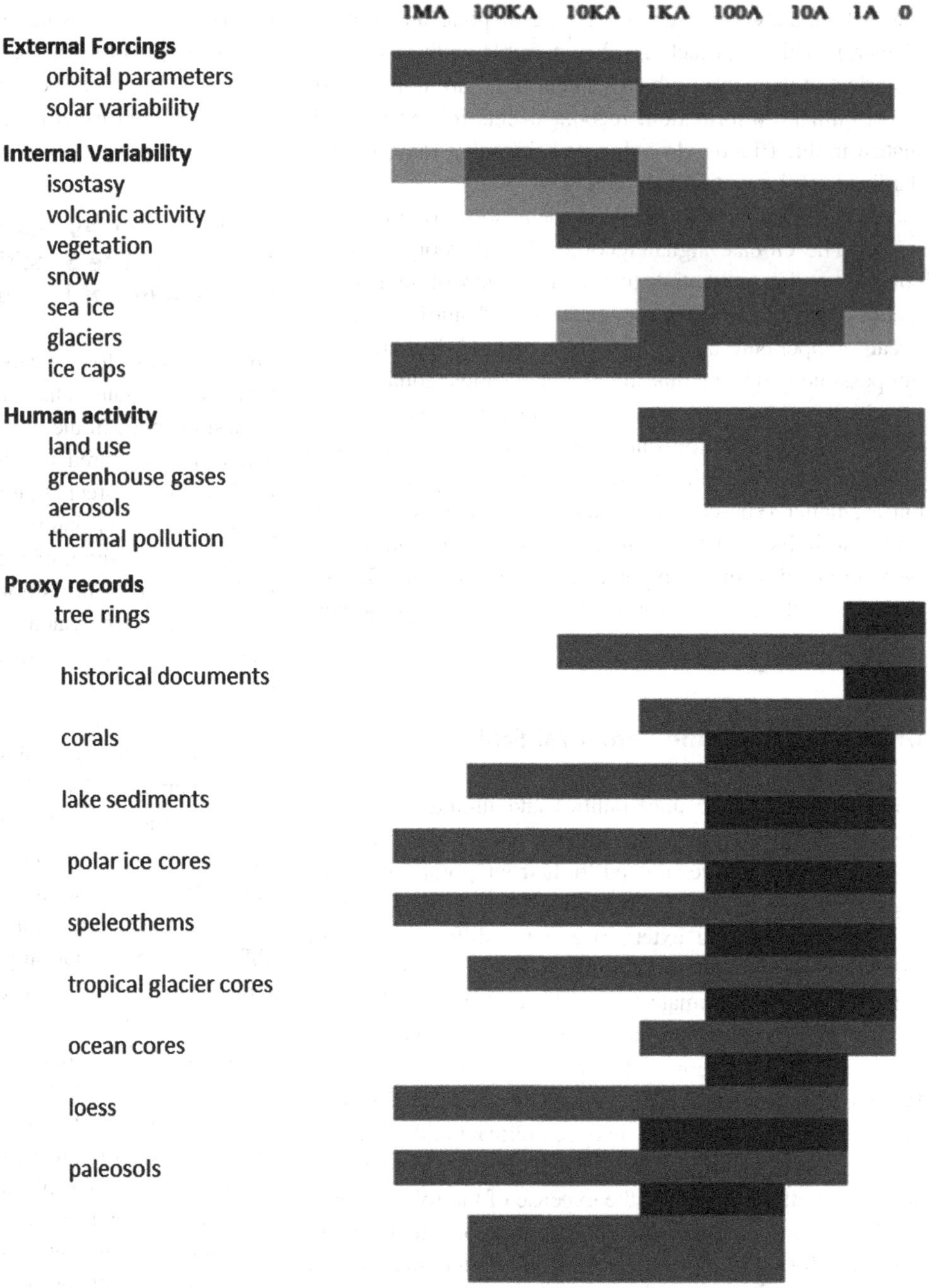

Fig. 10.1 Temporal characteristics of climate components and proxy supports used to reconstruct them. For climate components (above), the dark gray boxes represent the temporal range of the signal; these are extended by light gray boxes when the component becomes minor or occupies only part of the time slot. For proxies, black boxes indicate the maximum scope possible of the resolution, and gray boxes, the maximum range of time covered

External Forcings
  orbital parameters
  solar variability

Internal Variability
  isostasy
  volcanic activity
  vegetation
  snow
  sea ice
  glaciers
  ice caps

Human activity
  land use
  greenhouse gases
  aerosols
  thermal pollution

Proxy records
  tree rings
  historical documents
  corals
  lake sediments
  polar ice cores
  speleothems
  tropical glacier cores
  ocean cores
  loess
  paleosols

- uncertainty in the measurement of the isotopic content; this is generally low, with an impact of around a tenth of degree on estimates of temperatures;
- uncertainty in the processes able to modify the relationship between the isotopic composition of precipitations and the isotopic composition of the archive (ice, calcite, cellulose...). The physicochemical and biological processes controlling the transfer of information between the precipitation and the archive must be understood and quantified in order to assess the uncertainty in the reconstruction of the initial isotopic composition of the precipitation. Each archive has specific biases described in the following sections, which are not necessarily constant over time: these uncertainties can only be estimated through proper understanding and modeling of transfer processes between the water from precipitation and that of the archive (hydro-isotopic models of the functioning of trees, and of lake systems). Recent work has, for instance, shown that the isotopic composition of Greenland or Antarctic surface snow can evolve between snowfall events, possibly due to the interplay of snow metamorphism with changes in surface water vapor

Fig. 10.2 a Global network monitoring the isotopic composition of precipitations, stations with several years of measurements (International Atomic Energy Agency, IAEA, www.iaea.org/water). b The relationship between the content of ^{18}O in precipitations and the air temperature at ground level, based on network measurements by IAEA (filled-in circles) and from a synthesis of surface snow measurements in Antarctica (open circles). The slope of the isotope-temperature relationship (in ‰ by °C) is presented for Antarctica, for IAEA data with annual temperatures below 20 °C, and IAEA data with annual temperatures above 20 °C

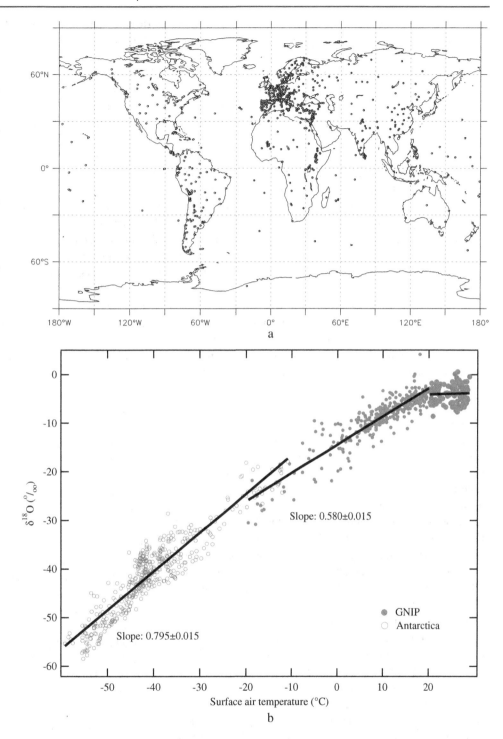

isotopic composition associated with different air mass trajectories. The exact effect of these uncertainties on temperature reconstructions remains difficult to quantify;

- uncertainty concerning the stability over time of the relationships between the isotopic abundances in precipitations and weather factors: even with an almost perfect archive of the isotopic composition of precipitations, how does this translate in terms of climate? Although condensation temperature is the key factor governing the isotopic distillation of an air mass, certain factors like the evaporation conditions, changes in the source of moisture or the vapor trajectory (including convective processes or not), continental recycling (precipitation/evaporation ratio), can affect the isotopic composition of the precipitations. Several critical aspects, such as the seasonality of the precipitations or the relationship between the

temperature of surface air and the temperature of condensation, can also affect the quality of reconstruction of paleoclimates. For Antarctica, for example, the uncertainty in the reconstruction of temperature is estimated to be of the order of 20–30% between glacial and interglacial periods, and recent work has produced evidence of spatio-temporal variations in this relationship over the last decades;
- Finally, it should be noted that in the tropics, it is not the temperature of surface air that determines the isotopic composition of rainfall. Indeed, both spatially and temporally (seasonal and inter-annual), the isotopic composition of rainfall is mainly related to its intensity ('mass effect'), because the intensity of the isotopic distillation depends on atmospheric convective activity, irrespective of the air temperature at the surface. The interpretation of isotopic signals of tropical glaciers is therefore fundamentally different from that of polar ice. Note that at time scales greater than a decade, it is still possible that tropical sea surface temperatures have a leading role on tropical precipitation isotopic composition through their impacts on atmospheric dynamics and convective activity (see Chap. 20).

Assessing the stability of the relationships between the isotopes of precipitation and meteorological parameters under different climate conditions requires the exploration of the processes associated with the three-dimensional atmospheric circulation and water cycle. This can be achieved using general or regional atmospheric circulation models implemented with the representation of the different isotopic forms of the water molecule and the associated fractionation processes. These modeling tools have been successfully used to explore the processes affecting the isotope-temperature relationship at glacial-interglacial scales. Current challenges are related to the ability to perform long (multi-centennial or longer) simulations using coupled ocean-atmosphere models equipped with water stable isotopes to quantify the climatic drivers of precipitation isotopic composition in different regions and over different time scales (seasonal, inter-annual, decennial, centennial etc.). Recently, new understanding has emerged from in situ and remote sensing monitoring of water vapor isotopic composition, which provides more continuous insights than the sampling of precipitation. These data are used to better understand the climatic drivers of water vapor isotopic composition, at the scale of weather events, but also to benchmark the ability of atmospheric models to correctly simulate the origin of atmospheric moisture.

Uncertainties Associated with Biological Indicators

The commonly used methods to reconstruct climate from biological assemblages are known under the term transfer function (Imbrie and Kipp 1971). Their principle is based on the expression of the relationship between the climate variable and the relative abundances of each taxon considered, as if the climate were dependent on the assemblage. This is an inverse approach, since the reality is that the assemblage depends on climatic conditions. The direct problem is called the response function. A few equations suffice to show the drawback of such an approach. Note X, the assemblage, C, all climate factors combined, D, all non-climatic factors that may also influence X (e.g. soil), and R, the response function of the assemblage to C and $X = R(C,D)$. If climate dominates over non-climatic factors, the relationship can be approximated as follows: $X = R_c(C)$. The transfer function may be obtained by inversion: $\hat{C} = \hat{R}_c^{-1}(X)$. But in general, the number of variables included in the vector C is far fewer than the number of X variables, and in this case, only a least squares method can solve the system of equations by minimizing the deviation between C and its estimation which expresses C as a function of X: $\hat{C} = \hat{T}(X)$ where $\hat{T}$ denotes the transfer function.

The 'transfer function' (TF) approach is based on several assumptions which should be kept in mind:

1. climate conditions are the ultimate cause of any changes observed in the data; human action which often modifies the landscape is assumed to be negligible;
2. the ecological properties of the studied species have not changed between the period of analysis and the present: the relationships between species and climate are constant through time;
3. current observations contain all the information necessary to interpret fossil data: so, it is necessary that the vegetation of the past, for example, survived somewhere in the world and that we have the corresponding information. This third hypothesis, added to the second, may be translated as the principle of uniformitarianism (the present is the key to the past), which is implicit in any paleontological approach.

It is clear that these three assumptions are quite strong. The differences found between the various approaches often stem from the fact that these assumptions are not always entirely verifiable.

The analog method (AM) does not operate by calculating a statistical relationship between climate and assemblages, but it is nevertheless based on the same assumptions, making it subject to the same biases when these assumptions are not met. However, this approach has its own peculiarities, because it is not based on a statistical calibration but on a calculation of similarity. The fossil pollen spectrum (or any other assemblage of fossils) for which we would like to know the climatic conditions is compared to all current spectra, and a measurement of each fit ('distance') is performed (see Chap. 12). The few current spectra with the lowest distance from the fossil one are considered as the best analogs. The reconstructed climatic conditions arise from the climatic conditions corresponding to these analogs, weighted according to the inverse of the distance from each analog to the fossil assemblage.

Figure 10.3 illustrates three marginal cases where the two approaches (transfer function and analog) behave quite differently. In this figure, the horizontal axis represents the climate space (this space has several dimensions but here it is simplified into one). Similarly, the vertical axis represents the space of the assemblages (in reality one axis per taxon). The gray circles represent the current data and the empty circles represent the fossil data. The line represents the transfer function (TF). Once we know the abundance of each taxon, on the horizontal axis, then the ordinate can be found by projection along the line, and the climate conditions thereby deduced (see Example A whose climate is T_A). The three cases are represented by three different letters:

- The fossil assemblage A falls in an area without modern equivalent assemblage, but the TF allows the climate to be easily inferred by T_A even though this value does not exist in the current data. The closest analog of A is A_o whose climate is $C(A_o)$. This shows that the AM is unable to provide a climate different from that which exists in the data.
- The fossil assemblage B also falls in an area without current data. While the TF provides an estimate of T_B completely outside the realm of current data—which may not be realistic, the AM provides the climate $C(B_o)$ that may underestimate the reality, but which has the advantage of being realistic.
- The assemblage C has a very close analog (C_o) which is isolated from the other points. The AM will naturally take climate $C(C_o)$ as an estimate, but the TF will provide an estimate T_c which is very far from reality. The TF therefore follows the dominant gradient of the data and is unable to provide a reliable estimate for rare assemblages.

This illustration shows that there is probably no perfect method and that the most effective way to validate the results (apart from comparing them to reconstructions from other proxies) is to try several methods, such as advocated by Kucera et al. (2005). About ten techniques can be identified in the literature, with the two TF and AM families (Guiot and de Vernal 2007). It is therefore possible to select a few of them and compare the reconstructions. Consistent results are a clear indication of their robustness.

Another problem arises from the fact that the climate variables to be reconstructed are often inter-correlated. If assemblages are available from only either wet and cold climates, or dry, hot climates, it would be impossible to reconstruct wet and warm, or dry and cold climates. If it is possible to collect some assemblages which differ from the dominant gradient, the analysis of point C in Fig. 10.3 shows that the AM was then more efficient than the TF.

A climate reconstruction is calibrated on current data and applied to fossil data. There is often a gap between the two situations. For example, in a continental environment, human activities act as a disruptive factor in the reference sample, and direct application of this to past data may cause biased reconstructions. This problem can only be minimized by selecting current data with limited anthropogenic influence.

Another problem inherent in any calibration is the risk of overestimation. In principle, if the number of parameters to be estimated (here, the weighting coefficients of each of the taxa in the TF) is high compared to the number of reference assemblies, it is possible to adjust a TF so that it passes through almost all the points (in Fig. 10.3, all the points are located on the line). Unfortunately, this line cannot provide a reliable forecast. Following principles of statistics, a good model is based on the lowest number possible of parameters to be estimated. An effective way to limit this problem is to

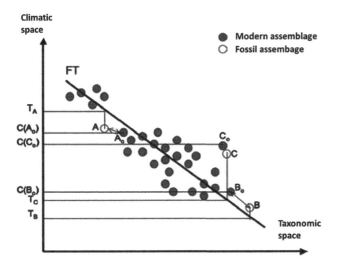

Fig. 10.3 Schematic representation of the main uncertainties related to transfer functions

divide the current database into two parts, to calibrate the TF on the first part, to apply it to the second, called the validation sample, and to deduce the average validation error. If this error remains reasonable, being inevitably greater than the average calibration error, then the TF is considered to be applicable. The same procedure is repeated after exchanging the calibration and the validation subsets. If the two resulting TFs obtained are acceptable, the TF can then be calibrated on the full data set and used for fossil assemblies.

Other statistical techniques exist (bootstrap, jackknife, permutations, etc.), but the basic paradigm is always to verify the quality of the estimates on data independent from those used for the calibration. Another important point is to systematically provide confidence intervals associated with climate reconstructions. This can be achieved for both the TF and the AM. These confidence intervals mean that the tolerance of the biological assemblies for a more or less wide range of climatic conditions can be assessed and the imperfection of the model to adjust all the current reference points can be taken into account.

References

Guiot, J., & de Vernal, A. (2007). Transfer functions: Methods for quantitative paleoceanography based on microfossils. In C. Hillaire-Marcel & A. de Vernal (Eds.), *Developments in marine geology* (Vol. 1, pp. 548–588). Dordrecht: Elsevier.

Imbrie, J., & Kipp, N. G. (1971). A new micropaleontological method for quantitative paleoclimatology: Application to a Late Pleistocene Caribbean core. In K. K. Turekian (Ed.), *The Late Cenozoic glacial ages* (pp. 71–181). USA: Yale University Press.

Kucera, M., Weinelt, M., Kiefer, T., Pflaumann, U., Hayes, A., Weinelt, M., et al. (2005). Reconstruction of sea-surface temperatures from assemblages of planktonic foraminifera: Multi-technique approach based on geographically constrained calibration data sets and its application to glacial atlantic and Pacific Oceans. *Quaternary Science Reviews, 24*, 951–998.

Pfister, C. (1980). The climate of Switzerland in the last 450 years. *Geographica Helvetica* (numéro spécial), 15–20.

Air-Ice Interface: Polar Ice

Valérie Masson-Delmotte and Jean Jouzel

The deposition and preservation of layers of snow, year after year, enable many climate and environmental parameters to be recorded in the structure and composition of the ice, as well as in its gas inclusions and impurities. This section is devoted specifically to information on climate variables; the composition of the atmosphere and biogeochemical cycles are discussed in Chap. 1 of this volume and Chap. 1 of Volume 2.

Polar ice contains records of past changes in numerous climate variables, some specific to that particular site, such as temperature and accumulation with others relevant to a larger geographical scale, such as atmospheric circulation and the monsoon regimes. In a unique way, reconstructions of local variables, temperature and accumulation, are drawn from physical processes. Estimates of the accumulation of snow can be derived from the dating of ice (see volume 1, Chap. 9), for example, from the identification of seasonal cycles or reference horizons. Variations in accumulation throughout the ages are also estimated from changes in temperature in the past, through the dependent relationship between the saturation vapor pressure in the air and temperature. This chapter focuses on the various methods used to quantify temperature variations. We will discuss the exploitation of freeze-back layers caused by summer melt in some polar regions, the inversion of temperature profiles measured in the boreholes, the analysis and modeling of the stable isotope composition of the ice, and finally, the analysis of the isotopic composition of nitrogen and argon in the air trapped in the ice.

V. Masson-Delmotte · J. Jouzel (✉)
Laboratoire des Sciences du Climat et de l'Environnement, LSCE/IPSL, CEA-CNRS-UVSQ, Université Paris-Saclay, 91190 Gif-sur-Yvette, France
e-mail: jean.jouzel@lsce.ipsl.fr

© Springer Nature Switzerland AG 2021
G. Ramstein et al. (eds.), *Paleoclimatology*, Frontiers in Earth Sciences,
https://doi.org/10.1007/978-3-030-24982-3_11

Melt Index and Borehole Temperatures

Certain glaciers and ice caps from the Arctic coastal areas form a first collection of records where the structure of the ice allows the changes in summer temperatures to be estimated. Many of these locations are characterized by a summer temperature that can exceed 0 °C. In this case, the surface snow melts, percolates into the deeper colder layers and then refreezes. Identifying the freeze-back layers in the physical structure of the ice allows the indices of summer melting intensity to be reconstructed. This method was used to estimate changes in summer temperatures over time scales of a few centuries, such as in Spitsbergen (Svalbard) or in the Russian Arctic, and during the Holocene, in the Canadian Arctic (Koerner and Fisher 2002). In Greenland and Western Antarctica, some low-altitude sites (Siple Dome, Antarctica peninsula) are also characterized by the regular or occasional occurrence of summer melting. However, the vast majority of deep core samples taken from the Antarctica and Greenland ice are from sites where the temperature remains below 0 °C throughout the year and where there are no freeze-back layers in the ice. In these cases, several methods have been used to estimate past changes in the local temperature.

The diffusion of heat through the structure of the ice causes fluctuations in the vertical temperature profile, which, once drilling operations have been completed, can be measured with an accuracy of one thousandth of a degree in the liquid in the drill holes. The numerical inversion of the temperature profile allows, in principle, the large variations of past surface temperatures to be estimated. However, this problem is poorly constrained and requires general assumptions to be made about the shape of the function sought. There is a broad uncertainty associated with the temperature estimates both in terms of amplitude and chronology. Through this method, estimates of temperature variations over the last century, and even over the last millennia, have been made in some sites with high accumulation (Dahl-Jensen et al. 1999) as well as estimates of the glacial-interglacial amplitude in the center of Greenland (Dahl-Jensen et al. 1998) and Antarctica (Salamatin et al. 1998).

Stable Isotopes of Water and Temperature

The most commonly used method to reconstruct variations in past temperatures at the center of Antarctica and Greenland is based on analyzing the isotopic composition of the ice. The study of the abundance of the isotopic forms of water molecules in precipitations, initiated in the 1950s (Dansgaard 1953), helped to highlight a spatial relationship between depletion in heavy isotopes and site temperature, a relationship on which the concept of the 'isotopic thermometer' is based. Natural waters, formed mainly of $H_2^{16}O$ molecules (99.7%), also present some rarer stable isotopic forms, including 0.2% of $H_2^{18}O$ and 0.03% of $HD^{16}O$ (D represents deuterium ^2H). The isotopic concentrations are expressed as the deviation in permil in δ notation (δD and $\delta^{18}O$) against an international standard, the V-SMOW. At temperate and polar latitudes, a linear relationship is observed between the isotopic ratios in precipitations today, δD or $\delta^{18}O$, and the temperature of the site. Figure 11.1 illustrates the 'isotopic thermometer' in Antarctica, where more than 900 sites were sampled. The spatial gradients observed are of the order of 6‰/ °C for δD and 0.8‰/ °C for $\delta^{18}O$.

Modeling the isotopic composition of precipitations has been developed using conceptual distillation models (Ciais and Jouzel 1994) and atmospheric general circulation models incorporating the representation of the cycles of the different isotopic forms of the water molecule (Joussaume et al. 1984). These digital tools take into account the effect of different fractionations related to the differences between the saturation vapor pressure (equilibrium effect) and the diffusivity in the air (kinetic effect) of the relevant molecules, and allow the distillation process behind this spatial relationship to be understood (Fig. 11.2).

The reconstruction of past temperatures is based on measuring the isotopic ratio of a thin strip of ice taken along the length of the cores. The isotope-temperature relationship is then applied to this isotopic measurement. The estimate of changes in past temperatures relies on the assumption that the current spatial relationship is applicable to an estimate of the difference in temperature between any two given periods at the drilling site; it assumes that this 'temporal' slope is equal to the spatial slope. For changes at the glacial-interglacial scale, a correction linked to variations in the isotopic composition of the ocean (Jouzel et al. 2003) needs to be taken into account. In the best of cases, the accuracy of measurements by mass spectrometry is ± 0.5‰ for δD and ± 0.05‰ for $\delta^{18}O$. Temporal resolution is very variable. In sites where the accumulation rate is high (more than 10 cm per year), it is possible to find a sub-annual (seasonal) resolution. However, the diffusion of water vapor in the upper layers of the firn quickly brings about a 'smoothing' of the isotopic composition and a loss of information with each snowfall. In low-accumulation sites, the redistribution of surface snow by the winds makes climate reconstruction on a time scale of less than twenty years impossible.

The uncertainty in the estimation of changes in past temperatures is not dependent on the accuracy of the measurements but rather on the different parameters that can influence the isotope-temperature relationship. Through the

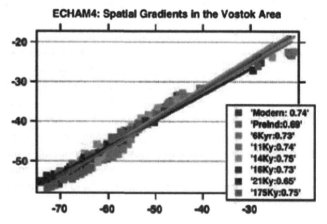

Fig. 11.2 Modeling of the isotopic composition of precipitations ($\delta^{18}O$) in the Vostok region (central plateau of Eastern Antarctica) using the ECHAM atmospheric general circulation model which includes explicit modeling of the cycle of the stable isotopes of water. This model was forced by boundary conditions (sunlight, ice caps, surface sea temperatures, sea ice, and atmospheric composition) estimated for different time intervals (current, pre-industrial, 6, 11, 14, 16, 21, 175 ky BP). The squares represent the results obtained for each simulation for the Vostok region: the temporal slope between $\delta^{18}O$ and temperature can then be estimated, and it can be seen that it is very close to the modern spatial slope

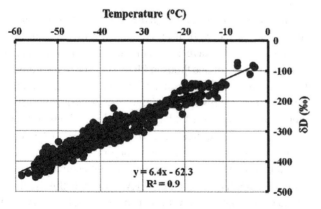

Fig. 11.1 Spatial relationship found between δD of surface snow and average annual temperature for 900 sites in Antarctica where both variables were measured (see Masson-Delmotte et al, 2008)

ages, changes in the seasonality of precipitation, in the altitude of condensation, in the trajectory and origin of precipitation, can all have significant effects on the relationship between isotopic composition and local surface temperature. Isotopic distillation models and atmospheric general circulation models, equipped with an explicit representation of stable water isotopes allow the impact of these factors to be estimated through sensitivity studies. A set of simulations conducted using the ECHAM-iso model, (Werner et al. 2001; Jouzel et al. 2007a; Werner et al. 2017) has helped to highlight the stability of the temperature-isotope relationship for the Antarctic central plateau, between the glacial and current climates, and to show that the temporal slope is very similar to the spatial slope. This supports the 'isotopic thermometer' approach for sites on the Antarctic plateau with an accuracy of between −10 and +30%, according to Jouzel et al. (2003).

However, this approach is debatable for climates warmer than the present. Sime et al. (2008) examined a scenario showing an increase in atmospheric concentration of CO_2 with the HadAM3-iso model. This simulation suggests a reduction in the isotope-temperature relationship in the Dome C region of Antarctica, in the context of a warmer global climate. This result remains difficult to apply to measurements from ice cores, because of a lack of analogy between climate changes caused by modifications in the Earth's orbit ('warm' interglacial periods) and those caused by an increase in the greenhouse effect.

Similarly, climate-isotope modeling and the comparison of isotopic analysis with other paleothermometry methods suggest that in central Greenland, the two slopes differ significantly, by up to a factor of 2 (Dahl-Jensen et al. 1998; Masson-Delmotte et al. 2005); the conventional approach underestimates variations in temperature by this amount.

Currently, the oldest isotopic composition profiles go back to 800,000 years in Antarctica on the Dome C site (Jouzel et al. 2007a) (Fig. 11.3); in Greenland, the deepest core, drilled at NorthGRIP, provides about 123,000 years of archives of the isotopic composition of ice (NorthGRIP-community-members 2004).

Cores from Summit, in Greenland, have revealed even older but discontinuous ice segments, identified by comparing the composition of the air to reference series obtained in Antarctica. The study of the isotopic composition of ice has thus allowed reconstructions of changes in local temperatures in the past to be proposed (see Masson-Delmotte et al. 2006), with a high level of consistency between cores from Eastern Antarctica, Vostok, Dome C (Watanabe et al. 2003) and Dome Fuji, where a new core now covers the last 720,000 years (Dome Fuji Ice Core Project Members 2017). The combined study of the different isotopic forms of water, also gave rise to a parameter of the second order, deuterium excess defined as $d = \delta D - 8\delta^{18}O$ (Dansgaard 1964). This parameter is strongly conditioned by the evaporation conditions of atmospheric water vapor masses, and it has been used to estimate the changes in the origin of polar precipitation over time (Jouzel et al. 2007b; Vimeux et al. 2001). Changes in 'source temperatures' calculated in this way, are difficult to compare with the reconstructions of surface temperature of the oceans (Chap. 10), because they

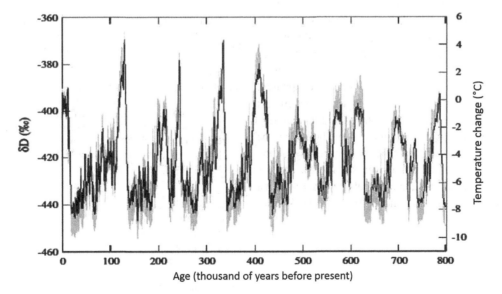

Fig. 11.3 Isotopic recordings of the EPICA Dome C drilling in Antarctica (Jouzel et al. 2007b). Data are shown as a function of time (x-axis, in thousands of years before the present, i.e. before the year 1950). The isotopic composition of the ice samples is indicated by the black line, on the left axis (δD, in ‰). The estimate of the corresponding temperature change (relative to the current temperature, in °C), calculated using the current spatial gradient (see Fig. 11.1), and adjusted for changes in isotopic composition of seawater, is indicated in gray (on the right axis)

correspond to a source of precipitations which may vary geographically over time. In addition, the isotopic composition of the ice may be affected by changes in altitude and run-off, bringing to a drill site on the side of a dome, old ice previously formed in a distant area, effects which need to be taken into consideration (Masson et al. 2000).

Stable Isotopes of Air and Temperature

New methods to quantify abrupt changes of temperature have been implemented, taking advantage of the thermal fractionation of nitrogen or argon (whose isotopic composition is stable in the atmosphere over these time scales) that occurs when air is trapped in the firn (Severinghaus et al. 1998). This approach reveals temperature change markers in the gas phase, and also allows a more detailed characterization of the phase shifts between changes in atmospheric composition (concentrations of greenhouse gases measured in the gas phase) and changes in polar temperature.

This method does not allow an accurate estimate of the temperature changes in the Antarctic, where gravitational fractionation effects dominate, because changes in temperature are slower and less intense. However, it was used to determine the phase shift between Antarctic warming and increasing atmospheric concentrations of carbon dioxide during deglaciations (Caillon et al. 2003).

In Greenland, study of the thermal fractionation of gases was conducted on a series of rapid events, recorded in the ice cores from GISP2, GRIP and NorthGRIP (Capron et al. 2010). Estimates of temperature changes through thermal fractionation of the air have led to an upward revision of the intensity of temperature changes during the warm Dansgaard-Oeschger events, reaching an amplitude of 8 to 16 ± 3 °C, and have called into question the presumed stability of the cold phases as they appear in the continuous recordings of the stable isotopes of water. The apparent discrepancy between the quantifications from the current spatial relationship between water isotopes and temperature and this alternative paleothermometry method can be explained by significant changes in the seasonality of the deposition of snow in Greenland over the ages, a process highlighted by the climate-isotope models between the Last Glacial Maximum and present times (Krinner 1997; Werner et al. 2001).

Conclusions

It is important to note that a number of methods may be applied to polar ice cores to quantify temperature changes. New avenues are being explored to improve quantifications of temperature change: isotopic measurements at very high resolution in order to quantify isotopic diffusion, a process which depends on the temperature; analysis of oxygen-17 in water to more accurately estimate the evaporation conditions (temperature, relative humidity) at the surface of the ocean; continuous analysis of the isotopic composition of argon, nitrogen and noble gases in order to characterize the thermal and gravitational fractionations of the firn. Much remains to be learned about the spatial and temporal variability of temperatures in Greenland and Antarctica, both over recent centuries and in the ancient climate cycles recorded in the polar ice.

References

Caillon, N., Severinghaus, J. P., Jouzel, J., Barnola, J. M., Kang, J. and Lipenkov, V. Y. (2003). Timing of atmospheric CO$_2$ and Antarctic temperature changes across termination III. *Science, 299,* 1 728–1 731.

Capron, E., et al. (2010). Millennial and sub-millennial scale climatic variations recorded in polar ice cores over the last glacial period. *Climate of the Past, 6,* 345–365.

Ciais, P., Jouzel, J. (1994). Deuterium and Oxygen 18 in Precipitation: An isotopic model including mixed cloud processes. *Journal of Geophysics Research, 99,* 16 793–16 803.

Dahl-Jensen, D., Morgan, V. I., & Elcheikh, A. (1999). Monte carlo inverse modeling of the law dome (Antarctica) temperature profile. *Annals of Glaciology, 29,* 145–150.

Dahl-Jensen, D., Mosegaard, K., Gundestrup, N., Clow, G. D., Johnsen, S. J., Hansen, A. W., et al. (1998). Past temperatures directly from the greenland ice sheet. *Science, 282,* 268–271.

Dansgaard, W. (1953). The abundance of ^{18}O in atmospheric water and water vapour. *Tellus, 5,* 461–469.

Dansgaard, W. (1964). Stable isotopes in precipitation. *Tellus, 16,* 436–468.

Dome Fuji Ice Core Project Members. (2017). State dependence of climatic instability over the past 720,000 years from Antarctic ice cores and and climate modeling. *Science Advances, 3,* e1600446.

Joussaume, S., Jouzel, J., & Sadourny, R. (1984). A general circulation model of water isotope cycles in the atmosphere. *Nature, 311,* 24–29.

Jouzel, J., Vimeux, F., Caillon, N., Delaygue, G., Hoffmann, G., Masson-Delmotte, V., & Parrenin, F. (2003). Magnitude of the isotope-temperature scaling for interpretation of central antarctic ice cores. *Journal of Geophysical Research, 108,* 1029–1046.

Jouzel, J., Masson-Delmotte, V., Cattani, O., Dreyfus, G., Falourd, S., Hoffmann, G., et al. (2007a). Orbital and millennial Antarctic climate variability over the past 800,000 years. *Science, 317,* 793–797. https://doi.org/10.1126/science.1141038.

Jouzel, J., Stiévenard, M., Johnsen, S. J., Landais, A., Masson-Delmotte, V., Sveinbjornsdottir, A., et al. (2007b). The GRIP deuterium-excess record. *Quaternary Science Reviews, 26,* 1–17.

Koerner, R. M., & Fisher, D. A. (2002). Ice-Core evidence for widespread arctic glacier retreat in the last interglacial and the early Holocene. *Annals of Glaciology, 35,* 19–24.

Krinner, G., C. Genthon, and J. Jouzel (1997). GCM analysis of local influences on ice core δ signals. *Geophysical Research Letters, 24,* 2 825–2 828.

Masson, V., Vimeux, F., Jouzel, J., Morgan, V., Delmotte, M., Hammer, C., et al. (2000). Holocene Variability in Antarctica based on 11 ice core isotopic records. *Quaternary Research, 54,* 348–358.

Masson-Delmotte, V., Jouzel, J., Landais, A., Stievenard, M., Johnsen, S. J., White, A., et al. (2005). Deuterium excess reveals millennial and orbital scale fluctuations of greenland moisture origin. *Science, 309*, 118–121.

Masson-Delmotte, V., Braconnot, P., Dreyfus, G., Johnsen, S., Jouzel, J., Kageyama, M., et al. (2006). Past temperature reconstructions from deep ice cores: relevance for future climate change. *Climate of the Past, 2*(2), 145–165.

Masson-Delmotte, V., Hou, S., Ekaykin, A., Jouzel, J., Aristarain, A., Bernardo, R. T., Bromwhich, D., Cattani, O., Delmotte, M., Falourd, S., Frezzotti, M., Gallée, H., Genoni, L., Isaksson, E., Landais, A., Helsen, M., Hoffmann, G., Lopez, J., Morgan, V., Motoyama, H., Noone, D., Oerter, H., Petit, J. R., Royer, A., Uemura, R., Schmidt, G. A., Schlosser, E., Simões, J. C., Steig, E., Stenni, B., Stievenard, M., van den Broeke, M., van de Wal, R., van den Berg, W. J., Vimeux and F., White, J. W. C. (2008). A review of antarctic surface snow isotopic composition: observations, atmospheric circulation and isotopic modelling. *Journal of Climate, 21*, 13, 3 359–3 387.

NorthGRIP-community-members. (2004). High resolution climate record of the northern hemisphere reaching into last interglacial period. *Nature, 431*, 147–151.

Salamatin, A. N., Lipenkov, V. Y., Barkov, N. I., Jouzel, J., Petit, J. R. & Raynaud, D. (1998). Ice-core age dating and paleothermometer calibration on the basis of isotope and temperature profiles from deep boreholes at Vostok station (East Antarctica). *Journal of Geophysical Research, 103*, 8 963–8 977.

Severinghaus, J., Sowers, T., Brook, E. J., Alley, R. B., & Bender, M. (1998). Timing of abrupt climate change at the end of the younger dryas interval from thermally fractionated gases in polar ice. *Nature, 391*, 141–146.

Sime, L. C., Tindall, J. C., Wolff, E. W., Connolly, W. M., & Valdes, P. J. (2008). Antarctic isotopic thermometer during a CO_2 forced warming event. *Journal Geophysical Research, 113*, D24119.

Vimeux, F., Masson, V., Delaygue, G., Jouzel, J., Petit, J.-R., &Stievenard, M. (2001). A 420,000 year deuterium excess record from East Antarctica: Information on Past Changes in the Origin of Precipitation at Vostok. *Journal of Geophysical Research, 106*, 31 863–31 873.

Watanabe, O., Jouzel, J., Johnsen, S., Parrenin, F., Shoji, H., & Yoshida, N. (2003). Homogeneous climate variability across East Antarctica over the past three glacial cycles. *Nature, 422*, 509–512.

Werner, M., Heimann, M., & Hoffmann, G. (2001). Isotopic composition and origin of polar precipitation in present and glacial climate simulations. *Tellus, 53B*, 53–71.

Werner, M., Jouzel, J., Masson-Delmotte, V., & Lohmann, G. (2018). Reconciling glacial Antarctic water stable isotopes with ice sheet topography and the isotopic paleothermometer. *Nature Communications, 9*, 3537. https://doi.org/10.1038/s41467-018-05430-y

Air-Vegetation Interface: Pollen

Joël Guiot

From the Production of Pollen to Sediment

In order to persist in difficult conditions, most plants and terrestrial animals contain hard parts that are preserved in the sediments after death. For higher-level terrestrial plants, these parts consist mainly of pollen grains and spores which provide a widely-used tool in paleoclimatology, thanks to their abundance in wet sediments. These grains, which are dispersed over variable distances depending on their shape and size (from 5 to 100 microns), are an essential factor in the reproduction of higher-level plants. Their outer envelope (exine) is composed of sporopollenin, a highly resistant substance once it is protected from oxidation. Lakes and bogs are particularly good environments for the conservation of these plant remains. These grains are scattered by wind, insects, birds, water. In temperate regions, wind is the predominant vector, and because of its relative inefficiency, lots of pollen grains and spores are found not far from their source, in continental sediments and marine sediments near the coast. In rainforests, animals play a much more important role, so that many species are underrepresented in sediments.

After choosing a site representative of the surrounding vegetation, with good conservation of pollen and sufficient accumulation rates to permit studies over the desired time scale, cores are extracted, usually at the center of the lake or bog. These cores are studied stratigraphically and samples are dated in order to establish an absolute chronology of climate events. Samples are taken at regular intervals along the core. They are subjected to physical and chemical treatments to make the pollen grains clearly visible for examination under a microscope. The grains are then identified on the basis of their exine which have different morphologies depending on the plant type. The palynologist counts each pollen type to work out the relative abundance of the species of trees or grasses that produced the grains. It is not always possible to determine each species, and many plants can only be recognized at the level of the genus or even family. Because of this heterogeneity in the classification, the term 'pollen taxon' is used to characterize the type of plant that produced it. The total number of grains counted varies depending on the diversity of the vegetation: tropical vegetation is more diverse than temperate vegetation and therefore requires a much higher total number, sometimes more than a thousand grains compared with a few hundred in a temperate vegetation, and this ensures good statistical significance of the various fluctuations detected. The relative abundance of each taxon makes up a pollen assembly or spectrum. It provides information on the relative composition of the surrounding vegetation, but this signal is influenced by the abundance of pollen productivity, its mode and capacity of dispersion. Statistical methods are needed to reliably decode the pollen spectrum (Moore et al. 1991).

The Pollen Diagram

The set of pollen spectra along the core is presented graphically to provide a pollen diagram whose complexity depends on the number of taxa counted. It can be simplified either by grouping together similar taxa (i.e. taxa co-evolving in similar environments), or by representing only the most important ones. Figure 12.1 shows an example of a simplified diagram. The interpretation of the diagram is complex because of the large number of processes occurring between the pollen production by the vegetation and its record in the sediment. As in most paleontological disciplines, interpretation is done by comparison with modern data. This supposes that the principle of uniformity, that is, where the present is the key to the past, applies. This is generally the case for data from the Quaternary, and in

J. Guiot (✉)
European Centre for Research and Teaching in Environmental Geosciences CEREGE, Aix-Marseille University, CNRS, IRD, INRAE, Collège de France, BP 80, 13545, Aix-en-Provence, Cedex 04, France
e-mail: guiot@cerege.fr

© Springer Nature Switzerland AG 2021
G. Ramstein et al. (eds.), *Paleoclimatology*, Frontiers in Earth Sciences,
https://doi.org/10.1007/978-3-030-24982-3_12

particular for continuous sequences that have been correlated over 500 000 years (Tzedakis et al. 1997). To verify this assumption, pollen samples are taken from modern or recent moss and their pollen spectra are compared with current vegetation. From a qualitative point of view, we can already sketch the relationship between the abundances of pollen and the composition of the vegetation. For example, the study of an altitudinal transect provides an impression of what happened during a cooling climate. Many published articles are based on this approach. But if current pollen spectra are collected from vegetation as diverse as possible, the problem can be treated quantitatively and objective keys to interpretation can be established. We develop this perspective below.

Figure 12.1 shows that the vegetation around Lake Rotsee, Switzerland (Lotter and Zbinden 1989) was dominated by herbaceous vegetation before 11,000 years BP: sagebrush (*Artemisia*) and grasses (*Poaceae*), indicating an open environment with few trees. Around 11,500 years BP, birch (*Betula*) and, a century later, pine (*Pinus*) flourished until they occupied nearly 80% of the assemblage. This pioneer arboreal vegetation thrived due to a warmer climate. Around

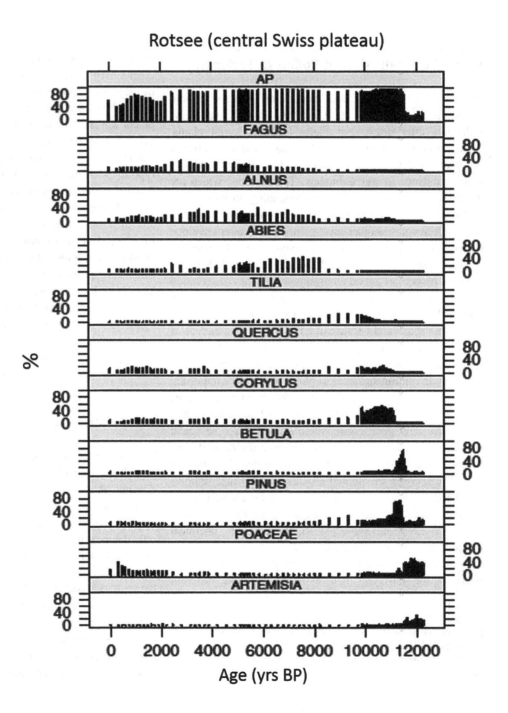

Fig. 12.1 Simplified pollen diagram of Lake Rotsee (8.33° E, 47.08° N, 419 m) in the Swiss Alps (Lotter and Zbinden 1989). The time period covered is from 12 000 years before now (BP) in calendar age to the present day. AP is the sum of all the arboreal pollen abundances

11,100 years BP, strong warming encouraged the proliferation of the hazel (*Corylus*), the oak (*Quercus*) three centuries later, and the lime tree (*Tilia*) five centuries later again, replacing these pioneers. This succession over less than a millennium is not necessarily exclusively due to climate: species expand geographically from refuges more or less distant from the studied area at a speed that is specific to them. Another, less dramatic change in climate (probably a wetter and cooler climate) around 8200 years BP enabled the fir (*Abies*) and alder (*Alnus*) to become established. Around 6300 years BP, it was the beech (*Fagus*) which progressed in the region. Around 2000 years BP, during the Roman period, the proportion of grasses increased and trees decreased (AP), a sign of anthropogenic deforestation.

Comparison of the different graphs in Fig. 12.1 shows that some taxa evolve together while others are diachronous. Taxa that are present at the same time on the same site will likely thrive in the same climatic conditions: they will be temperate or boreal, they will be resistant to drought or only survive in wet conditions. If, to this, we add characteristics related to the size of the plant (tree, shrub, grass), phenology (evergreen plant or deciduous), type of leaf (needles or broad leaves), a reasonable classification can be made. Prentice et al. (1996) have proposed one for Europe. This was then applied to other continents (Jolly 1998; Tarasov 1998). The types of plants defined in this way, known as plant functional types (PFTs), can be directly compared to simulations by vegetation models based on the same typology. These PFTs are used to define the vegetation of a site in the form of a biome: a bio-geographic area characterized by the species of plants (and animals) that live there. In Fig. 12.2 four such PFTs have been reproduced. The biome is determined by comparing them. Before 11,500 years BP, herbaceous plants dominated, as is currently the case in the arctic tundra and alpine grasslands. This period is called the Younger Dryas. Then, the presence of boreal deciduous trees, followed by conifers, indicated a warming sufficient for the taïga, as the forest in Northern Europe is called today, to develop. From 11,100 years BP, at the beginning of the Holocene, the temperate forest became established. The arrival of conifers around 8200 years BP, probably due to a well-known abrupt cooling (Tinner and Lotter 2006) transformed the landscape from a deciduous forest to a mixed forest which lasted until about 5000 years BP. The consequent growth of deciduous trees was disturbed around 4000 years BP and even more so around 2000 years BP, as a result of widespread deforestation by man. The reconstructed biome then became the steppe, although it was not exactly a proper one, being a mix of grasslands and forests.

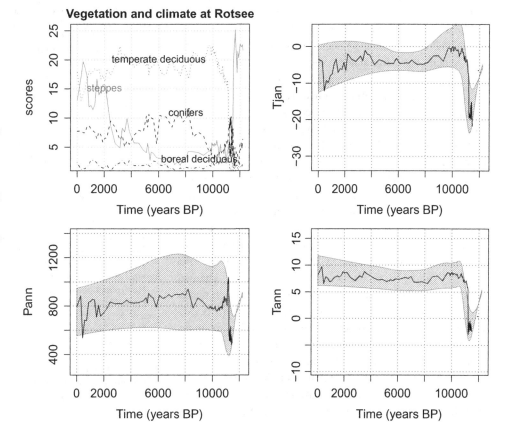

Fig. 12.2 Evolution of four functional plant types at Lake Rotsee. Continuous line gray: herbaceous steppe and tundra, dashed dotted line: boreal deciduous trees; dashed line: conifers; dotted line: temperate deciduous trees. On the same figure, the biomes that can be inferred from this are represented: steppe, temperate forest, mixed forest, taïga, tundra. The other three panels represent the three climate parameters reconstructed with their error bars: Pann (annual precipitation in mm/year), Tjan (January temperatures in °C), Tann (annual average temperature in °C). The error bar is given by the total variability between the analogs

Reconstruction of Climate

These interpretations are essentially qualitative. Transfer functions and the analog method (see Chap. 10, Sect. 10.1.3) provide quantitative information on climate. Here, we use the analog method, not on the percentages of taxa, but on the PFT scores (Fig. 12.2.), which has many advantages: it reduces the number of variables and groups together the taxa with similar behavior, making the approach more robust. Twelve analogs for each fossil spectrum were retained.

Figure 12.2 shows these reconstructions with the shaded area representing the range of variability between the analogs. The quality of the reconstructions is controlled by applying the same method to modern data: for each modern spectrum, the best analogs are determined (obviously excluding the spectrum itself) and the present climate is reconstructed so that it can be compared to direct observations. In this case, we find a coefficient of determination (r^2) of 0.64 for the annual precipitation (Pann), 0.89 for temperatures in January (Tjan) and 0.93 for annual average temperatures (Tann). The estimates obtained for precipitation will therefore contain larger errors than the other variables. This is confirmed by Fig. 12.2. In terms of climate, we see that the Younger Dryas was very cold (14 $\pm$ 7 °C colder than now) and dry (400 $\pm$ 400 mm/year less rainfall than now over the annual average), but that the uncertainties are large (in this method, this is the variability between analogs rather than an actual error bar). The temperature maximum (4 $\pm$ 2 °C more than now) occurred at around 10,000 years BP, when oak dominated, and the rainfall maximum (100 $\pm$ 250 mm more) occurred around 8200 years BP, with the arrival of the mixed forest. However, the 8200 BP event does not appear to have been short-lived since this forest continued for several millennia longer.

Several other methods have been proposed to reconstruct climate from pollen data (see Brewer et al. 2007; Birks 2011; Juggins 2013). They all have their strengths and weaknesses. It is not always easy to find the optimal method. It is recommended to try several and compare the results. The convergence of estimates is an indication of the robustness of the reconstruction, and their divergence is often a sign that the initial assumptions used for the reconstruction were not entirely valid. In particular, when the climate changes rapidly, vegetation adapts with a certain delay, which makes it difficult to find current analogs. Another problem is estimating the impact of non-climate constraints. For example, atmospheric CO_2 concentration during the Quaternary glaciations was lower than the current concentration (about 200 ppm instead of more than 280 ppm). We know that this level is an activator of photosynthesis and so it is unlikely that the principle of uniformity applies: vegetation will not respond to climate change in the same way when atmospheric CO_2 levels change. This problem was solved by the use of mechanistic models of vegetation (Guiot et al. 2000). Another avenue takes pollen dispersal and associated biases into account when reconstructing the landscape (Sugita 2007; Trondman et al. 2015). The relative abundance of pollen of a taxon depends on the height and distance of the productive plant, its pollen productivity, the weight and shape of the pollen grain, the dimensions of the lake into which it falls.

To improve climate reconstructions without resorting to overly complex models, the 'multiproxy' approach is recommended. This involves taking several simultaneous climate indicators into account: pollen, macro remains of plants or animals, carbon isotopes, lake levels, sedimentological parameters etc. as was done by Cheddadi et al. (1997). This multiproxy approach is also supported by mechanistic models (Rousseau et al. 2006. Guiot et al. 2009). These avenues of research are being explored both to reconstruct the climate of the past and to understand how it has influenced the vegetation.

References

Birks, H. J. B. (2011). Strengths and weaknesses of quantitative climate reconstructions based on late-quaternary biological proxies. *The Open Ecology Journal, 3*, 68–110. https://doi.org/10.2174/1874213001003020068.

Brewer, S., Guiot, J., Barboni, D. (2007). Pollen Data as Climate Proxies. In *Encyclopedia of Quaternary Sciences*, Elsevier.

Cheddadi, R., Yu, G., Guiot, J., Harrison, S. P., & Prentice, I. C. (1997). The Climate of Europe 6000 years ago. *Climate Dynamics, 13*, 1–9.

Guiot, J., Torre, F., Jolly, D., Peyron, O., Boreux, J. J., & Cheddadi, R. (2000). Inverse vegetation modeling by monte carlo sampling to reconstruct palaeoclimates under changed precipitation seasonality and CO_2 conditions: Application to glacial climate in mediterranean region. *Ecological Modelling, 127*, 119–140.

Guiot, J., Wu, H., Garreta, V., Hatti, L., & Magny, M. (2009). A few prospective ideas on climate reconstruction from a statistical single proxy approach towards a multi-proxy and dynamical approach. *Climate of the Past, 5*, 571–583.

Jolly, D. et al. (1998). Biome reconstruction from pollen and Plant Macrofossil data for Africa and the Arabian Peninsula at 0 and 6 ka. *Journal of Biogeography, 25*, 1 007–1 028.

Juggins, S. (2013). Quantitative reconstructions in palaeolimnology: New paradigm or sick science? *Quaternary Science Review, 64*. https://doi.org/10.1016/j.quascirev.2012.12.014.

Lotter, A. F., & Zbinden, H. (1989). 'Late-glacial pollen analysis, Oxygen-Isotope and Radiocarbon stratigraphy from Rotsee (Lucerne), Central Swiss Plateau. *Ecologae Gologicae Helvetiae, 82*, 191–202.

Moore, P. D., Webb, J. A., & Collinson, M. E. (1991). *Pollen Analysis*. Oxford: Blackwell Scientific Publications.

Prentice, I. C., Guiot, J., Huntley, B., Jolly, D., & Cheddadi, R. (1996). Reconstructing biomes from Palaeoecological data: A general method and its application to European pollen data at 0 and 6 ka. *Climate Dynamics, 12,* 185–194.

Rousseau, D., Hatté, C., Guiot, J., Duzer, D., Schevin, P. and Kukla, G. (2006). Reconstruction of the Grande Pile Eemian Using Inverse Modeling of Biomes and d13C. *Quaternary Science Reviews, 25,* 2 806–2 819.

Sugita, S. (2007). Theory of quantitative reconstruction of vegetation I: Pollen from Large Sites REVEALS regional vegetation composition. *The Holocene, 17,* 229–241.

Tarasov, P. E. et al. (1998). Present-Day and Mid-Holocene Biomes reconstructed from pollen and plant macrofossil data from the Former Soviet Union and Mongolia. *Journal of Biogeography, 25,* 1 029–1 054.

Tinner, W., & Lotter, A. (2006). Holocene expansions of Fagus silvatica and Abies alba in Central Europe: where are we after eight decades of debate? *Quaternary Science Reviews, 25*(5–6), 526–549. https://doi.org/10.1016/j.quascirev.2005.03.017.

Trondman, A. K., Gaillard, M. J., Mazier, F., Sugita, S., Fyfe, R., Nielsen, A. B., et al. (2015). Pollen-based quantitative reconstructions of Holocene regional vegetation cover (plant-functional types and land-cover types) in Europe suitable for climate modelling. Glob. Chang. Biol. 21. https://doi.org/10.1111/gcb.12737.

Tzedakis, P. C., Andrieu, V., de Beaulieu, J.-L., Crowhurst, S., Follieri, M., Hooghiemstra, H., et al. (1997). Comparison of terrestrial and marine record of changing climate of the last 500,000 years. *Earth and Planetary Science Letters, 50,* 171–176.

Ground-Air Interface: The Loess Sequences, Markers of Atmospheric Circulation

Denis-Didier Rousseau and Christine Hatté

Abstract

Atmospheric circulation is responsible for the rapid distribution of heat and moisture across the Earth and hence determines our weather and regional climate, today and in the past. During past climate cycles, the atmosphere has been much more dustier, except during the interglacials, inducing uncertainties about the impact of mineral aerosols on the past climate dynamics. There are abundant traces of the combination of past atmospheric dynamics and paleodust cycle such as eolian mineral material transported and deposited in terrestrial archives as loess. Records from loess deposits consistently suggest that atmospheric dynamics was highly variable, during the past climate cycles, much more than presently where the only sources of dust are the major deserts. In this chapter we explained the four main categories of parameters allowing to identify loess deposits as reliable markers for past air circulation.

Overview of Loess

Loess is an aeolian sediment which is relatively common in continental areas (Fig. 13.1). In Europe and North America, it is dispersed near former polar deserts or near the frontal moraines of the gigantic ice sheets that developed on these continents during the ice ages. In Asia, however, the situation is different, since the Loess Plateau is located south of the deserts of northern China and southern Mongolia. The loess sediments thus cover vast regions of the northern hemisphere, at latitudes where very few other records of the glacial paleoclimates are available, while in more southern latitudes, lakes, peat bogs, speleothems and marine sediments collected close to the coast provide continental records of a high quality. In Southern Hemisphere, loess is mostly associated with fluvial and piedmont deposits in the Pampean plains in South America, and fluvial abrasion and glacial grinding in New Zealand.

Loess is a fine sediment transported mainly by wind to various altitudes depending on the particle grain size, but also on the state of the substratum and on the environmental conditions in the region of origin of the material (Fig. 13.2). Dust emission is favored by a low rate of vegetation development and extension of river-emerged banks, source of fine mineral dust. Loess deposits are thus specific to glacial times with low sea-levels thus high river incision and weak vegetation development.

To simplify, coarse material is carried over distances of less than hundred kilometers and transport is mainly by saltation, that is to say, by a series of jumps. However, fine material (clay), which may be of local or remote origins, may equally have been transported over hundreds of kilometers to altitudes of several kilometers in the form of mineral dust and may have been redeposited through wet or dry deposition (depending on whether or not they are associated with precipitations). Loess is therefore usually a marker for air circulation in the past.

To illustrate our words and the research potential of loess sequences, most of our examples will be drawn from the extensively studied Nussloch loess sequence (Rhine Valley, Germany).

D.-D. Rousseau (✉)
École Normale Supérieure de Paris, Laboratoire de Météorologie Dynamique, UMR CNRS 8539, Université Paris Sciences et Lettres, 24 Rue Lhomond, 75231 Paris, France
e-mail: denis-didier.rousseau@lmd.ens.fr

Lamont-Doherty Earth Observatory of Columbia University, Palisades, NY 10964, USA

C. Hatté
Laboratoire des Sciences du Climat et de l'Environnement, LSCE/IPSL, Université Paris-Saclay, 91190 Gif-Sur-Yvette, France

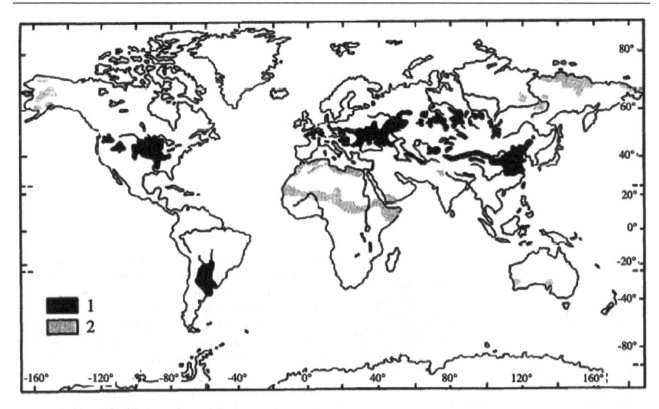

Fig. 13.1 Global distribution of loess sediments and equivalents. 1: loess sediments; 2: loess derivatives (according to Pécsi (1990))

Paleoclimate Indicators

Different paleoenvironmental indicators can be identified for this type of material, deposited at medium and high latitudes with climate conditions unfavorable for biological development (glacial periods). These parameters allow their interpretation in term of atmospheric circulation. Four major categories are presented below.

Sedimentological Indicators

The conceptual view

Even the observation of a loess sequence is instructive. Indeed, when a record is of one or more climate cycles, this is never made up of a single stratigraphic unit.

While loess sediment is characteristic of cold and quite arid periods, other warmer periods are marked by paleosols. Thus, if at low resolution, the loess-paleosol sequence corresponds to one or more climatic cycles; at higher resolution, the identification of more precise events is achieved by observing the sequence of soils or a particular hierarchy of pedo-sedimentary units. (Kukla and An (1989)). In Europe, a leached brown paleosol (Bt), at the base of a brown soil, indicates an interglacial level, while a humus-rich forest soil or a tundra gley (hydromorphic soil) are indicative of more temperate interstadial intervals (Antoine (2009)).

Since they present detailed records of past climate, reproducing the same pedostratigraphic and sedimentological units with the same succession of these units, the pedosedimentary record of loess sequences can be used as a powerful tool for chronostratigraphical correlation (Antoine, et al, (1999); Rousseau et al. (2007a, b)).

The grain size index

The study of the grain size, calculated through determination of the dominant classes, their mode distribution and relationships between them, can characterize the relative strengths of the winds that created the deposits (Rousseau et al. (2007a, b)). These wind dynamics are linked to changes in the general atmospheric circulation (Fig. 13.3). On a finer scale, Nussloch shows a progressive coarsening of the loess deposits between ca 30 and 22 ka (Rousseau et al. (2007b)). This coarsening trend ends with a short but major decrease in grain size, followed by an increase to a new maximum at 20 ± 2 ka ("W" shape). Correlation between the loess grain-size index and the Greenland ice-core dust records suggests a global connection between North Atlantic and Western European global atmospheric circulation and wind regimes (Antoine (2009); Rousseau et al. (2007b)).

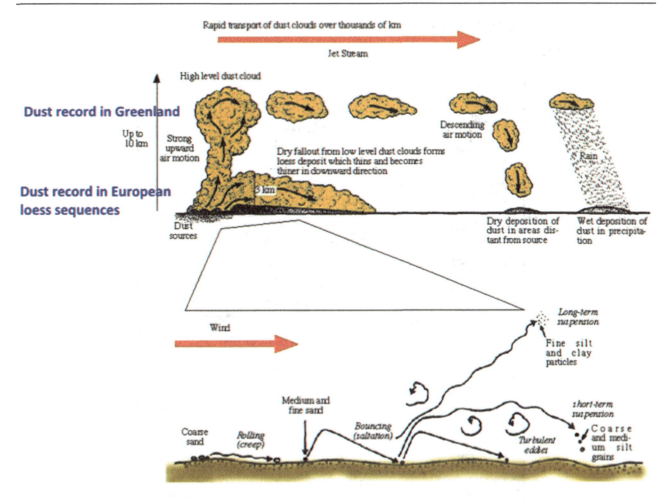

Fig. 13.2 Representation of the different types of dust transportation contributing to the formation of loess deposits. A—Two modes of transport and deposition of wind-blown dust from the northwestern deserts of China to the Loess Plateau and the North Pacific Ocean during the Quaternary (from Pye and Zhou (1989) modified). B—Diagram showing the various modes of transport of wind sediment particularly to the lower levels of the atmosphere (from Pye (1995) modified)

The loess deposit thickness

The thickness of the units is another characteristic which also allows the link with wind dynamics to be established. Indeed, monitoring a particular unit in a given territory allows the characterization of gradients which will be oriented according to the prevailing winds, the thickest part being upwind (Rousseau et al. (2007a)) (Figure 13.4).

Furthermore, the cyclic variation of the sedimentation rate has been shown to be a potential response to the North-Atlantic rapid climate changes, i.e. the Greenland stadial/interstadial cycles and the Heinrich events. This hypothesis has been tested by modeling the impact of North-Atlantic climate variations on dust emissions. This study clearly highlighted that, besides wind, precipitation, soil moisture and snow cover showing some differences in the dust emission intensity, vegetation cover is the main impacting parameter (Sima et al. (2009)). Dust fluxes for the cold climate states (Greenland stadial and Heinrich event) generally become more than twice as high as those for the relatively warmer Greenland interstadial, in agreement with the observed loess data (Sima et al. (2009)) (Fig. 13.4).

The mineralogy

The mineralogy, in particular its composition of heavy minerals, also helps to trace the origin of certain deposits and to deduce the prevailing winds responsible for their transport (Lautridou (1985)).

Biological Indicators

These are relatively diverse. Although the remains of micro-mammals: bones, teeth or skulls, or of larger mammals are identified quite sporadically, other fossils are more

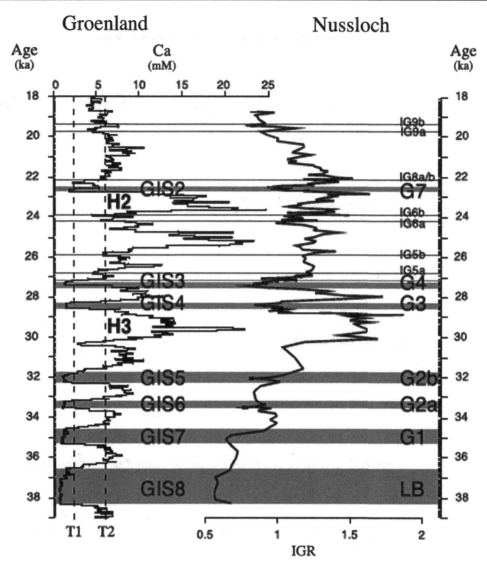

Fig. 13.3 Correlation of variations in particle size index (IGR) as defined at Nussloch with variations in calcium concentration (representing transported dust) from the GRIP survey. The T1 threshold applied to Greenland data highlights the main intervals with low dust concentration, which correspond to Dansgaard-Oeschger (DOI) interstadials 8 to 2. These are correlated to intervals where the IGR is low in Nussloch, corresponding to a brown boreal soil called Lohne Boden (LB) or to well-developed tundra gleys (G1, G2a, G2b, G3, G4 and G7). The T2 threshold defines intervals with very high dust concentration in Greenland, corresponding to the DO stadials, as well as some less significant peaks in dust (according to Rousseau et al. (2007b))

frequent. Among these, terrestrial mollusks form populations typical of diverse environments (Wu (2001); Rousseau (1987)).

Mollusk assemblages

Usually identified at the species level, these organisms have the advantage of persisting to the present day. In accordance with the principle of actualism, it is possible to apply modern requirements and ecological characteristics to individual fossils. Multivariate statistical analyses are used to reconstruct the environment in which a fossil community developed (Rousseau (1987)). Two examples follow: the first, in China, illustrates the long-term variations that can be correlated to terrestrial orbital frequencies and the second shows the short-term variation relating to the internal variability of the climate.

The long-term variability

Using the ecological requirements of taxa identified in the Chinese loess, it was possible to define environmental groups that have proven to be reliable indicators of the summer and winter monsoons through the ages (Fig. 13.5). Transfer

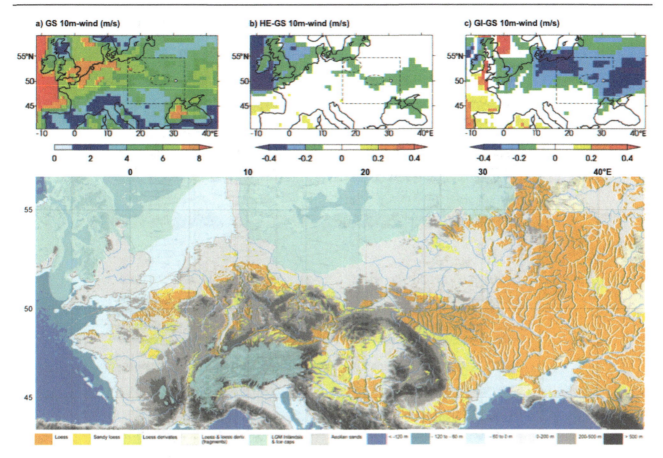

Fig. 13.4 Map of European loess deposits indicating the maximum expansion of the Fennoscandian and British ice sheets during the last glacial maximum (map by Antoine in [10] modified), with wind speed reconstructed for three climate conditions corresponding to northern oceanic influences: GS, HE-GS and GI-GS with GS = Greenland stadial, HE = Heinrich stadial, GI—Greenland interstadial (according to Sima et al. (2009) modified)

functions from terrestrial mollusks and using the principle of modern analogs, were developed to reconstruct the seasonal temperatures from the European loess sequences over three climate cycles (Rousseau (1991); Moine et al. (2002)).

The short-term variability

The high-resolution study (1 sample every 10 cm) of malacological assemblages from the Nussloch loess sequence allows vegetation change to be described along the 70 to 34 kyr cal BP period, recorded in 6 meters of sediment (Moine et al. (2005)). The mollusk changes reflect three short phases of vegetation development and climatic improvement related to soils of the interstadials. A steppe to herb/shrub tundra shift characterizes the Lower-Middle Pleniglacial transition and is followed by a decline in vegetation and humidity increase ending with a new increase in temperature and vegetation cover (Moine et al. (2005)).

Many other methods have been developed using micro-mammals or beetles, with the current distribution of species and the average ranges of associated climate acting as a reference (Liu et al. (1985)). Terrestrial mollusks have also been studied from their signature left in amino acids. This approach helps to distinguish important differences between one climate cycle and another (Oches and McCoy (1995)).

Another paleoclimate index also studied in loess sediments are phytoliths, siliceous concretions present in superficial plant tissue. In some cases, they are well preserved and identified, particularly in the Chinese loess sequences, where they helped in the reconstruction of the temperature and precipitation of the last climatic cycle (Lu et al. (2007)). Pollen, in contrast, are poorly preserved in loess sequences and are the subject of very few studies (Gerasimenko and Rousseau (2008); Rousseau et al. (2001)).

Recently, *earthworm granules* have been established as a new biological proxy (Prud'homme et al. (2015)). This proxy is not based on species recognition but is based on abundance. Counts of earthworm granules reveal a link between their abundance and the nature of the stratigraphic units and their associated climate conditions. They are in very high abundances in tundra gley and boreal brown soil horizons, i.e. during Greenland interstadial intervals and are

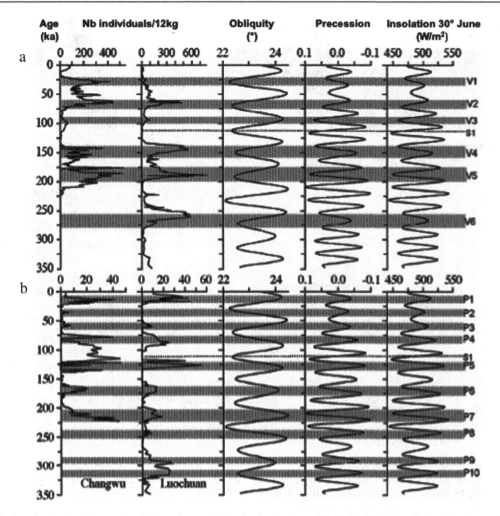

Fig. 13.5 Variations in the abundance of mollusks, characteristic of ancient summer and winter monsoons over the last three climate cycles in two sequences of the Chinese Loess Plateau (Changwu and Luochuan) and their relationship to astronomical parameters and insolation at 30° N. A—Variations in the abundance of *Vallonia tenera* and *Pupilla aeoli*, indicators of ancient winter monsoons. B—Variations in the abundance of *Punctum orphana*, indicator of ancient summer monsoons (according to Wu (2001))

almost absent in typical calcareous loess, associated with Greenland and Heinrich stadials (Prud'homme et al. (2015)).

Geophysical Indicators

This is mainly the magnetic sensitivity in low fields which allows, firstly, the different lithological units present in the same sequence to be characterized, and secondly, to identify the source of the matter through the magnetic particle size (Lagroix and Banerjee (2002)). In general, typical loess units have relatively low field magnetic susceptibility, unlike paleosols, which are characterized by significantly higher values. Widely used for the Chinese sequences, low-frequency magnetic sensitivity was interpreted by Kukla et al. (1990) as corresponding to a relatively constant supply of fine matter through the ages. According to this theory, the formation of soil in the various Chinese interglacial paleosols caused the magnetic grains to become concentrated. This led Kukla to propose a chronological model, independent of any astronomical calibration, while assuming continuous and complete sequences. However, the discovery in the paleosols of bacteria secreting magnetic grains called this debatable assumption into question (Zhou et al. (1990)). On the other hand, research on modern Chinese soils along gradients, reflecting the impact of the summer monsoon, has enabled the calibration of the sensitivity signal and the establishment of a transfer function for the region to reconstruct annual paleoprecipitation associated with the variations in the East Asian monsoon (Maher and Thompson (1995)). A new method of characterizing the origin of loess is to work directly taking a quartz grain and to study both the crystallinity index and the intensity of the spin resonance signal. This new technique makes it possible to differentiate between the origins of the grains and thus to follow the variations in the source of the transported material. Applied

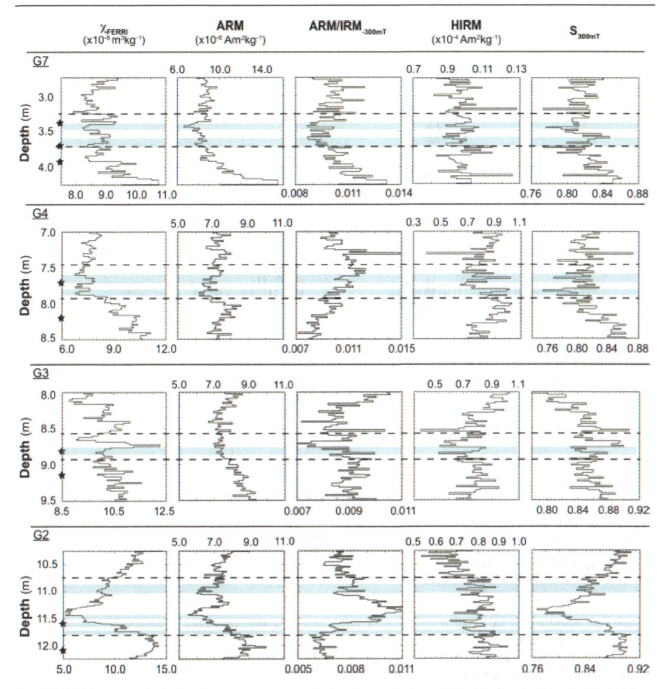

Fig. 13.6 Variation versus depth of rock magnetic parameters allowing the waterlogging in different tundra gleys from the Nussloch loess sequence to be characterized. From Taylor et al. (2014)

for the first time in the Chinese loess sequences, this method has helped to clarify the origin of the quartz grains transported to the Loess Plateau (the Gobi desert during stadial periods, deserts of northern China during the warmer periods) (Sun (2008)). More recently, the study of the mineral magnetic composition completed the classical interpretation of the magnetic susceptibility record by allowing waterlogging processes in tundra gley supporting the correlation of these paleosols with Greenland interstadials to be characterized (Taylor et al. (2014)) (Fig. 13.6).

Geochemical Indicators

The fourth category of indicators concerns geochemical indices and, in particular, isotopic parameters which constitute a powerful tool for interesting research perspectives.

Tracing paleodust sources

As representatives of the age of the geological formation, the dust particles derived from Sr and Pb isotopes are commonly

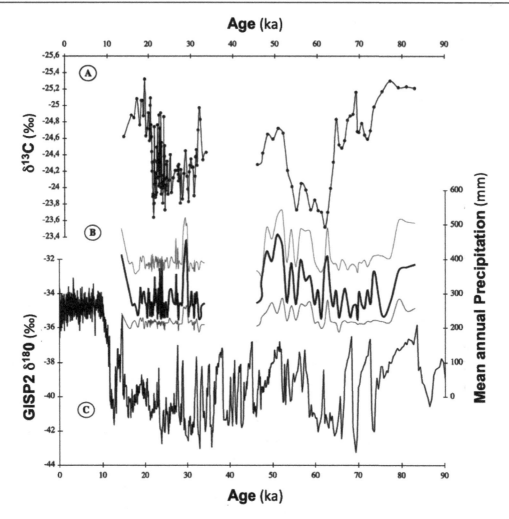

Fig. 13.7 Reconstructing the paleoprecipitation at Nussloch over the last 80 ka. A—$\delta^{13}C$ over time. The range of values obtained corresponds to plants of C3 photosynthetic type. B—Reconstruction of paleoprecipitation by inverse modeling of the isotopic signal (according to Hatté and Guiot (2005)). C—Comparison with $\delta^{18}O$ in Greenland ice (GISP2). During periods of high sea level, the warm phases of DO events result in a net increase in rainfall (+30 to 40%), while during periods of low sea level, the distance from the coastline insulates Nussloch from any climatic improvement that might result from a warm episode

regarded as source tracers. A recent study has been performed on loess samples, dated from the Last Glacial Maximum, located along a 50° N transect (from English Channel to Ukraine), chosen to represent the geographic and petrographic variability of the European loess belt. Geochemical results combined with dust emission simulations revealed that the dust was transported only over a few hundred kilometers. Moreover, the results highlighted that the sources were regional and that distinct sources prevailed within the European continent (Rousseau et al. (2014)).

Tracing past precipitation and seasonality

The analysis of $\delta^{13}C$ from organic matter preserved in very small quantities in the loess sediment indicates the presence of vegetation; this can also be observed through the presence of carbonated concretions developed around herbaceous rootlets (for example, Wang and Follmer (1998); Hatté et al. (2001a)). The values of this isotopic ratio also allow the characterization of the type of vegetation which trapped the dust during its deposition, and therefore the associated environmental and climatic conditions. Distinctions may be made at the level of the photosynthetic cycle (plants in C3 compared to those in C4) or within a similar photosynthetic type through the definition of seasonal variations in temperature or precipitation.

The inversion of vegetation models incorporating an isotopic fractionation module allows the reconstruction of fluctuations in the annual rainfall patterns occurring during the deposition of the sequence (Hatté and Guiot (2005)) (Fig. 13.7). Analysis of other chemical elements, such as rare earth elements, or other isotopes, contributes to the

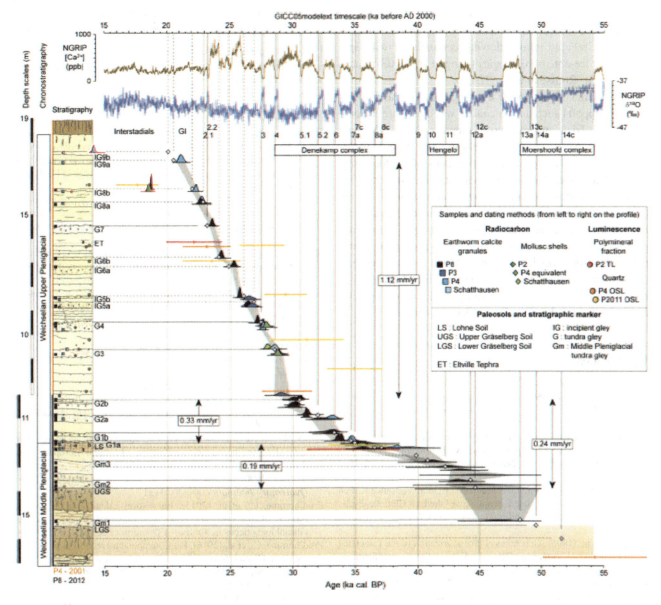

Fig. 13.8 ^{14}C dates obtained from earthworm granules collected in the different paleosols from the Nussloch sequence. Correlation with the NGRIP climate changes described by the dust and the δ^{18}O variations from Moine et al. (2017)

search for the origin of the transported matter, using samples taken from potential source areas, and thus, to estimate the transport-related mechanisms (Rousseau et al. (2014); Gallet et al. (1996); Guo et al. (2002)).

Tracing past precipitation and past temperature

At the same time as the abundance count of earthworm calcitic granules emerged as a paleoclimate proxy, study of their isotopic properties has also developed (Prud'homme et al. (2016), (2018)). The δ^{18}O granules and interlinked transfer functions between water cycle, air and soil temperatures allow the estimation of air temperatures. In Nussloch, the mean summer air temperature during Greenland interstadial has been estimated to have been 10–12 ± 4 °C (Prud'homme et al. (2016)). In line with loess organic δ^{13}C interpretation, δ^{13}C of earthworm granules has been interpreted as a proxy of paleoprecipitation. Thanks to transfer functions, the past precipitation in Nussloch during the Greenland interstadial has been estimated at about 159–574 mm/yr (Prud'homme et al. (2018)).

Loess Chronology

Loess chronology remains a major challenge since OSL returns a wide range of value depending on the choice of the mineral (type and grain size), the choice of light ray (IRSL,

OSL,), and the analytical choice to reconstruct the evolution through time of the ambient dose rate. Nevertheless, it provides a reliable frame allowing comparisons with speleothem or ice-core chronologies (Lang et al. (2003); Fuchs et al. (2013); Rousseau et al. (2013)).

^{14}C is limited by its 50–0 kyr range and by the fact that reliable supports such as charcoal or wood are rare in loess sequences. Trials on ubiquitous supports (loess OM bulk (Hatté et al. (2001b)) and alkane (Häggi (2014)) were not that conclusive, highly dependent on the status of the loess section studied and on the cleaning method of the outcrop. Even in the best conditions, they show poor reliability around (paleo) soil. However, the ^{14}C dates obtained from the calcitic earthworm granules allow perfect dating of the various paleosols (embryonic soils, tundra gleys and arctic brown paleosols) from which they are collected, as was recently demonstrated in the Nussloch sequence (Fig. 13.8) (Moine et al. (2017)). This new approach opens the door to precise and independent chronologies of loess sequences.

References

Antoine, P., et al. (1999). Last interglacial-glacial climatic cycle in loess-palaeosol successions of north-western France. *Boreas, 28*, 551–563.

Antoine, P., et al. (2009). Rapid and Cyclic Aeolian Deposition During the Last Glacial in European Loess: A High-Resolution Records from Nussloch, Germany. *Quaternary Science Reviews, 28*, 2 955–2 973.

Fuchs, M., et al. (2013). The loess sequence of Dolni Vestonice, Czech Republic: A new OSL based chronology of the Last Climatic Cycle. *Boreas, 42*, 664–677. https://doi.org/10.1111/j.1502-3885.2012.00299.x.

Gallet, S., et al. (1996). Geochemical characterization of the Luochuan Loess-Paleosol sequence, China, and Paleoclimatic Implications. *Chemical Geology, 133*, 67–88.

Gerasimenko, N., & Rousseau, D.-D. (2008). Stratigraphy and Paleoenvironments of the Last Pleniglacial in the Kyiv Loess Region (Ukraine). *Quaternaire, 19*(4), 293–307.

Guo, Z. T., et al. (2002). Onset of Asian desertification by 22 Myr ago inferred from loess deposits in China. *Nature, 416*, 159–163.

Häggi, C., et al. (2014). On the stratigraphic integrity of leaf-wax biomarkers in loess paleosols. *Biogeosciences, 11*, 2455–2463.

Pécsi, M. (1990). Loess Is not just the accumulation of dust. *Quaternary International, 7/8*, 1–21.

Hatté, C., et al. (2001a). $\delta^{13}C$ Variation of loess organic matter as a potential proxy for paleoprecipitation. *Quaternary Research, 55*, 33–38.

Hatté, C., et al. (2001b). Development of an accurate and reliable ^{14}C chronology for loess sequences. Application to the loess sequence of Nußloch (Rhine valley, Germany). *Radiocarbon, 43*, 611–618.

Hatté, C., & Guiot, J. (2005). Palaeoprecipitation reconstruction by inverse modelling using the isotopic signal of loess organic matter: application to the Nussloch Loess Sequence (Rhine Valley, Germany). *Climate Dynamics, 25*, 315–327.

Kukla, G., et al. (1990). Magnetic susceptibility record of Chinese Loess. *Transactions of the Royal Society Edinburgh: Earth Sciences, 81*, 263–288.

Kukla, G. J., & An, Z. S. (1989). Loess stratigraphy in central China. *Palaeogeography, Palaeoclimatology, Palaeoecology, 72*, 203–225.

Lagroix, F., & Banerjee, S. K. (2002). Paleowind directions from the magnetic fabric of loess profiles in Central Alsaka. *Earth and Planetary Science Letters, 195*, 99–112.

Lang, A., et al. (2003). High-resolution chronologies for loess: Comparing AMS ^{14}C and optical dating results. *Quaternary Science Reviews, 22*, 953–959.

Lautridou, J. P. (1985). *Le Cycle Périglaciaire Pléistocène en Europe du Nord-Ouest et plus particulièrement en Normandie* (pp. 908). Centre Géomorphologie Caen: Thèse Etat, Université Caen, Caen.

Liu, T. S., et al. (1985). *Loess and the environment* (pp. 251). Beijing: China Ocean Press.

Lu, H. Y., et al. (2007). Phytoliths as quantitative indicators for the reconstruction of past environmental conditions in China II: Palaeoenvironmental reconstruction in the loess plateau. *Quaternary Science Reviews, 26*, 759–772.

Maher, B. A., & Thompson, R. (1995). Paleorainfall reconstructions from pedogenic magnetic susceptibility variations in the Chinese Loess and Paleosols. *Quaternary Research, 44*, 383–391.

Moine, O., et al. (2002). Paleoclimatic reconstruction using mutual climatic range on terrestrial mollusks. *Quaternary Research, 57*, 162–172.

Moine, O., et al. (2005). Terrestrial molluscan records of Weichselian Lower to Middle Pleniglacial climatic changes from the Nussloch loess series (Rhine Valley, Germany): the impact of local factors. *Boreas, 34*, 363–380.

Moine, O., et al. (2017). The impact of Last Glacial climate variability in west-European loess revealed by radiocarbon dating of fossil earthworm granules. *Proceedings of the National Academy of Sciences, 114*, 6209–6214.

Oches, E. A., & McCoy, W. (1995). Amino Acid geochronology applied to the correlation and dating of central European loess deposits. *Quaternary Science Review, 14*, 767–782.

Pécsi, M. (1990). Loess is not just the accumulation of dust. *Quaternary International, 7*(8), 1–21.

Prud'homme, C., et al. (2015). Earthworm calcite granules: a new tracker of millennial-timescale environmental changes in Last Glacial loess deposits. *Journal of Quaternary Science, 30*, 529–536.

Prud'homme, C., et al. (2016). Palaeotemperature reconstruction during the Last Glacial from $\delta^{18}O$ of earthworm calcite granules from Nussloch loess sequence, Germany. *Earth and Planetary Science Letters, 442*, 13–20.

Prud'homme, C., et al. (2018). $\delta^{13}C$ of earthworm calcite granules: a new proxy for palaeoprecipitation reconstructions during the Last Glacial in Western Europe. *Quaternary Science Reviews, 179*, 158–166.

Pye, K. (1995). The nature, origin and accumulation of loess. *Quaternary Science Reviews, 14*, 653–667.

Pye, K., & Zhou, L. P. (1989). Late Pleistocene and Holocene Aeolian Dust Deposition in North China and the Northwest Pacific-Ocean. *Palaeogeography, Palaeoclimatology, Palaeoecology, 73*, 11–23.

Rousseau, D.-D. (1987). Paleoclimatology of the Achenheim Series (Middle and Upper Pleistocene, Alsace, France). A Malacological Analysis. *Palaeogeography, Palaeoclimatology, Palaeoecology, 59*, 293–314.

Rousseau, D.-D. (1991). Climatic transfer function from quaternary mollusks in european loess deposits. *Quaternary Research, 36*, 195–209.

Rousseau, D.-D., et al. (2001). Late Pleistocene environments of the central Ukraine. *Quaternary Research, 56*, 349–356.

Rousseau, D.-D. et al. (2007a). Evidence of cyclic dust deposition in the US great plains during the last deglaciation from the high-resolution analysis of the peoria loess in the eustis sequence

(Nebraska, USA). *Earth and Planetary Science Letters, 262*, 159–174.

Rousseau, D.-D. et al. (2007b). Link Between European and North Atlantic Abrupt Climate Changes over the Last Glaciation. *Geophysical Research Letters, 34*, https://doi.org/10.1029/2007gl031716.

Rousseau, D. D., et al. (2013). Major dust events in Europe during marine isotope stage 5 (130–74 ka): A climatic interpretation of the 'markers. *Climate of the Past, 9*, 2213–2230.

Rousseau, D.-D., et al. (2014). European Glacial Dust Deposits: Geochemical Constraints on Atmospheric Dust Cycle Modeling. *Geophysical Research Lettters, 41*, 7666–7674. https://doi.org/10.1002/2014GL061382.

Sima, A., et al. (2009). Imprint of North-Atlantic millennial-timescale variability on Western European loess deposits as viewed in a dust emission model. *Quaternary Science Reviews, 28*, 2851–2866.

Sun, Y. et al. (2008). Tracing the provenance of fine-grained dust deposited on the central Chinese loess plateau. *Geophysical Research Letters, 35*. https://doi.org/10.1029/2007GL031672.

Taylor, S. N., et al. (2014). Mineral magnetic characterization of the Upper Pleniglacial Nussloch loess sequence (Germany): an insight into local environmental processes. *Geophysical Journal International, 199*, 1463–1480.

Wang, H., & Follmer, L. R. (1998). Proxy of monsoon seasonality in carbon isotopes from paleosols of the Southern Chinese loess plateau. *Geology, 26*, 987–990.

Wu, N. Q. et al., (2001). Orbital forcing of terrestrial mollusks and climatic changes from the loess plateau of China during the past 350 ka. *Journal of Geophysical Research—Atmospheres, 106*, 20 045–20 054.

Zhou, L. P., et al. (1990). Partly Pedogenic Origin of Magnetic Variations in Chinese Loess. *Nature, 346*, 737–739.

Air-Ground Interface: Reconstruction of Paleoclimates Using Speleothems

Dominique Genty and Ana Moreno

Speleothems: Description, Distribution, Formation and Preservation

The term speleothem refers to carbonate deposits in caves: mainly stalactites, stalagmites and stalagmite floors. It is taken from the English (*speleothem*) which has its roots in Greek signifying 'subject' (*thema*) and 'cave' (*spelaion*). Composed of calcium carbonate, speleothems are most commonly made of calcite (ones in aragonite are rarer and less studied). Carbonate massifs which contain caves, within which speleothems are found, are widely scattered around the globe, and are found at all latitudes and in all continents, although less frequently in the southern hemisphere (Fig. 14.1).

Speleothems acquire their geochemical and structural characteristics as a result of the infiltration of rainwater through a limestone or dolomitic environment. The formation process is basically comprised of three steps: (1) at ground level, CO_2 (produced by the roots of plants and by microorganisms in the soil) is dissolved in rainwater; (2) the dissolution of the surrounding rock (limestone, dolomitic limestone, calcareous dolomites) either at ground level (known as dissolution in an open system – at the meeting of the three elements: air, water and rock), or at the level of the many tiny fissures in the surrounding carbonate (closed system); (3) once it arrives in the underground gallery, there is degassing of CO_2 and precipitation of $CaCO_3$.

Drops of water, emerging from micro-fissures, form stalactites at the ceiling of the gallery, and stalagmites on the ground. It is the latter which is most frequently studied because of its simple structure (resembling inverted interlocking calcite cups). The physical and chemical parameters that allow us to reconstruct paleoclimates from speleothems (stable isotopes, trace elements, growth rate) depend partly on the precipitation conditions of the calcite (the temperature in the cave, flow rate, moisture level) and, partly on the geochemical characteristics of the water supplying the speleothems (McDermott 2004; Fairchild and Baker 2012). Speleothems, once formed, seldom undergo subsequent changes (no erosion, no internal recirculation, some rare examples of diagenesis, such as, for example, the transformation of aragonite into calcite). The calcite in speleothems is a material that can be dated as far back as 500 ka using radiometric methods (U-Th series dating) or even as far back as several million years depending on their geochemical characteristics (U-Pb dating).

Growth and Chronology of Speleothems

The growth of speleothems is determined by the presence of infiltrating water. As a result, extremely dry or cold climates (water freezing above ground), causes a halt in growth (hiatus) except in certain exceptional circumstances found in the mountains (Luetscher et al. 2015). The speed of growth of a stalagmite depends not only on the amount of water infiltrating into the karst, but also on a number of environmental factors such as the Ca^{2+} concentration of the infiltrating water, the thickness of the film of water on the surface of the speleothem, the temperature and the partial pressure of CO_2 in the cave's atmosphere. Geochemical modeling of growth shows that the dominant factors are the flow rate and the Ca^{2+} content of the water (Dreybrodt

D. Genty (✉)
Laboratoire des Sciences du Climat et de l'Environnement, LSCE/IPSL, CEA-CNRS-UVSQ, Université Paris-Saclay, 91190 Gif-sur-Yvette, France
e-mail: dominique.genty@lsce.ipsl.fr;
dominique.genty@u-bordeaux.fr

EPOC (Environnements et Paléoenvironnements Océaniques et Continentaux), Université de Bordeaux, bat. B18N, Allée Geoffroy Saint-Hilaire, 33615 Pessac Cedex, France

A. Moreno
Instituto Pirenaico de Ecología—CSIC, Avda. Montañana 1005, 50059 Saragossa, Spain

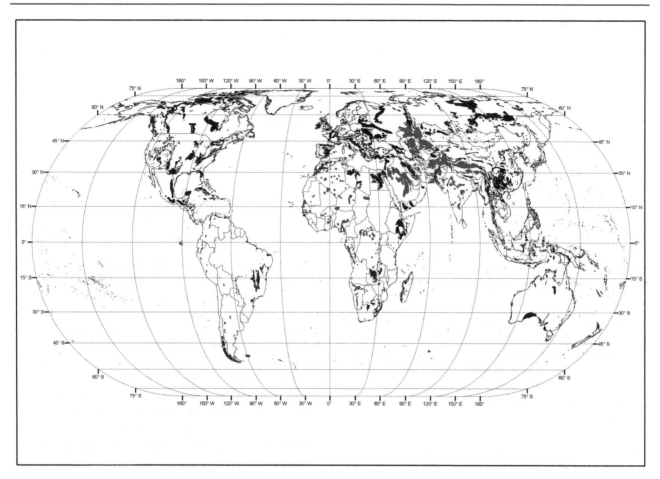

Fig. 14.1 Map showing the distribution of carbonate massifs (in black) where speleothems can be found (according to Ford and Williams (2007) and the University of Auckland http://web.env.auckland.ac.nz/our_research/Karst/)

1988). This model was verified by comparing the theoretical growth rates with the actual growth rates measured on modern calcite deposits at various latitudes. In a cold and humid climate (Scotland), the average vertical growth of a stalagmite is only 20 microns/year, while it can reach 1 mm/year in the caves of southern France (Baker et al. 1998). The sensitivity of the growth rate to environmental conditions provides an indicator of paleoclimatic conditions: more speleothems with a faster growth rate are produced in warm and humid periods than in cold, dry periods. It appears that below a certain rainfall threshold, the growth of speleothems may slow down due to under-saturation of the infiltration water.

The temperature of a cave is generally stable throughout the year and is close to the average annual exterior temperature. Depending on the depth of the cave, the exterior thermal wave which determines the cave temperature may take between several months and several years to travel through the thickness of the rock. However, there are marked seasonal variations in the water infiltration rate and in the concentrations of the different chemical elements (Ca, Sr, Mg, U, MO, etc.) even if the average infiltration time can be several years (Genty and Deflandre 1998; Genty et al. 2014). These have resulted in a seasonal variation in the growth rate and are probably responsible for the formation of visible or luminescent annual growth laminae (Genty and Quinif 1996). When such laminae are present, their identification (a clear and dark lamina is deposited each year) can provide a precise timeline and allows the study of the evolution of an isotopic or geochemical parameter to the nearest year. But annual lamination of stalagmites is not systematic and is often broken; examples rarely go back further than 1000 years (Baker et al. 2015).

The most common way to date speleothems is to measure the isotope series of uranium: ^{234}U, ^{238}U, ^{230}Th (See Chap. 6). Uranium, which is soluble, enters the karst system, while thorium, insoluble, remains above ground. The ^{230}Th measured in the stalagmites is therefore, theoretically, the result of the disintegration of ^{234}U. Sometimes, thorium is also brought by infiltrated water, often along with some detrital clay; a correction must then be made to take account of ^{232}Th (linked to detrital contamination) in the light of the initial ^{230}Th/^{232}Th ratio (Hellstrom 2006). Finally, uncertainty on the age of a calcite fragment is linked to the size of the correction for detritus, and to the uranium concentration in the speleothem (varying from ~30 ppb to ~1 ppm). This is, on average, between less than 1% and 5% for the last three climate cycles, as far back as ~500 ka. The most common current technique to measure these isotopes is by MC-ICP-MS; this requires between a few milligrams to tens of milligrams of calcite to make a measurement. Moreover, thanks to the technical progress in ICP-MS equipment and an improved understanding of half-life constants of radioactive isotopes in recent years, the error in U-Th dates can be even lower (Cheng et al. 2013). Other methods are used incidentally, such as ^{14}C, whose concentration at the moment of deposition of the calcite depends on the proportions of atmospheric CO_2 and of CO_2 without ^{14}C, coming from the dissolution of limestone (Genty et al. 1999), ^{226}Ra for recent millennia (Ghaleb et al. 2004), and the U-Pb method which can trace back more than a million years (Woodhead et al. 2006).

Paleoclimate Reconstruction: A Qualitative Approach

The proxies used in speleothems to reconstruct past climates are isotopes of calcite ($\delta^{18}O_c$, $\delta^{13}C_c$), isotopes of fluid inclusions trapped in the calcite (δD, $\delta^{18}O$), trace elements (Sr, Ba, Mg, U), organic matter (lipids, amino acids), petrography and the growth rate.

The most common crystalline fabrics (layout and shape of the calcite crystals) in stalagmites is the palisading columnar one: large elongated crystals perpendicular to the growth laminae. However, the addition of detrital components, variations in humidity, in temperature or in pCO_2 over the course of climate variations can produce a variety of crystalline fabrics (dendritic, fibrous, microcrystalline, etc.). Even if we could establish links between crystal structures and environmental conditions through the study of modern calcite deposits (Frisia et al. 2000), this type of relationship is complex and should be examined with caution. Recently, a new approach to stalagmite characterization consists of precisely describing the petrography and categorizing the stalagmites and their bounding surfaces into six different classes (from individual crystallites to major nonconformities) (Martín-Chivelet et al. 2017).

Speleothem calcite contains, either in its crystalline defects, or by substitution of Ca in the crystal lattice, minor or trace elements, which provide information on the paleoenvironment (Fairchild et al. 2000). Interpretation of this is complex because it involves several factors: composition of the soil and of the surrounding limestone, the intensity of dissolution and the precipitation conditions of the calcite (temperature, supersaturation, growth rate, etc.). At the annual level, analysis of these trace elements contributes to an understanding of the dynamics of infiltration, especially when the elements measured are linked to the residence time of the infiltrating water. A new model called I-Stal was recently developed to investigate the factors involved in the interpretation of trace elements in speleothems (Stoll et al. 2012). To measure these elements, precise techniques are used to analyze points of a few micrometers in diameter on a polished section of stalagmite (LA-ICP-MS, XRF, X fluorescence by synchrotron, ionic microprobe). During the transfer of these trace elements, organic colloids play a major role. The sensitivity of each one to environmental conditions is different; for example, Ba, Na and Sr may be sensitive to the speleothem growth rate while Mg and U may reflect paleohydrology. Some, such as P, Zn and Cu, are related to the vegetation above ground. Trace elements are also chronological markers: when the annual laminae are not visible in the calcite structure, then the analysis of trace elements can reveal seasonal variations, thereby providing a relative or absolute chronology accurate to the nearest year. This type of analysis is used to determine the duration of a rapid climate event or of a transition (Bourdin et al. 2011).

The stable isotopes of calcite are the most commonly used to reconstruct climate variations, even though interpreting them into temperature and precipitation terms is not easy. The $\delta^{18}O_c$ of the calcite, when it precipitates at isotopic equilibrium (see below), depends on the temperature of the precipitation of calcite (temperature in the cave and therefore the average annual exterior temperature) and on the $\delta^{18}O_w$ of the infiltrating water. The latter is linked to the exterior temperature above the site, to the amount of water extracted from the cloud masses between the source of evaporation and the site and to the isotopic composition of the source, usually the ocean (isotope distillation process). During the precipitation of calcite, there is an inverse relationship between $\delta^{18}O_c$ and temperature (~ −0.24‰/°C), whereas the relationship is direct between the rainfall $\delta^{18}O$ and outside temperature of (e.g. 0.3–0.7‰/°C). In summary:

$$\delta^{18}O_c = f(\delta^{18}O_w, T_{cave}, \text{isotopic equilibrium})$$

with:

$$\delta^{18}O_w = f(\delta^{18}O_{rain}, \text{evapotranspiration in some cases})$$

and

$$\delta^{18}O_{rain} = f(T_{ext.}, \text{quantity of rain, trajectory of cloud masses}, \delta^{18}O_{source}).$$

Consequently, depending on the location of the site relative to the main source of evaporation and on the prevailing conditions when the water masses were transported, the relationship between $\delta^{18}O_c$ and climate will be more or less marked. Thus, in the stalagmites of Southeast Asia the variations in intensity of the monsoon over the last two climate cycles can be seen remarkably well: the $\delta^{18}O_c$ is systematically depleted by ~4‰ during periods of strong summer monsoon, mainly due to the effect of mass (the $\delta^{18}O$ of the rain is inversely proportional to the volume of rainfall) but also due to changes in the source (Wang et al. 2008). Thanks to the many U-Th datings carried out on several stalagmites in caves in Sanbao, Hulu and Dongge (25° N to 32° N, China), it has been shown that the $\delta^{18}O_c$ of the calcite had a periodicity of 23 ka and was directly correlated with changes in insolation at 65° N, demonstrating that variations in the monsoons were caused by orbital changes (Fig. 14.2) (Wang et al. 2008). Superimposed on these large climate variations, millennial climate events were also recorded and linked to Dansgaard-Oeschger events (D/O) detected in the Greenland ice, which shows the strong connection between the climate systems of South Asia and the North Atlantic. This interconnection between the monsoon regime and the North Atlantic region was confirmed by one of the longest and most precise paleoclimate recordings produced by the study of Chinese speleothems (Cheng et al. 2016). The variations in monsoon intensity were reconstructed for the past 640,000 years with an unmatched precision. The multiplicity of samples, precise U-Th dating and the very high resolution of isotopic analyses suggest new theories on the causes of the major glacial-interglacial climate cycles and millennial variability. One of these theories posits that the period between terminations (glacial-interglacial transitions) is equal to a multiple of the length of the precession cycles; moreover, a very strong teleconnection between the dynamics of the ice caps of the northern hemisphere, the circulation of the Atlantic Ocean and the Asian monsoon is implied. Similarly, the variation in monsoon intensity in South America was recorded in stalagmites from southern Brazil (24° S to 27° S), at the orbital and the millennial scale (Cruz et al. 2005). In this case, the latitudinal changes in the position of the ITCZ (intertropical convergence zone) seem to be the root cause of these variations.

At higher latitudes, in Europe for example, the $\delta^{18}O_c$ of calcite has a less pronounced response to changes in climate, probably because of the opposite effects controlling the $\delta^{18}O_c$ of calcite (e.g. external temperature impacts on the $\delta^{18}O$ of the rain while cave temperature impacts on $\delta^{18}O_c$ of calcite). However, the study of several European samples has revealed a certain logic within the paleoclimate chronicles of varying appearance: there is a variable gradient in the $\delta^{18}O_c$ speleothems during the Holocene along a west-east transect (McDermott et al. 2011). This is related to differences in warming during the Holocene depending on longitude and also on different atmospheric circulations. This logic shows that even if speleothems do not precipitate at isotopic equilibrium, they contain valuable information on atmospheric paleocirculation. As a result of the influence of these multiple climatic, hydrological and kinetic factors, the evolution of the $\delta^{18}O_c$ of the speleothem calcite is not uniform from one region to another: in general the $\delta^{18}O$ decreases when the climate becomes warm and humid (in temperate regions for example), while this trend may be reversed ($\delta^{18}O$ increases as the temperature increases) at certain altitudes as has been observed in the Alps (Boch et al. 2011; Moseley et al. 2014) and in eastern Europe on the coast of the Black Sea (Fleitmann et al. 2009).

The $\delta^{13}C_c$ of calcite may also react to climate changes, and sometimes in a more obvious way than $\delta^{18}O_c$. The carbon atoms of the calcite molecule, $CaCO_3$, which makes up the speleothems come from two main sources: CO_2 from the soil and $CaCO_3$ from the surrounding rock. Soil CO_2, produced by plant roots and microbial activity, has a $\delta^{13}C_c$ close to −24‰ (for C3 type plants, most frequently found in temperate areas), $CaCO_3$ from marine limestone has a $\delta^{13}C_c$ of between ~−2 and +2‰. It has been shown that the main source of carbon in speleothems is the CO_2 from the soil which can represent up to 90% of C contained in the $CaCO_3$ of speleothems (Genty et al. 1998). In several sites in the South of France, the carbon from the dissolution of limestone (also called dead carbon because it contains no ^{14}C) represents, in this particular case, only 15 ± 5% of the C in the speleothems. Consequently, any change in the vegetation above a cave brought about by climate change, such as proportion of type C3 to C4 or vegetation density, will have an impact on the $\delta^{13}C$ of the CO_2 in the soil and thus on the $\delta^{13}C_c$ of the speleothems. In summary:

$\delta^{13}C_c = f$ (type of vegetation, density of vegetation, hydrology, isotopic equilibrium).

The $\delta^{13}C_c$ in the stalagmites in the Villars cave (South-West France) shows abrupt changes from −2‰ to −5‰ over the last 80 ka. These have been linked to Dansgaard-Oeschger events recorded in Greenland ice cores and to temperature reconstructions using analyzes of pollen from lakes and marine cores (Genty et al. 2003).

The value of this comparison is, firstly, to test the chronology from other archives against the absolute chronology provided by speleothems, and if necessary, to adjust it. Refining the age of an abrupt climatic transition, such as the one which occurred at the beginning of D/O 12 (Fig. 14.3), is important in order to find out its cause by comparing a sequence of climate events with the external forcings and with archives from other latitudes. Comparison with pollen reconstructions shows the close link between changes in vegetation brought on by variations in temperature and humidity, and the $\delta^{13}C_c$ recorded in speleothems. Other examples show that the $\delta^{13}C_c$ recorded the last deglaciation, in New Zealand as well as in Europe (Genty et al. 2006; Moreno et al. 2010). Thus, a stalagmite from the Chauvet cave (Ardèche, France) records the climate events (e.g. Bølling-Allerød, Intra Allerød Cold Period, Younger-Dryas) that punctuated the last deglaciation with a

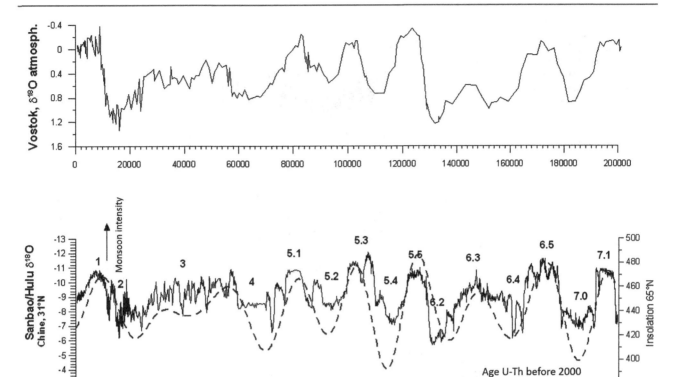

Fig. 14.2 After Wang et al, 2008—Example of a recording of the variations in intensity of the Asian monsoon using the $\delta^{18}O_c$ of Chinese stalagmites from Sanbao and Hulu caves (bottom graph). Comparison with July insolation at 65°N and the $\delta^{18}O_{atm}$ of Vostok ice core, Antarctica (dashed line). There is a good correlation between $\delta^{18}O_c$ and the insolation, dominated here by the precession (cycles of 23 ka). The $\delta^{18}O_{atm}$ from Vostok reflects the impact of the precession on low latitude water cycle and productivity of the biosphere (See Chap. 11 on polar ice)

resolution comparable to that obtained in ice cores from Greenland, with, however, some differences in the trends caused by different climate gradients.

Paleoclimate Reconstruction: A Quantitative Approach

During its growth, the speleothem calcite traps water in the form of microscopic fluid inclusions (1–10 microns large) or, more rarely, macroscopic ones (several mm). This water comes from the rainwater contemporaneous to the calcite deposition and can therefore be dated indirectly by dating the surrounding calcite. With an average composition of only a few nL of water per gram of calcite, the technical difficulties of extracting and analyzing this water have only recently been resolved (Verheyden et al. 2008; Vonhof et al. 2006; Affolter et al. 2014; Arienzo et al. 2013). There are many reasons for measuring the isotopic composition (δD, $\delta^{18}O$) of the fluid inclusions of speleothems: (1) the value of the $\delta^{18}O$ of the inclusion water is close to that of the rainwater and so is a tracer of atmospheric circulation; (2) in conjunction with the $\delta^{18}O_c$ of the calcite, it is possible to calculate the temperature of calcite formation in the cave, which is equivalent to the average annual exterior temperature. For this second case, it is necessary for the calcite precipitation to have occurred at thermodynamic equilibrium (isotopic, by extension), in other words, that the exchanges between the different carbon species (e.g. HCO_3^-, CO_3^{2-}, CO_2 gas) have been completed.

Examples using this new technique are still rare, but are among the only ones, on land, to express the evolution of temperature (from direct measurements) over a precise absolute time scale. A stalagmite from Peru has thus shown that the isotopic composition of the rainwater followed the local winter insolation (6° S) due to changes in the intensity of convective rainfall during the Holocene, themselves linked to latitudinal variations in the ITCZ (Van Breukelen et al. 2008). The temperature, calculated using the above method, varied little (±2 °C) over the last 13.5 ka, unlike at higher latitudes, as is shown by another example from Vancouver Island (Canada 49° N), where the temperature has varied by more than 10°C between 6 ka and 10 ka (Zhang et al. 2008).

However, this method cannot be applied to all speleothems. Indeed, conditions of low humidity or low pCO_2

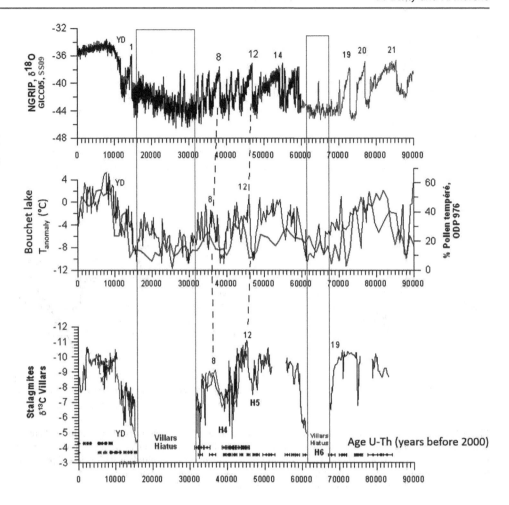

Fig. 14.3 Comparison between the $\delta^{13}C$ from Villars Cave stalagmites, the $\delta^{18}O$ from NGRIP (Greenland), the temperature reconstructions from Bouchet Lake (Massif Central) (Guiot et al. 1989) and marine core ODP976 from the south of Spain (Combourieu Nebout et al. 2002). The dots at the bottom of the graph show the U-Th datings with 2σ error bars

inside the cave can cause rapid degassing of CO_2, and thus lead to a kinetic effect causing a thermodynamic disequilibrium. There are several equations linking the isotopic compositions of calcite and of water with temperature, allowing this equilibrium to be checked (Kim and O'Neil 1997; Tremaine et al. 2011). However, the study of deposits of modern calcite shows that many of them are not deposited at isotopic equilibrium, but rather reflect a kinetic effect (Genty 2008; Mickler et al. 2006). In this case, calculation of the temperature is not valid (the $\delta^{18}O_c$ measured from calcite is generally too high, causing the calculated temperature to be too low).

There is a recent, promising method to reconstruct paleotemperatures using only isotopes from the mineral phase of the calcite, thus overcoming the difficulty of extracting water from the fluid inclusions (Daeron et al. 2008, 2011). It involves the clumped isotopes of the CO_3^{2-} molecules and is based on the thermo-dependence of the isotopic exchanges between these different molecules:

$$^{13}C^{16}O_3^{2-} + {}^{12}C^{18}O^{16}O_2^2 <=> {}^{13}C^{18}O^{16}O_2^{2-} + {}^{12}C^{16} + {}^{12}C^{16}O_3^2$$

Called the Delta 47 method, this method reflects the statistical overabundance of mass 47 clumped isotopes in CO_2 ($^{13}C^{18}O^{16}O$) produced by acidc attack of carbonate minerals. When applied to speleothems which were deposited at thermodynamic equilibrium, precipitation temperatures can be reconstructed independently of the previous method which uses the $\delta^{18}O$ of the calcite and of the fluid inclusions. However, tests on modern deposits from caves in the South of France and in vitro experiments show the existence of a kinetic effect during precipitation (Daeron et al. 2011, 2008). In this case, the temperature can only be calculated by using both the Delta 47 measurement and the measurement of the $\delta^{18}O$ in calcite and fluid inclusions. The value of this is twofold: in addition to calculating the precipitation temperature, this method quantifies the state of thermodynamic disequilibrium, often difficult to detect, and which is also a reflection of the conditions of the paleoenvironment.

Another way of quantifying climate parameters is by calibrating proxies from the growth laminae of fast-growing modern stalagmites (Baker et al. 2007; Domínguez-Villar et al. 2018). The method consists of finding the best correlation between the signal measured on the stalagmite (e.g. the thickness of the annual growth laminae) and the climate signal from instrumental measurements outside (e.g., rainfall). However, the water arriving at the end of a stalactite is, in fact, a mixture of water from a variety of networks of large and small micro-fissures which are more or less complex, with a variable transfer time from the exterior to the inner gallery (ranging from a few hours to few weeks, months or years). This mixture was modeled by A. Baker in a cave in Ethiopia, with components of different ages: a small proportion (10–30%) of water flowing quickly through the rock (from a few days to a few weeks) and a greater proportion (70–90%), of the water passing through slowly (months to a few years). The variation in the amount of water arriving at the particular stalagmite being studied (which largely determines the thickness of the annual laminae) is calculated by applying this mixture model to the rainfall provided by meteorological data. The proxy (the thicknesses, in this case) is then compared with the amount of water supplying the stalagmite (modeled using this simple mixture model). The timeframe is given by the annual growth laminae. This new method, applied to stalagmites in Ethiopia, has helped to reconstruct summer rainfall over the last 80 years (Baker et al. 2007). Other work is being developed around the modeling of the $\delta^{18}O$ of infiltration water involving reservoirs in the unsaturated zone (Fairchild and Baker 2012) or by empirical adjustment with modern climate data (Domínguez-Villar et al. 2018).

Conclusion

Speleothems constitute continental paleoclimate records that allow the reconstruction of climate changes on a precise and absolute chronological timescale over the last half-million years or more. They help to improve the accuracy of chronologies derived from other climate archives and, depending on the site, can be used to reconstruct variations in rain intensity, rain source and vegetation density on the continent. However, to correctly interpret the proxies measured in the speleothems, it is essential to know the present day environment of the studied site: climate, seasonality, hydrology, origin of the precipitations. Recent methods of analysis of fluid inclusions and of Delta 47 create opportunities to establish long series of continental temperature changes and of the isotopic composition of rainwater. The recent proliferation of studies of speleothems and the increasingly structured organization of the scientific community working on this archive (Karst Record symposiums for example) have led to the discovery of important results on climate change and their causes. Interactions between increasingly large datasets (INTIMATE-Moreno et al. 2014, SISAL database being constructed, Comas Bru et al. 2017) and climate models are multiplying so as to better understand the processes and causes of climate variations.

References

Affolter, S., Fleitmann, D., & Leuenberger, M. (2014). New online method for water isotope analysis of speleothem fluid inclusions using laser absorption spectroscopy (WS-CRDS). *Climate of the Past, 10*, 1291–1304. https://doi.org/10.5194/cp-10-1291-2014.

Arienzo, M. M., Swart, P. K., & Vonhof, H. B. (2013). Measurement of $\delta18O$ and $\delta2H$ values of fluid inclusion water in speleothems using cavity ring-down spectroscopy compared with isotope ratio mass spectrometry. *Rapid Communications in Mass Spectrometry, 27*, 2616–2624. https://doi.org/10.1002/rcm.6723.

Baker, A., Asrat, A., Fairchild, I. J., Leng, M. J., Wynn, P. M., Bryant, C., & Umer, M. (2007). Analysis of the climate signal contained within delta O-18 and growth rate parameters in two ethiopian stalagmites. *Geochimica et Cosmochimica Acta, 71*, 2975–2988.

Baker, A., Genty, D., Dreybrodt, W., Barnes, W. L., Mockler, N. J., & Grapes, J. (1998). Testing theoretically predicted stalagmite growth rate with recent annually laminated samples: Implications for past stalagmite deposition. *Geochimica et Cosmochimica Acta, 62*, 393–404.

Baker, A., Hellstrom, J. C., Kelly, B. F. J., Mariethoz, G., & Trouet, V. (2015). A composite annual-resolution stalagmite record of North Atlantic climate over the last three millennia. *Scientific Reports, 5*, 10307. https://doi.org/10.1038/srep10307.

Boch, R., Cheng, H., Spötl, C., Edwards, R. L., Wang, X., & Häuselmann, P. (2011). NALPS: A precisely dated European climate record 120–60 ka. *Climate of the Past Discussions, 7*, 1049–1072. https://doi.org/10.5194/cpd-7-1049-2011.

Bourdin, C., Douville, E., & Genty, D. (2011). Alkaline-earth metal and rare-earth element incorporation control by ionic radius and growth rate on a stalagmite from the Chauvet Cave, Southeastern France. *Chemical Geology, 290*, 1–11.

Cheng, H., Edwards, R. L., Sinha, A., Spötl, C., Yi, L., Chen, S., et al. (2016). The Asian monsoon over the past 640,000 years and ice age terminations. *Nature, 534*, 640–646. https://doi.org/10.1038/nature18591.

Cheng, H., Lawrence Edwards, R., Shen, C.-C., Polyak, V. J., Asmerom, Y., Woodhead, J., et al. (2013). Improvements in 230Th dating, 230Th and 234U half-life values, and U-Th isotopic measurements by multi-collector inductively coupled plasma mass spectrometry. *Earth and Planetary Science Letters, 371–372*, 82–91. https://doi.org/10.1016/j.epsl.2013.04.006.

Comas Bru, L., Burstyn, Y., & Scroxton, N. (2017). From caves to climate: Creating the SISAL global speleothem database. *Past Global Change Magazine, 25*, 156–156. https://doi.org/10.22498/pages.25.3.156.

Combourieu Nebout, N., Turon, J. -L., Zahn, R., Capotondi, L., Londeix, L., & Pahnke, K. (2002). Enhanced aridity and atmospheric high-pressure stability over the western Mediterranean during the North Atlantic cold events of the past 50 ky. *Geology, 30* (10), 863–866.

Cruz Jr, F. W., Burns, S. J., Karmann, I., Sharp, W. D., Vuille, M., Cardoso, et al. (2005). Insolation-driven changes in atmospheric circulation over the past 116,000 years in subtropical Brazil. *Nature, 434*, 63–66.

Daeron, M., Guo, W., Eiler, J., Genty, D., Blamart, D., Boch, R., et al. (2011). (13)C(18)O clumping in speleothems: Observations from natural caves and precipitation experiments. *Geochimica et Cosmochimica Acta, 75*, 3303–3317.

Daeron, M., et al. (2008). Absolute speleo-thermometry, using clumped isotope measurements to correct for kinetic isotope fractionations induced by CO2 degassing. *Geochimica et Cosmochimica Acta, 72*, A193–a193.

Domínguez-Villar, D., Lojen, S., Krklec, K., Kozdon, R., Edwards, R. L., & Cheng, H. (2018). Ion microprobe $\delta18O$ analyses to calibrate slow growth rate speleothem records with regional δ18O records of precipitation. *Earth and Planetary Science Letters, 482,* 367–376. https://doi.org/10.1016/j.epsl.2017.11.012.

Dreybrodt, W. (1988). *Processes in karst systems.* Springer Verlag, 288p.

Fairchild, I. J., & Baker, A. (2012). *Speleothem science: From process to past environments.* Wiley.

Fairchild, I. J., Borsato, A., Tooth, A. F., Frisia, S., Hawkesworth, C. J., Huang, Y., et al. (2000). Controls on Trace Element (Sr-Mg) compositions of carbonate cave waters: Implications for speleothem climatic records. *Chemical Geology, 166,* 255–269.

Fleitmann, D., Cheng, H., Badertscher, S., Edwards, R. L., Mudelsee, M., Gokturk, O. M., et al. (2009). Timing and climatic impact of Greenland interstadials recorded in stalagmites from northern Turkey. *Geophysical Research Letters, 36.*

Ford, D. C. & Williams, P. (2007). *Karst Geomorphology and Hydrology,* Wiley, 562p.

Frisia, S., Borsato, A., Fairchild, I. J. & McDermott, F. (2000). Calcite Fabrics, Growth Mechanisms, and Environments of formation in Speleothems from the Italian Alps and Southwestern Ireland. *Journal of Sedimentary Research, 70,* 1183–1196.

Genty, D. (2008). Paleoclimate research in Villars Cave (Dordogne, SW-France). *International Journal of Speleology, 37,* 3.

Genty, D., Blamart, D., Ghaleb, B., Plagnes, V., Causse, C., Bakalowicz, M., et al. (2006). Timing and dynamics of the last deglaciation from European and North African δ13C stalagmite profiles—Comparison with Chinese and South Hemisphere stalagmites. *Quaternary Science Reviews, 25,* 2118–2142. https://doi.org/10.1016/j.quascirev.2006.01.030.

Genty, D., Blamart, D., Ouahdi, R., Gilmour, M., Baker, A., Jouzel, J., et al. (2003). Precise dating of Dansgaard-Oeschger climate oscillations in western Europe from stalagmite data. *Nature, 421,* 833–837.

Genty, D., & Deflandre, G. (1999). Drip flow variations under a stalactite of the Père Noël cave (Belgium). *Evidence of seasonal variations and air pressure constraints, Journal of Hydrology, 211,* 208–232.

Genty, D., Labuhn, I., Hoffmann, G., Danis, P. A., Mestre, O., Bourges, F., et al. (2014). Rainfall and cave water isotopic relationships in two South-France sites. *Geochimica et Cosmochimica Acta, 131,* 323–343. https://doi.org/10.1016/j.gca.2014.01.043.

Genty, D., Massault, M., Gilmour, M., Baker, A., Verheyden, S., & Kepens, E. (1998). Calculation of past dead carbon proportion and variability by the comparison of AMS ^{14}C and TIMS U/Th ages on two Holocene stalagmites. *Radiocarbon, 41,* 251–270.

Genty, D., & Quinif, Y. (1996). Annually laminated sequences in the internal structure of some Belgian stalagmites—Importance for paleoclimatology. *Journal of Sedimentary Research, 66,* 275–288.

Ghaleb, B., Pons-Branchu, E., & Deschamps, P. (2004). Improved method for radium extraction from environmental samples and its analysis by thermal ionization mass spectrometry. *Journal of Analytical Atomic Spectrometry, 19,* 906–910.

Guiot, J., Pons, A., de Beaulieu, J. L., & Reille, M. (1989). A 140,000-year continental climate reconstruction from two European pollen records. *Nature, 338,* 309–313.

Hellstrom, J. (2006). U-Th dating of speleothems with high initial Th-230 using stratigraphical constraint. *Quaternary Geochronology, 1,* 289–295.

Kim, S.-T., & O'Neil, J. R. (1997). Equilibrium and nonequilibrium oxygen isotope effects in synthetic carbonates. *Geochimica et Cosmochimica Acta, 61,* 3461–3475.

Luetscher, M., Boch, R., Sodemann, H., Spötl, C., Cheng, H., Edwards, R. L., et al. (2015). North Atlantic storm track changes during the Last Glacial Maximum recorded by Alpine speleothems. *Nature Communications, 6,* 7344. https://doi.org/10.1038/ncomms7344.

Martín-Chivelet, J., Muñoz-García, M. B., Cruz, J. A., Ortega, A. I., & Turrero, M. J. (2017). Speleothem Architectural Analysis: Integrated approach for stalagmite-based paleoclimate research. *Sedimentary Geology, 353,* 28–45. https://doi.org/10.1016/j.sedgeo.2017.03.003.

McDermott, F. (2004). Paleo-climate reconstruction from stable isotope variations in speleothems: A review. *Quaternary Science Reviews, 23,* 901–918.

McDermott, F., Atkinson, T. C., Fairchild, I. J., Baldini, L. M., & Mattey, D. P. (2011). A first evaluation of the spatial gradients in δ18O recorded by European Holocene speleothems. *Global and Planetary Change, 79,* 275–287. https://doi.org/10.1016/j.gloplacha.2011.01.005.

Mickler, P. J., Stern, L. A., & Banner, J. L. (2006). Large kinetic isotope effects in modern speleothems. *Geological Society of America Bulletin, 118,* 65–81.

Moreno, A., Stoll, H. M., Jiménez-Sánchez, M., Cacho, I., Valero-Garcés, B., Ito, E., & Edwards, L. R. (2010). A speleothem record of rapid climatic shifts during last glacial period from Northern Iberian Peninsula. *Global and Planetary Change, 71,* 218–231. https://doi.org/10.1016/j.gloplacha.2009.10.002.

Moreno, A., Svensson, A., Brooks, S. J., Connor, S., Engels, S., Fletcher, W., et al.(2014). A compilation of Western European terrestrial records 60–8 ka BP: towards an understanding of latitudinal climatic gradients. Quaternary Science Reviews, Dating, Synthesis, and Interpretation of Paleoclimatic Records and Model-data Integration: Advances of the INTIMATE project (INTegration of Ice core, Marine and Terrestrial records, COST Action ES0907) *106,* 167–185. https://doi.org/10.1016/j.quascirev.2014.06.030.

Moseley, G. E., Spötl, C., Svensson, A., Cheng, H., Brandstätter, S., & Edwards, R. L. (2014). Multi-speleothem record reveals tightly coupled climate between central Europe and Greenland during Marine Isotope Stage 3. *Geology, 42,* 1043–1046. https://doi.org/10.1130/G36063.1.

Stoll, H. M., Müller, W., & Prieto, M. (2012). I-STAL, a model for interpretation of Mg/Ca, Sr/Ca and Ba/Ca variations in speleothems and its forward and inverse application on seasonal to millennial scales: I-STAL SPELEOTHEM TRACE ELEMENT MODEL. Geochem. Geophys Geosystems 13, n/a-n/a. https://doi.org/10.1029/2012GC004183.

Tremaine, D. M., Froelich, P. N., & Wang, Y. (2011). Speleothem calcite farmed in situ: Modern calibration of δ18O and δ13C paleoclimate proxies in a continuously-monitored natural cave system. *Geochimica et Cosmochimica Acta, 75,* 4929–4950. https://doi.org/10.1016/j.gca.2011.06.005.

Van Breukelen, M. R., Vonhof, H. B., Hellstrom, J. C., Wester, W. C. G., & Kroon, D. (2008). Fossil dripwater in stalagmites reveals Holocene temperature and rainfall variation in Amazonia. *Earth and Planetary Science Letters, 275*(1–2), 54–60.

Verheyden, S., Genty, D., Cattani, O., & van Breukelen, M. R. (2008). Water release patterns of heated speleothem calcite and hydrogen isotope composition of fluid inclusions. *Chemical Geology, 247,* 266–281.

Vonhof, H. B., van Breukelen, M. R., Postma, O., Rowe, P. J., Atkinson, T. C., & Kroon, D. (2006). A continuous-flow crushing device for on-line delta H-2 analysis of fluid inclusion water in speleothems. *Rapid Communications in Mass Spectrometry, 20,* 2553–2558.

Wang, Y., Cheng, H., Edwards, R. L., Kong, X., Shao, X., Chen, S., & An, Z. (2008). Millennial- and orbital-scale changes in the East Asian monsoon over the past 224,000 years. *Nature, 451,* 1090–1093.

Woodhead, J., Hellstrom, J., Maas, R., Drysdale, R., Zanchetta, G., Devine, P., & Taylor, E. (2006). U-Pb geochronology of speleothems by MC-ICPMS. *Quaternary Geochronology, 1,* 208–221.

Zhang, R., Schwarcz, H. P., Ford, D. C., Schroeder, F. S., & Beddows, P. A. (2008). An absolute paleotemperature record from 10 to 6 Ka inferred from fluid inclusion D/H ratios of a stalagmite from Vancouver Island. *British Columbia, Canada, Geochimica Cosmochimica Acta, 72,* 1014–1026.

Air-Interface: $\delta^{18}O$ Records of Past Meteoric Water Using Benthic Ostracods from Deep Lakes

Ulrich von Grafenstein and Inga Labuhn

Introduction

Oxygen-isotope records offer an alternative way to quantitatively reconstruct paleoclimate, which, at first view, is independent of an ecosystem's reaction to climate change. The ice core records from high latitude inland ice, where the former precipitation and its isotopic composition ($\delta^{18}O_P$) are preserved in an almost original state, are widely accepted as valuable sources of paleoclimate information. Worldwide systematic observation of $\delta^{18}O_P$ (Rozanski et al. 1992), starting in the late sixties, together with the incorporation of the water isotopes in several generations of general circulation models (Hoffmann et al. 1998; Jouzel et al. 1987) and models of intermediate complexity have, not only increased the confidence in $\delta^{18}O_P$ as a powerful paleotemperature indicator in a number of key regions, but have also demonstrated its importance as a primary hydrometeorological parameter in regions where temperature dependence is less evident or absent.

In non-polar regions, investigations have been conducted on lake sediment (carbonate, bulk organic matter, cellulose, diatoms), speleothems, tree rings and soil carbonate. While all these terrestrial records respond to changes in the isotopic composition of past precipitation ($\delta^{18}O_P$), a number of secondary effects can alter the original atmospheric signal and thus limit or exclude a quantitative interpretation. Detecting past changes of $\delta^{18}O_P$ therefore not only provides a valuable paleoclimate proxy, but also largely facilitates the quantification of secondary effects in complex isotopic records. The problem is to find material, like polar ice, which contains quantitative information about the isotopic composition of past precipitation and from which secondary effects can either be excluded or quantified. Those include the alteration of the primary signal ($\delta^{18}O_P$) on its way to through the water cycle (hydrological effects) and during the formation of material in which it will be preserved (isotope fractionation effects).

Lake sediments very often cover reasonably long periods with relatively high accumulation rates, allowing sampling resolutions from individual years to decades. Ostracod valves preserved in those sediments can, with respect to their dominance and habitat, be considered as the lacustrine equivalent to benthic foraminifera in marine sediments (De Deckker 2002). As such, they are ideally suited for the geochemical characterization of former lake water, including stable isotope studies (Holmes and Chivas 2002). A major advantage they have compared to other materials, such as bulk carbonate or organic matter, is that ostracod valves can be relatively easily separated, cleaned (Danielopol et al. 2002), and assigned to different species for which biological and behavioural information is available from the study of modern materials (Horne et al. 2002). Moreover, the specific fractionation can be calibrated using the valves of individuals that grew and calcified under known conditions (Chivas et al. 2002; von Grafenstein et al. 1999a, b; Xia et al. 1997). The same applies to benthic molluscs belonging to the family of sphaeriides, which frequently can be found together with ostracods.

The focus of this chapter is on the special situation of ostracods and molluscs in the profundal sediments of deep lakes, where seasonal and long-term temperature changes are almost absent and where the $\delta^{18}O$ of the local water regularly approaches the average of the entire water column due to the cold season overturn. The fossil valves of these animals can be used to reconstruct precisely the isotopic composition of the lake water ($\delta^{18}O_L$) and, depending on the lake's hydrologic setting, the isotopic composition of former precipitation ($\delta^{18}O_P$). The first part of this chapter provides an overview of published deep-lake oxygen-isotope records. The following two parts more specifically address the processes implicated in the signal transfer from the atmospheric

---

U. von Grafenstein (✉) · I. Labuhn
Laboratoire des Sciences du Climat et de l'Environnement, LSCE/IPSL, CEA-CNRS-UVSQ, Université Paris-Saclay, 91190 Gif-sur-Yvette, France
e-mail: Ulrich.von-Grafensstein.fr@lsce.ipsl.fr

precipitation to the ostracod valve, i.e. (1) the major hydrological processes controlling the present and past links between $\delta^{18}O_L$ and $\delta^{18}O_P$, and (2) the effects related to the valve formation (temperature-dependent fractionation and vital effects). Most of the data used to illustrate those processes are from lakes situated within a small region of southern Germany and are the author's published or unpublished material. The data was collected to support and refine the quantitative interpretation of the $\delta^{18}O$ records from sub-recent and late Glacial to Holocene mainly from one of those lakes (Ammersee), which is presented in the fourth part of this chapter. The reason for this regional concentration is the lack of case studies as comprehensive from other regions, combining investigation of the modern isotope hydrology, the physical limnology, and the ostracod geochemistry within the settings of a fossil record. The last part of the chapter aims to show how $\delta^{18}O_P$ reconstruction from deep lake ostracods could be further developed and exploited to address secondary effects in more complex archives. This chapter is an update of a previously published book section (von Grafenstein 2002). It incorporates major advances concerning our understanding of the oxygen isotope signal preserved in ostracod valves and gives an overview of new and upcoming records in Europe.

Existing Deep-Lake Oxygen-Isotope Records

The first published ostracod stable isotope record was from Lake Erie (Fritz et al. 1975). This showed a 4‰ shift at the late Wisconsin to Holocene transition, which was interpreted as a temperature controlled change in $\delta^{18}O_P$. Later, deep-lake oxygen-isotope ostracod records from the Great Lakes demonstrated that the Late Wisconsin and Holocene re-arrangements of the hydrological pathways and episodic drainage from pro-glacial lakes largely controlled the isotopic composition of the lakes' water (Colman et al. 1994; Dettman et al. 1995; Forester et al. 1994; Lewis and Anderson 1992; Rea et al. 1994). Although it is almost impossible to extract information on $\delta^{18}O_P$ from those ostracod records, they are still excellent examples of successful and very useful reconstructions of relative $\delta^{18}O_L$ changes. Data for the isotopic composition of modern ostracod valves have been reported for Lake Huron (Dettman et al. 1995). However, whilst these data suggested that the most important taxa showed vital offsets, they were insufficient for these offsets to be quantified.

Lister et al. (1991) presented an oxygen-isotope record from Lake Qinghai (Qinghai-Tibetan Plateau) based on measurements of *Limnocythere inopinata* and *Eucypris inflata*, which alternately dominate the lake's benthic ostracod assemblage throughout the past 15,000 years. The record shows an overall increase of 6‰ from 8000 ^{14}C-years BP to 3000 ^{14}C-years BP, which the authors attribute to the slow accumulative evaporative concentration following a major reduction in humidity in the region. This hypothesis is strengthened by the fact that the reconstructed oxygen-isotope values of the lake water, prevailing for the last 3000 years, are close to the endpoint of isotopic enrichment under modern conditions. Short term excursions from this trend and a substantial variability before 8000 ^{14}C-years BP are interpreted as episodic changes in the lake's water balance and water level variations. However, it cannot be excluded that shifts of $^{18}O_P$ and/or $\delta^{18}O_A$ (which are documented in the $\delta^{18}O$ record from the Guliya ice cap (Thompson et al. 1997) or changes of the bottom water temperature (the lake today is only 23 m deep) might have contributed to both the long-term shifts and to the small-scale fluctuations of $\delta^{18}O_L$.

In Europe, early monospecific deep-lake records exist from Lake Zürich (Lister 1988) and Lake Lugano (Niessen and Kelts 1989), both have a strong glacial to interglacial shift but lack resolution of abrupt events during the transitions. Two moderately better-resolved $\delta^{18}O_L$ histories of two neighbouring lakes in the northern alpine foreland (Ammersee and Starnberger See, southern Germany) (von Grafenstein et al. 1992) provided evidence of climate-induced changes of $\delta^{18}O_P$ consistent with the pollen-inferred local climate history, including a strong negative excursion during the Younger Dryas. They also show systematic and constant offsets between the $\delta^{18}O$ values of different taxa from the same sediment layer, indicative of physiologically controlled fractionation in addition to temperature-dependent fractionation. $\delta^{18}O$ records of monospecific ostracod samples from a deep-lake core from Lac Neuchâtel (Switzerland), depicted similar millennial-scale, late-glacial variations, in addition to sudden changes in the lake's water balance and shifts of the mean isotopic composition of the input, due to the episodic connection to the Aare river system (Schwalb et al. 1994). Shallow-water, late-glacial records from the Ammersee (von Grafenstein et al. 1994), from southern Sweden (Hammarlund et al. 1999), and Switzerland (von Grafenstein et al. 2000, 2013) give evidence that the large shifts of $\delta^{18}O$, bracketing the Younger Dryas cold period, were accompanied by relative changes in summer water temperatures, consistent with air-temperature controlled shifts of $\delta^{18}O_P$. Quantitative reconstruction of $\delta^{18}O_P$ from those records remains biased by the temperature effects, by seasonal variation of shallow water $\delta^{18}O$, and often by unknown changes in the lakes' water balances. The best-resolved European oxygen-isotope record from deep-lake ostracod valves is from Ammersee (von Grafenstein et al. 1996, 1998, 1999a, b), and the accompanying hydrological and isotope-geochemical calibration (von Grafenstein et al. 1999a, b) will be used below as an example to discuss the possible effects in more detail.

A lower-resolved record from Lake Constance (Schwalb 2003) is in good agreement with the Ammersee record. A low-resolution record also exists from Lake Geneva covering the Holocene. More recently, the first results from a long-term project aiming to produce a new high-resolution record using a new sediment core from Mondsee (Austria) show a remarkable synchrony for the negative excursion during the 8.2 ky-event (Andersen et al. 2017) and over the entire glacial-interglacial transition (Lauterbach et al. 2011) Currently, a multidisciplinary team continues to work on the complete high-resolution $\delta^{18}O_P$ records from Ammersee (Germany), Mondsee (Austria) and Lac d'Annecy (France) for the last 15,000 years.

Hydro-Meteorological Effects

The Isotopic Composition of Atmospheric Precipitation

The isotopic composition of atmospheric precipitation at a given place and time is controlled by various fractionation effects, including (1) the moisture formation in the (mainly oceanic) source regions, (2) loss through precipitation along the moisture transport pathways, (3) re-evaporation and (4) the condensation and precipitation process at the sampling site. Thirty years of monthly observation at meteorological stations around the world show that, in high to mid latitudes, losses of the original moisture content along horizontal temperature gradients seem to prevail, leading to the existence of an overall positive correlation between $\delta^{18}O_P$ and local air temperatures on seasonal and inter-annual time scales as well as geographically (Dansgaard 1964; Rozanski et al. 1992). In low latitudes, a negative correlation between $\delta^{18}O_P$ and the amount of precipitation is observed and is explained by the vertical temperature gradients. These climate-$\delta^{18}O_P$ relations are, however, overlain by noise due to high short-term variability, reflecting the effects of the different condensation processes and the admixture of vapor from different marine or continental sources during a single precipitation event, and through spatial variability, due to relief effects. Evidently, the quantitative relation between $\delta^{18}O_P$ and climatic parameters, like temperature and precipitation, has to be checked for each basin before conclusions in terms of paleoclimate can be drawn from a record. However, even if such a relation is weak or absent, the reconstructed $\delta^{18}O_P$ itself remains a valuable information source for paleoclimate analyses and for the quantitative interpretation of isotopic records with strong secondary effects.

It is important to be aware of the seasonal variations of $\delta^{18}O_P$, which are large even with respect to the greater shifts recognised in paleo-records. In many cases, the retention times of water within the catchment and in the lake are long enough, so that the effective input into the lake carries an average isotope signal over several years. This average is often close to the average of $\delta^{18}O_P$, but might be offset by preferential losses of a part of the precipitation by evaporation and transpiration (see below). In the special case of short catchment retention times combined with a short residence time in the lake, the isotopic composition of the lake water ($\delta^{18}O_L$) can respond with oscillations either in or out of phase with the seasonal $\delta^{18}O_P$ cycle.

Catchment Effects

The term 'catchment effect' has been introduced and discussed in detail by Gat et al. (1995). It can be defined as the sum of effects from all hydrological processes occurring in the drainage basin that might alter the relation between long term $\delta^{18}O_P$ and $\delta^{18}O_L$. Besides the 'catchment effect' sensu stricto, which is the measurable difference between the isotopic compositions of the drainage basin runoff $\delta^{18}O_I$ and $\delta^{18}O_P$ (Δ_{P-I}), we also have to consider those processes that control the amount of runoff (I) and the spatial and temporal variability of both I and $\delta^{18}O_I$.

Potentially responsible for a significant Δ_{P-I} are losses via evaporation or sublimation, which can lead to an isotopic enrichment of the remaining water. The importance of these losses can be estimated by comparing the deuterium excess of the runoff with that of the mean precipitation. Usually they are small in humid climates and they depend on the proportion of open-water surfaces within a drainage basin. Alternatively, non-fractionating losses such as the uptake and consequent transpiration and evaporation of water by plants, or the recharge of deep groundwater, can be selective against a portion of the annual precipitation and thus can lead to a relative shift of $\delta^{18}O_P$. This effect theoretically could be greater than 1‰, if the evapotranspiration is restricted to the summer period and only summer precipitation is taken up. In practice, the largest part of water for evapotranspiration comes from the soils, which retain significant amounts of water over the year and where the seasonal variation of $\delta^{18}O_P$ and, in consequence, the effect on Δ_{P-I} is considerably reduced.

Stronger effects are to be expected in high-altitude drainage basins, including those in high alpine areas, where build-up of glaciers can retain significant amounts of isotopically-depleted winter precipitation and thus shift $\delta^{18}O_I$ to more positive values. Glacier growth is a relatively long-term process, which takes place over tens to hundreds of years during relatively cold periods, whereas ablation can be almost spontaneous with a consequent discharge of isotope-light water easily exceeding the amount of yearly retention during glacier growth. The isotopic composition of runoff from partially glaciated areas may therefore be biased to more positive values compared to $\delta^{18}O_P$ during cold phases with glacier growth, and marked by strong negative spikes during subsequent warm periods.

Environmental changes including climate shifts can increase or decrease the runoff without significantly affecting Δ_{P-I}. The most evident is a change in precipitation, but evapotranspiration and groundwater recharge and discharge can also change either in concert with precipitation, or independently, as a response to climate change, vegetation change and human activity. Such variations of the runoff will alter the lake water balance and thus the link between $\delta^{18}O_P$ and the isotopic composition of lake water $\delta^{18}O_L$ by an amount dependent on the hydrological sensitivity of the lake (see below). Changes in the seasonal runoff characteristics can affect the link between $\delta^{18}O_P$ and $\delta^{18}O_L$. These may occur without any measurable changes in the annual runoff and mean annual $\delta^{18}O_I$, by moderation of the short-term reservoirs (soil, snow cover). Such effects are important in lakes with water residence times of close to or less than one year, but may also be visible in lakes with longer response times.

Probably the most efficient way to characterize the catchment effects in a given basin is through a survey of the isotopic composition of both the precipitation and the river runoff. Figure 15.1 gives an example for the Ammer River (southern Germany), the main contributor to Lake Ammersee. River water was sampled just upstream of the lake in two-weekly intervals for more than two years and on a daily base for the second of these two years (von Grafenstein et al. 1996). Atmospheric precipitation is collected routinely at one meteorological station (Hohenpeissenberg) within the drainage basin and at two additional stations (Neuherberg and Garmisch), about 50 km to the north and south, respectively, providing monthly values for $\delta^{18}O_P$ since 1970 AD. The oxygen isotope composition of the total two-years of runoff is −10.3‰, with a tendency to increase from −10.5‰ to −10.1‰ during this period. Interestingly, the two-year precipitation mean for the same period is −10.1‰, indicating that at least a part of the runoff is from longer term (groundwater) reservoirs. Indeed, the daily record of the second year allows the spontaneous, soil water and groundwater components to be separated out with a mean retention time of 9 months for the soil water component and of at least several years for the groundwater component. In the long term, the total contribution of soil water and spontaneous runoff is about 30%. It is therefore impossible to define an exact estimate of Δ_{P-I} without considering the history of $\delta^{18}O_P$ for the past few decades. The best explanation for the tendency of $\delta^{18}O_I$ over more than 4 years (including lower frequency measurements from 1989 on) is a mean age of three years for the groundwater component. However, even if we consider a much longer retention in the groundwater, Δ_{P-I} is less than +0.3‰ for the present-day conditions in the Ammersee drainage basin. A small decrease in the deuterium excess of the mean runoff (8‰) compared to that of the precipitation over the last decade (10‰) might indicate that this probable catchment effect in the case of the Ammer river is due to surface evaporation (from some smaller lakes in the basin).

In summary, catchment effects, even if they are seemingly small as in our example, have to be at least considered as a possible reason for changes of $\delta^{18}O_L$ in the past. Quantification is however rather complicated and has to be based on a proper description of the modern geomorphologic conditions and the vegetation cover of the drainage basin.

Lake Water Balance Effects

The isotopic composition of lake water is controlled by the amounts and the isotopic compositions of the inflowing water (I, $_I$) and loss via evaporation (E, $_E$). In steady state, i.e. if the environmental conditions are considered as stable for some time, δ_L approaches a constant value representing a mixture between δ_I (weighted by I) and δ^* (weighted by the evaporation and the atmospheric water vapor deficit):

$$\delta_L = (\delta_P I + \delta^* Eh/(1-h))/(I + Eh/(1-h)) \quad (1)$$

δ^* is the isotopic composition of water being in isotopic equilibrium with the atmospheric water vapor and thus the maximum that can be reached by evaporation:

$$\delta^* = (h\,\delta_A + \varepsilon)/(h - \varepsilon) \quad (2)$$

where h is the relative humidity, δ_A the isotopic composition of the atmospheric water vapour and ε the sum of the equilibrium and kinetic fractionation between water and vapour (Gat et al. 1994; Gibson et al. 1993) Several important facts with respect to the reconstruction of δ_P can be derived from these relations: (1) the isotopic enrichment due to evaporation is independent of the residence time of water (I/V), but depends, in addition to the atmospheric moisture conditions, on I (which is the product of P-ET, Precipitation-EvapoTranspiration, and the surface of the catchment basin) and E (which is the product of evaporation per surface unit and the lake's surface). Lakes with differing ratios between the lake surface A_L and drainage basin area A_C will therefore have significantly differing δ_L, even if the atmospheric conditions including δ_P, δ_A and the evaporative flux are equal. (2) Any change of P-ET or E will be weighted by this factor A_L/A_C, i.e. will induce a stronger reaction of δ_L for lakes with a larger A_L/A_C. The smaller the lake surface compared to the drainage basin, the smaller will be the influence of changing hydrology on the quantitative link between δ_I and δ_L (and consequently the link between δ_P and δ_L). (3) Comparison of

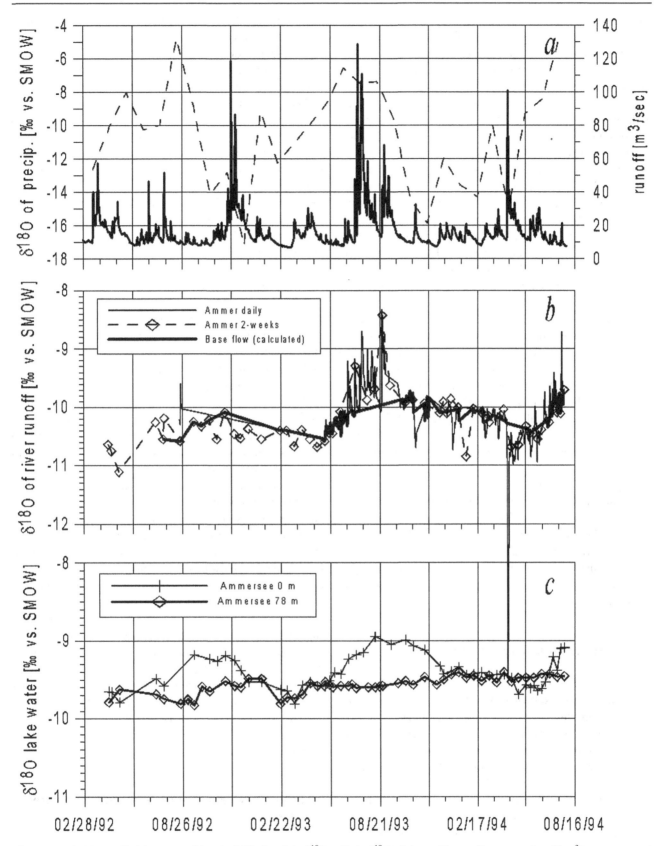

Fig. 15.1 a Daily runoff of the Ammer River (solid line) and the $\delta^{18}O$ of monthly precipitation at the meteorological station Hohenpeissenberg (stippled line); **b** $\delta^{18}O$ of the Ammer River sampled daily (continuous line) and in two-week intervals (diamonds and stippled line). $\delta^{18}O$ of days with runoff not exceeding 12 m^3/s are connected with the thick line and supposed to represent $\delta^{18}O$ of the base-flow; **c** $\delta^{18}O$ of surface and deep water in Ammersee (redrawn from von Grafenstein 2002)

δ_L from three neighbouring lakes with differing A_L/A_C should allow quantification of the evaporative enrichment and, in consequence, of Δ_{P-I}, P − ET, E, and h, if one stable water isotope is used ($\delta^{18}O$ or δD), and from two lakes if both $\delta^{18}O_L$ and δD_L are available.

Figure 15.2 gives an example of such a set of lakes (which in part are also used below for δ_P reconstruction). These four lakes are fed today by surface water of almost identical composition and are sufficiently close to assume that P − ET, h, δ_A, and evaporation from a surface unit are equal. Individual water samples, taken within the period from 1989 to 1994 in a δD–$\delta^{18}O$ diagram scatter around mean values from a lake, are clearly separated from the local meteoric precipitation and from lake to lake. Averages of the samples taken in the winter, when the lakes are well mixed, are plotted on a line with a slope of 4.9, indicating that the distance from the long-term average of the precipitation is due to the relative importance of evaporation from the individual lake surface. Using Eqs. 1 and 2, $\delta^{18}O_L$ measurements of the four lakes can be reproduced using mean climatic conditions for the region and assuming that the isotopic composition of atmospheric water vapor (δ_A) is in equilibrium with the long term mean of the isotopic composition of the precipitation ($\delta^{18}O_P$). Evaporation from a surface unit, as calculated from the isotope balances, is 535 mm, close to independent estimates from the water balances (ca. 600 mm) and from energy balance modelling (580 mm).

Figure 15.3 gives an example of a sensitivity test exploiting how the δ_L of the lakes in Fig. 15.2 will change if hydrologic conditions change (without a change of δ_P), such as a doubling of the input (I) or a doubling of the evaporation (E). Evidently, the lake with the lowest modern evaporative offset is also the least sensitive to changing hydrology, and therefore should give the most accurate estimate of δ_P from reconstructed δ_L.

Transient Changes and Dynamic Effects

The scatter of $\delta^{18}O_L$ of individual lake water samples in Fig. 15.2 is the result of temperature stratification, which leads to short-term deviations from the 'steady state.' In summer, the warm epilimnic water body, which is efficiently separated from the much larger hypolimnion, is both evaporatively enriched and fed by isotopically-enriched summer precipitation and river runoff. The first effect leads to systematic shifts parallel to the local evaporation line, whereas the second leads to less systematic scatter along lines connection the prevailing $_L$ with the respective isotopic composition of the precipitation and river water. Under present day conditions, these seasonal deviations of epilimnetic water in our examples can add up to at least 1.5‰ and are subject to significant inter-annual variability as shown in Fig. 15.1c. Hypolimnetic water, in consequence, is the better representation of the lake's reaction to the long-term development, being 'updated' once a year during the lake overturn in autumn, winter, and/or spring.

With respect to dynamic effects, the selection of the 'ideal' lake for δ_P reconstruction is a trade-off between a short residence time (I/V), which ensures an optimal response of δ_L to a change in δ_P, on the one hand, and a residence time that is long enough to efficiently suppress the seasonal variability of δ_I, on the other hand. A theoretical residence time of 2.7 years (as for the Ammersee) seems to be sufficiently long to suppress the present-day seasonal variability of δ_I. The reactivity of a lake to a change in $_P$ can be described by

$$\delta_{L(t)} = \delta_{L(0)}e^{-tI/V} + \delta_{L(\infty)}\left(1 - e^{-t/V}\right) \quad (3)$$

where $\delta_{L(0)}$ is the isotopic composition before a change, $\delta_{L(\infty)}$ is the new steady state after the change, and t is the elapsed time since the change. For example, after a change of 1‰ in $\delta^{18}O_P$, the isotopic composition of the lake will reach a value indistinguishable from the new equilibrium (±0.05‰) after t = −ln (0.05/$\delta_{L(0)}$ − $\delta_{L(\infty)}$) * V/I, which would be 8 years for the Ammersee against 63 years for the Starnberger See, the lake with the longest residence time in our selection. All variability of δ_P with higher frequency will have a response in δ_L with reduced amplitude.

In order to minimize the overall error of a $\delta^{18}O_P$ reconstruction from lake isotopic records, the 'ideal' lake should have a simple, well-defined drainage basin, which is large compared to the surface area of the lake to ensure a small evaporative isotopic enrichment of its water. The lake should be holomictic with a cold hypolimnion, i.e. a water depth exceeding 40 m, and should have a short residence time, but not much below 2 years. Most of these criteria can often be verified based on basic field observation and literature, but should be documented with isotopic determinations of river and lake water samples if a lake is considered to be a potential candidate for paleoclimatological or paleohydrological investigations. Such hydrological pre-site studies not only increase the significance of the interpretation of the isotopic records in terms of changing $\delta^{18}O_P$, but may also help to design studies on one or more distinct hydrological effects.

Isotope Geochemistry of Benthic Freshwater Ostracods

Theoretically, all benthic organisms living in the hypolimnion of lakes and producing identifiable fossil remains preserved in the sediments could be considered for the

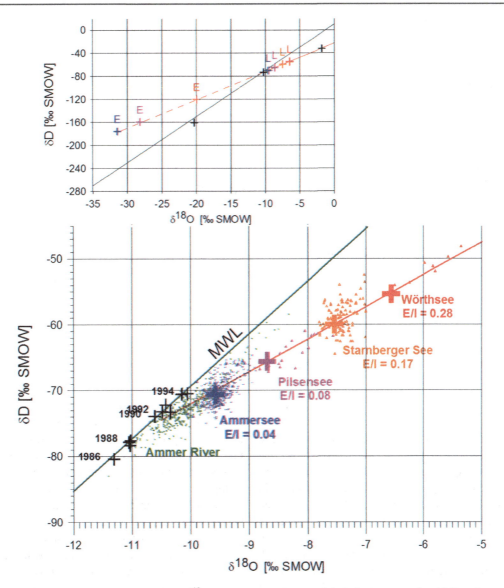

Fig. 15.2 Lower graph: water isotope ratios (δD versus $\delta^{18}O$) of lake water (Ammersee, blue; Pilsensee, magenta; Starnberger See, orange; Woerthsee, red), precipitation (IAEA-Station Hohenpeissenberg, large black crosses), and river water (Ammer, green) from the 'Fünf-Seen--Land' (Five Lakes District) close to Munich, Germany. Large symbols mark the average of the lake samples taken during the winter period, when the lakes are well mixed and effects from the thermal stratification are absent. Note the increasing distance of lake samples from the *Meteoric Water Line* (MWL) due to evaporative enrichment, proportional to the respective evaporation to input ratios (E/I). Scattering of the data from one lake is due to temporal variability of mean runoff (4 years of observation for Ammersee and Starnberger See) and to vertical differences in the water column in summer. All lakes have permanent outflow. The upper graph shows the calculated isotopic compositions of isotopically depleted water vapor produced from the lakes (δ_E, labelled as E), of lake water (δ_L, labelled as L), and of the limiting maximal enriched water (δ^*, labelled as *) governed by the mean local climatic conditions, by the mean isotopic compositions of the atmospheric water vapor (δ_A, labelled as A) and precipitation (δ_P, labelled as P). Colours for the different lakes are the same as for the lower graph

reconstruction of the isotopic composition of former lake water. However, at present, studies have been restricted to valves of ostracods and small bivalve molluscs belonging to the genus *Pisidium*. One reason for this choice is that these organisms build protective shells or valves from $CaCO_3$, which are relatively easy to separate from the rest of the sediment. In addition, we now have the capacity to measure oxygen and carbon isotope ratios in quantities of carbonate as small as 10 µg (Chivas et al. 1993; von Grafenstein et al. 1992). Contamination by littoral fauna are also detectable through the occurrence of species typically restricted to shallow water. The existence of fossil fauna representing the former life assemblage (De Deckker 2002) is itself an excellent paleolimnological indicator of sufficient oxygen in the hypolimnion and thus for a regular winter overturn of the water column, which is needed for the transfer of the $\delta^{18}O_P$ signal into the hypolimnion.

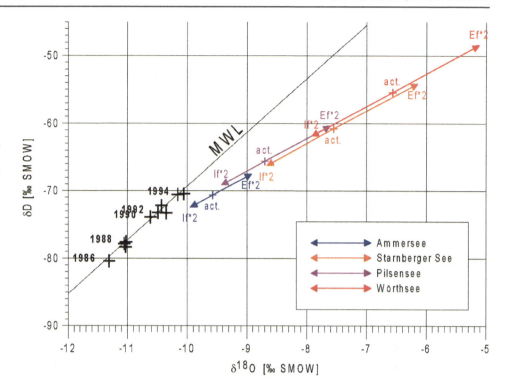

Fig. 15.3 Reaction of $\delta^{18}O_L$ to changes in the water balance (same colours as for Fig. 15.2). The modern isotopic compositions are labelled as 'act.', those resulting from to doubled input as 'I*2' and those from doubled evaporation as 'E*2'. Note that the $\delta^{18}O_L$ of Ammersee remains within a range of 1‰ for such hydrologic extremes, in contrast to the much stronger reaction of the $\delta^{18}O_L$ of Wörthsee (almost 3‰)

Vital Effects

However, the formation of the protective calcitic ostracod valves is a physiological process, which may affect the oxygen isotopic composition of the valves through systematic differences compared to an inorganic calcite formed in isotopic equilibrium. Indeed, such 'vital effects' have been postulated based on evidence from constant differences in valve $\delta^{18}O$ between different taxa in deep lake cores from Starnberger See and Ammersee (von Grafenstein et al. 1992). These effects have been quantified for some common European freshwater species by systematic field observation and collection (von Grafenstein et al. 1999a, b), and by laboratory cultivation for the North American species *Candona rawsoni* (Xia et al. 1997) and the Australian euryhaline nectic ostracod *Australocypris robusta* (Chivas et al. 2002). The result of the field observations in Ammersee and Starnberger See have been confirmed by a similar one-year observation in Lake Geneva (Decrouy et al. 2011). However, field and laboratory studies on ostracods from sites with variable salinity (Li and Liu 2010) and host water pH reporting (Chivas et al. 2002; Marco-Barba et al. 2012) indicate that the 'vital effect' might be influenced by host water conditions. Devriendt et al. (2017) therefore compiled all data from published 'calibration' studies (Chivas et al. 2002; Xia et al. 1997; von Grafenstein et al. 1999a, b; Decrouy et al. 2011; Li and Liu 2010; Didié and Bauch 2002; Keatings et al. 2002; Van der Meeren et al. 2011; Bornemann et al. 2012) and suggested that ostracod calcite reflects the oxygen isotopic composition of the sum of $[HCO_3^-]$ and $[CO_3^{2-}]$. For carbonate-dominated freshwater lakes with a pH of around 8.3 the contribution of $[CO_3^{2-}]$ to this sum is negligible. The isotopic composition of ostracods valves in these lakes therefore reflects almost exclusively the isotopic composition of $[HCO_3-]$ which in turn depends solely on $\delta^{18}O_L$ and the water temperatures, whereas, in waters with high and variable salinity, the variable contribution of $[CO_3^{2-}]$ leads to significant negative excursions in the $\delta^{18}O$ of the ostracod calcite, thus possibly influencing the reconstruction of the host water $\delta^{18}O$. This new calcification model nicely explains the apparent and up-to-now still enigmatic, positive offset of valve $\delta^{18}O$ with respect to an 'equilibrium calcite' in almost all freshwater settings. It could also be the key to understanding the significant differences of 'vital offsets' within taxonomic groups. For the deep freshwater lake situation, the correction for the empirical 'vital offset' and for the temperature-dependent fractionation between the $\delta^{18}O$ of ambient water and a calcite in isotopic equilibrium should still give an excellent measure of the $\delta^{18}O_L$, if the water temperature during the formation of the valve is known, or can be established.

Water Temperature Effects

Holomixis (the full overturn of the water column) of deep freshwater lakes occurs when the temperature of the entire water column approaches the density maximum of freshwater at 4 °C. The exact temperature for the overturn might, however, deviate slightly from 4 °C, if the wind forcing during the holomixis is strong enough to overcome the density gradients. Nevertheless, the possible temperature range of 3–5 °C is relatively small. In regions where air temperatures of the coldest month are equal to, or below, 4 °C, positive deviations from 4 °C are restricted to a short period during the onset of the holomixis, and negative deviations are limited to the winter and to cases of a continuing wind-forced overturn. Most probably, the deep water is at 4 °C at the moment of stabilisation due to spring warming. During summer, the hypolimnion is efficiently insulated from irradiative warming. Small temperature increases up to 6 °C are, however, observed in lakes in regions with high geothermal gradients. Warmer bottom-water temperatures occur in meromictic lakes, in which the winter overturn only reaches down to a limited depth and in which a deeper and denser water body (monimolimnion) exists. However, as mentioned above, no benthic in situ ostracod and mollusc fauna will be found there due to the absence of oxygen in the monimolimnion. In order to further minimize any error in the reconstruction of $\delta^{18}O_L$ from benthic carbonate fossils, bottom-water temperatures of the respective lakes should be followed over several seasons and compared with those simulated by energy- and water-balance models driven by observed climate conditions for the same period. The models, validated in this way, can then be used to estimate bottom water temperatures and thus provide error estimates for $\delta^{18}O_L$ for a large range of climate conditions. For Ammersee and Lac d'Annecy, two of the few lakes which should be excellently suited for quantitative reconstruction of $\delta^{18}O_P$, this modelling approach was used, confirming deep water temperatures astonishingly constant at 4 °C for a very wide range of climate conditions, colder than today for Ammersee, but with the risk of meromixis in warmer climates, especially when winter mean daily temperatures remain significantly above 4 °C. In contrast, Lac d'Annecy continues to be episodically meromictic during modelled warmer periods, because of its higher transparency during the warmest summer month, allowing deep water to reach the warmer winter minimal air temperatures within a couple of years (Danis et al. 2003, 2004)

In our example from the Ammersee, the average temperature at 80 m over the last 20 years (1980–2000, data provided by Dr. B. Lenhardt, WWA Weilheim) was 4.15 °C, with a standard deviation of 0.45 °C, and extremes of 3.2 °C and 5.0 °C. With such a narrow range of water temperatures and the relatively constant $\delta^{18}O_L$ of the hypolimnion (see Fig. 15.1c), the preferred moulting and calcification period of the ostracods and their instars is almost irrelevant. This is in strong contrast to sites within the epilimnion, where differences of up to 3‰ between the $\delta^{18}O$ of summer and winter produced valves (after correction for vital offsets) are common (von Grafenstein et al. 1994; 1999a, b, 2013; Dettman et al. 1995). However, a problem with such deep-lake studies is the low abundance of ostracods compared to littoral sites; this is most probably related to the combination of low population densities of the different species with relatively elevated sediment accumulation rates (in average ∼1 mm per year in core AS96-1 from Ammersee). High-resolution $\delta^{18}O$ records approaching the hydrologic resolution (ca. 8 years/sample) are therefore only possible if 10 to 20 valves of juveniles (instars A-5 to A-2) of the most abundant species *Fabaeformirscandona levanderi*, *F. tricicatricosa* or *Candona candida* are grouped together to produce samples of >10 μg calcite. This mixing of species and instars is not problematic, however, as the studied European *Candonidae* share identical vital offsets for $\delta^{18}O$.

The error of $\delta^{18}O_L$ reconstruction from Ammersee deep lake ostracod $\delta^{18}O$, is about ±0.2‰. The error of the $\delta^{18}O_P$ calculation, introduced by assuming that evaporative enrichment and catchment effects were as today, could, in times of extreme hydrological conditions, range from −0.3‰ to +0.6‰.

The Fossil Ostracod Record

Calibration Against the Instrumental Air Temperature Record

Despite the efforts to understand and quantify the transfer of an atmospheric $\delta^{18}O_P$ signal into a sedimentary archive, two points of significance with respect to paleoclimatic studies could not be addressed due to the short period of the relevant field observation. The first concerns the relationship between mean annual $\delta^{18}O_P$ and mean annual air temperature based on regional and European-wide inter-annual variability of the of the last three decades (Rozanski et al. 1992). This comparison had to be extended to at least the duration of existing instrumental records, to maximize the range of observed temperature changes and to meet the standards for the calibration of other paleotemperature proxies. The second open point was the overall reactivity of the coupled drainage basin-lake system to changes in $\delta^{18}O_P$, where, especially, the average retention time of the slowest runoff (groundwater) could only be roughly estimated to ∼3 years from a tentative match between a four-year shift of $\delta^{18}O_L$ and long-term averages of $\delta^{18}O_P$.

Both concerns could be addressed by establishing a 300-year long ostracod-derived record of $\delta^{18}O_L$ from the uppermost 120 cm of sediment in 80 m water depth in Ammersee (von Grafenstein et al. 1996). The age-depth model of the core was based on clearly identifiable annual lamination (Alefs et al. 1996; Czymzik et al. 2013) and the time represented by a sediment sample 1 cm thick was on average 2.5 years. A first, direct comparison of the youngest 200 years of the $\delta^{18}O_L$ with the adequately averaged mean annual air temperature (MAAT) record from Hohenpeissenberg (within the catchment of the lake) gave a reasonable correlation ($r^2 = 0.91$), but an apparent $\delta^{18}O_L$/MAAT sensitivity, which, with 0.38‰/°C, was significantly below the $\delta^{18}O_P$/MAAT sensitivity derived from three decades of observation in the region and for the rest of Europe (0.58‰/°C). In addition, all $\delta^{18}O_L$ values younger than 1920, while still showing the same temperature-dependence, were systematically displaced by −0.15‰ with respect to the older part of the record. The smaller-than-expected amplitude of $\delta^{18}O_L$ can clearly be attributed to oversampling of the record compared to the reactivity of the entire retention system and, in a next step, can be used to better quantify the catchment residence time. The abrupt relative shift of $\delta^{18}O_L$ is an indication of a change of the $\delta^{18}O_P$-$\delta^{18}O_L$ difference or of the catchment effect in its wider sense, related to the regulation of the Ammer River, which started in 1920 and ended in 1922. This regulation was designed to accelerate the runoff of storm-related flood events (which occur primarily in summer) and to allow cultivation of the extended river plains. Thus, after the regulation, a larger portion of isotopically-enriched summer water was transferred directly into the lake, was mixed into the epilimnion, and became partially lost via the outlet, whereas before regulation, those flood waters could infiltrate into the river plain aquifers and more efficiently change the longer term $\delta^{18}O_I$ and, in consequence, the $\delta^{18}O_L$.

After correction for this 'summer bypass effect', the entire $\delta^{18}O_L$ record was compared to the $\delta^{18}O_L$ calculated using a very basic lake model, by assuming a linear relation between $\delta^{18}O_P$ and the measured air temperature and by taking into account changes in the seasonal distribution of precipitation and the most obvious features of lake water mixing. Figure 15.4 shows the results of this modelling approach, which gives the best fit between calculated and observed $\delta^{18}O_L$, if the mean drainage basin retention time is set to 3 years and if the temperature dependence of $\delta^{18}O_P$ is assumed to be 0.58‰/°C, thus confirming the assumption that the correlation based on direct observation during the last decades was valid for the last two centuries.

The Record of $\delta^{18}O_P$ in Central Europe Over the Past 15,000 Years

The longer term history of $\delta^{18}O_P$ reconstructed from Ammersee deep-lake cores has been discussed in two publications, the first of which (von Grafenstein et al. 1998) concentrates on the synchrony of a cold event around 8200 years BP in Europe and Greenland and its probable forcing by the collapse of the Hudson Bay ice dome. The second (von Grafenstein et al. 1999a, b) makes a comparison between $\delta^{18}O_P$ in Europe and in Greenland over the period between 15,000 and 5500 years B.P. The most striking feature is probably the great similarity in the records, even at high frequencies, providing evidence that the climate of both regions (and most likely of the entire North Atlantic perimeter) experienced the same decadal variations, governed by the variability of the heat flux from the North Atlantic Ocean. Independent confirmation of the details of the $\delta^{18}O$ from ice in Central Greenland in a European record significantly increases confidence in the quantitative interpretation of changes of the regional air temperatures and helps to exclude alternative explanations such as significant changes in the prevailing water vapor sources for precipitation in both regions.

Like the 8.2-ka-event, the short, abrupt, cold oscillations during the relatively warm periods of the late Glacial and early Holocene were probably forced by cataclysmic fresh-water discharges into the Atlantic Ocean as a consequence of the disintegration of the continental ice sheets. Despite the high correlation at high frequencies, the quantitative comparison of both $\delta^{18}O_P$ records reveals periods of systematic change in the differences between Europe and Central Greenland around the Younger Dryas cold period. These might indicate slow systematic changes in the surface conditions and circulation in the Greenland-Norwegian Sea, governed by the persisting meltwater flux from the Scandinavian ice sheet during the relatively warm periods. The proposed mechanism (von Grafenstein et al. 1998) is also an attractive hypothesis for 'Dansgaard-Oeschger-events', occurring frequently during the period from ∼80,000 to ∼25,000 yr B.P.

Figure 15.5 shows an overview of the state of $\delta^{18}O_P$ reconstruction from Ammersee deep-lake ostracods. The lack of resolution for the last 5500 years is evident, even if the available data already show some similarity with the corresponding part of the $\delta^{18}O_P$ records from Greenland. The highest resolved parts of the record provide probably the best constrained and resolved $\delta^{18}O_P$ record existing in Europe. However, there is still uncertainty ranging between +0.6‰ and −0.3‰ from potential hydrological changes that have not been considered. In addition, the record might not be representative of the entire continent, at least during the late Glacial, when stronger and highly dynamic longitudinal climate gradients were very likely to have prevailed. Both the errors from hydrological change and those from regional European climatic gradients may partially explain the apparent cross North Atlantic differences. Another problem of the existing record from Ammersee is the quality of the age model which is based on the few radiocarbon measurements of macro plant and insect remains found in the

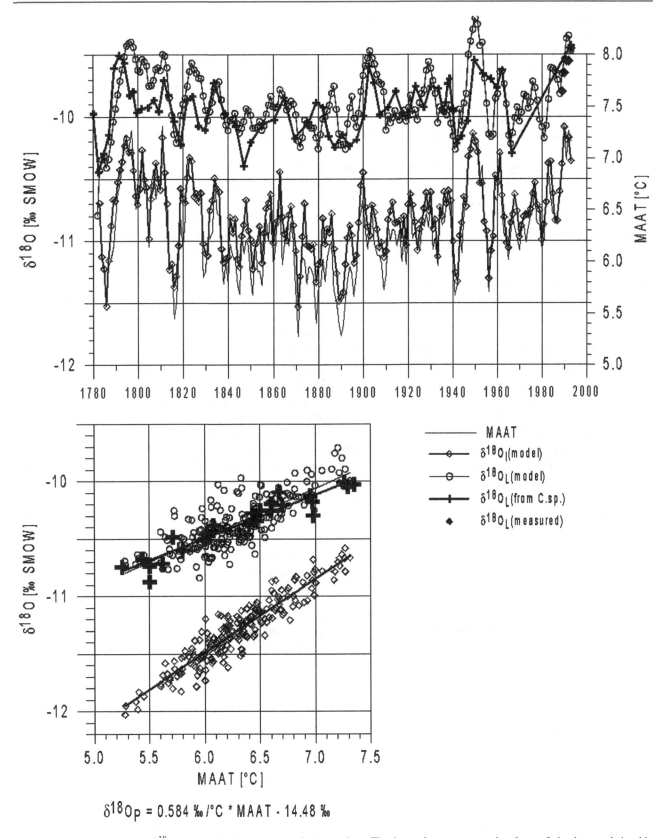

Fig. 15.4 Comparison of the $\delta^{18}O$ record of Lake Ammersee for the last two centuries derived from the $\delta^{18}O$ of juvenil Candona ($\delta^{18}O_L$(from C.sp), crosses) compared to the $\delta^{18}O_L$ as predicted by a two-box isotope hydrology model ($\delta^{18}O_L$ (model), open circles), if the lake was supplied with an inflow of river water with an isotopic composition ($\delta^{18}O_I$ (model), diamonds), controlled by the mean annual air temperature (MAAT) (using $\delta^{18}O_P = 0.58‰ * $ MAAT—14.48‰ as $\delta^{18}O_P$–MAAT relation). The upper graph shows the development over time. The lower box compares the slope of the input relationship (diamonds) with the slopes of the modelled (open circles) and the ostracod-derived $\delta^{18}O_L$-MAAT relation. Air-temperature variations dominate both the modelled and observed lake responses. The slopes of the relationship of both the modelled and observed lake response to air temperatures are decreased compared to the $\delta^{18}O_P$—temperature relation, due to the incomplete reaction of $\delta^{18}O_L$ to the fast climate changes (von Grafenstein et al. 1996)

sediment core. We therefore decided to take a new sediment core at a position close to that of core AS96-1, using the latest generation of the UWITEC coring system available in 2010 which now provides excellent material to improve the age model by combining radiocarbon dating and varve counting for large parts of the Holocene (Czymzik et al. 2013) and through radiocarbon dating and tephrochronology for the Late Glacial period. Sample treatment and the selection of monospecific shell material from the last 5500 years are ongoing. More highly-resolved and well-dated records will come from Mondsee (Austria) and Lac d'Annecy (France). For both records, all necessary monospecific carbonate samples have been prepared and are currently being analysed. A new sediment core from Lago d'Iseo (Italy) will provide a $\delta^{18}O_P$ record for the southern central Alps. However, this record will be discontinuous due to the lack of preserved carbonate material in parts of the sequence.

Late Glacial and Early Holocene Shallow-Water Temperatures

As shown above, the $\delta^{18}O$ of ostracods and molluscs living in shallow water carries combined information on the $\delta^{18}O_L$ and the water temperature at the moment of calcification. Some of the species and their instars have a seasonal preference for calcification. Therefore, it should be possible to calculate the water temperature of this time interval from $\delta^{18}O$ of fossil valves, if $\delta^{18}O_L$ for the given deposition time is known. Figure 15.6 shows, for example, the $\delta^{18}O$ records of *Pisidium* from three cores taken in Ammersee at 6, 7 and 11 m modern water depth, compared with the deep-lake $\delta^{18}O$ measured on juvenile *Candona*, all corrected for the respective vital offsets. The difference between littoral and profundal $\delta^{18}O$ values should give the mean water temperature for the shell growth period of *Pisidium*, i.e. summer (t_s) at the littoral sites according to $t_s = 4\ °C - (\delta^{18}O_{Cs} - {}^{18}O_{Pi})/(0.23‰/°C)$ (von Grafenstein et al. 1999a, b). For the sites at 6 and 7 m, summer water temperatures are rather similar and range between 6 and 12 °C for the period from 14.8 to 12.7 ky B.P. (Bølling/Allerød), i.e. on average 4 °C colder than today's 13 °C at this water depth in Ammersee. During the Younger Dryas they are 6 °C, i.e. −7 °C with respect to modern values. Summer water temperatures reach present-day values immediately after the end of the Younger Dryas. For the core from 11 m water depth, the early Holocene summer water temperatures are also close to the respective observed modern equivalent (9 ± 3 °C), in contrast to the 6 ± 2 °C during the Younger Dryas. Together, these results indicate a strong reduction in the stability of the thermal summer stratification of the water column due to considerably colder air temperatures during the Younger Dryas.

The summer water temperatures calculated from the differences between valves from littoral and profundal benthic organisms are an independent confirmation of the climatic significance of the changes in $\delta^{18}O_P$. They provide one of the rare temperature reconstructions for the late Glacial in Central Europe, which is derived from a physicochemical process, such as the temperature-dependent oxygen isotope fractionation between water and calcite. Theoretically,

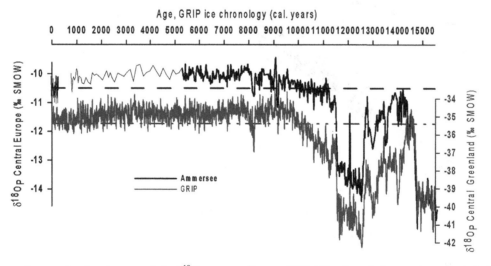

Fig. 15.5 Summary diagram showing the state of the $\delta^{18}O_P$ reconstruction from Ammersee deep-lake ostracods from AS Tmax (von Grafenstein 2002), AS 92-5 (von Grafenstein et al. 1998), and AS96-1 (von Grafenstein et al. 1998) cores. For sections sampled and measured with the optimal resolution (∼10 years/sample), $\delta^{18}O_P$ is traced by the thicker black line. The thin black line indicates the interval (5500 B. P. to ∼1700 AD) where further condensation is indicated. The high correlation with the $\delta^{18}O_P$ record from Central Greenland (grey line) that includes decadal events, provides strong evidence for a common control mechanism of the climate variability in both regions, probably via the North Atlantic thermohaline circulation

Fig. 15.6 Estimation of epilimnetic summer water temperatures in Ammersee by comparing the $\delta^{18}O$ of deep-lake ostracods (represented by $\delta^{18}O_L$, black line, data from (von Grafenstein et al. 1999a, b) with $\delta^{18}O$ of littoral *Pisidium* sp. in cores taken at water depths of 11 m (crosses), 7 m (filled circles), and 6 m (open circles) (von Grafenstein et al. 1994) The vital offsets and a systematic offset between epilimnetic summer $\delta^{18}O_L$ and hypolimnetic $\delta^{18}O_L$ (0.75‰) are taken into account

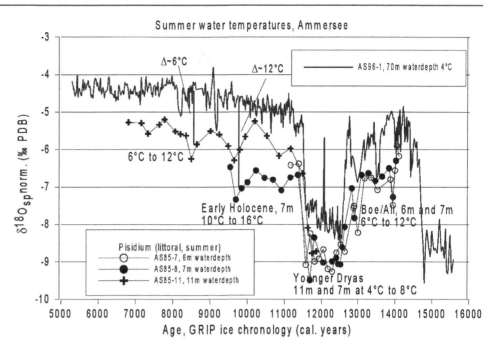

similar calculations could be made for all species and their instars present in shallow water cores and so provide water temperature information over the entire seasonal cycle and over a wide range of water depths. Because these temperature calculations are based on the differences alone, they are essentially independent of the respective $\delta^{18}O_L$ and therefore an attractive alternative among lake-based measurements, where the $\delta^{18}O_L$ itself is dominated by hydrological effects or in regions where the temperature-dependence of $\delta^{18}O_P$ is weak or absent (as for example in the north-American Great Lakes or in closed basins like Issyk-Kul).

Quantification of Hydrological Effects

Above, we have shown that the sensitivity of $\delta^{18}O_L$ to hydrological changes is dependent to a large extent on the ratio between the lake's surface and the drainage basin area (A_L/A_C), as this controls the relative importance of the evaporation (E/I). We also mentioned that the quantitative comparison of $\delta^{18}O_L$ records from neighbouring lakes with significantly differing A_L/A_C would allow detection of changes of the runoff (P-ET), the evaporation (E), and/or the relative humidity (h). These hydrological changes would lead, in contrast to a changing $\delta^{18}O_P$ and to deviations from the present day's $\delta^{18}O_L$ separation of the lakes.

Figure 15.7 shows an example of a comparison between the deep-lake records from Ammersee with a weak isotope-hydrological sensitivity (E/I = 0.04; $\Delta^{18}O_{L-P}$ = 0.75‰) and Starnberger See (E/I = 0.17; $\Delta^{18}O_{L-P}$ = 2.75‰). Similar to Ammersee, the core from Starnberger See comes from a water depth (88 m) sufficient to assume that calcification temperatures were very close to 4 °C. In addition, the species selected for isotope measurement are identical for both lakes, allowing direct comparison of their $\delta^{18}O$ values as representative of the $\delta^{18}O_L$ history of the lakes. The only restrictions for a direct comparison are the differences in the temporal resolution of the records and the much smaller reactivity of the Starnberger See to changes in $\delta^{18}O_P$. One sample from the Starnberger See record represents the average $\delta^{18}O_L$ over a period of several hundreds of years, compared to about ten years for an Ammersee sample. Similarly, the modern isotopic composition of Starnberger See water is integrated into the $\delta^{18}O_P$ history of at least the last 63 years, in contrast to about 8 years for the Ammersee. In order to get a meaningful 'modern' lake differential based on the same $\delta^{18}O_P$ reference, we calculated a longer term $\delta^{18}O_L$ average for the Ammersee since ~1930 from the ostracod record of the short gravity core TMAX and from the modelling of $\delta^{18}O_L$, which increase the reference lake offset to 2.5‰ compared to the apparent 2.0‰ offset in 1994. This is to a large extent due to the fact that Starnberger See water could not follow the rapid 0.5‰ increase of $\delta^{18}O_P$ between 1990 and 1994.

The 'modern' lake offset seems to be maintained or has slightly increased by up to 0.3‰ for the last 7000 years and for the Late Glacial (except for the Younger Dryas, which is not represented by ostracod valves in the core from Starnberger See). At the beginning of the Holocene, the difference was about 0.8‰ higher and approached the 'modern' offset between 9000 and 7000 years B.P. Lacking further evidence from another neighbouring lake or from δD_L records from both lakes, we can attribute the maximal increase of the difference to a reduction in the runoff (I = P − ET) of 50%,

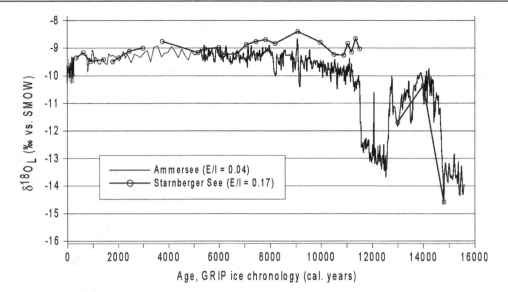

Fig. 15.7 Comparison of the deep-lake ostracod-derived $\delta^{18}O_L$ records of the Ammersee (black line) and the Starnberger See (open circles and line, corrected for the modern $\delta^{18}O_L$-difference between both lakes). The record from Starnberger See follows the low-frequency changes of $\delta^{18}O_L$ in Ammersee, with an offset very close to the modern difference for the last 7000 years and for the first warm period of the Late Glacial. The offset is increased by up to 0.8‰ during the Early Holocene (11,500 to 7000 years BP) indicating 'dryer than modern' conditions

an increase of the evaporation from the lakes' surfaces (E) of 100% or a reduction of the relative humidity from 0.74 to 0.5. Most probably, we need to distribute the overall effect to more moderate changes of all three parameters, which, in a natural world, are closely connected and often correlated. A reduction in I of 12%, together with an increase in E of 12%, and a relative humidity of 0.64, would, as would an infinitive number of other combinations, explain the increase in the lake offset of 0.8‰. However, all changes of these parameters had an effect not only on the difference of $\delta^{18}O_L$ in the lakes but also on the evaporative enrichment and $\delta^{18}O_L$–$\delta^{18}O_P$ of the Ammersee, which we considered as constant for the $\delta^{18}O_P$ reconstruction. The resulting correction of the $\delta^{18}O_P$ estimate for the Early Holocene would be −0.6‰ in the case of highly improbable isolated changes of either E or I, against only −0.25‰ for our example of combined h, I, and E changes. Besides important paleohydrological information, the lake-to-lake comparison also allows us to exclude errors in the $\delta^{18}O_P$ reconstruction arising from water balance changes for last 7000 years. For the early Holocene, the $\delta^{18}O_P$-record from Ammersee is probably up to 0.25‰ too high. A correction can be proposed using all available data points from the Starnberger See, which would then largely reduce the hydrological-induced uncertainties of the Ammersee $\delta^{18}O_P$-record. In future, a condensation of the Starnberger See record should be considered to increase confidence in the quantitative reconstruction of $\delta^{18}O_P$ from its neighbour.

Conclusions

Oxygen isotope records from deep-lake ostracods can provide high-resolution quantitative records of $\delta^{18}O_P$, if those lakes that are selected have (1) a small relative contribution from evaporation to the lake water balance, (2) a well-defined drainage basin without evidence of major re-organisation of its associated fluvial system, (3) a theoretical residence time between 2 and 10 years, (4) a water depth in excess of 50 m, (5) an annual complete overturn, and (6) proven preservation of in situ ostracod fauna among its profundal sediments. The temporal resolution of $\delta^{18}O_P$ from benthic deep-lake ostracods is limited to roughly ten years, the ε-folding time of lakes with a buffering volume large enough to suppress eventual bias from the seasonal variability of $\delta^{18}O$ of precipitation and river water. It was shown, using the example from Ammersee that a maximal theoretical resolution can be practically obtained. Detailed hydrological, limnological, and isotopic investigations can provide the basis for a realistic quality control of the $\delta^{18}O_P$ reconstruction. For the $\delta^{18}O_P$ record of the last 15,500 years, the maximal error from water temperature effects is ±0.2‰, to which must be added uncertainty from possible hydrological changes of −0.3‰ to +0.6‰. Correction of the record using the quantification of the hydrological effects through comparison with the more strongly affected Starnberger See further reduces the overall error to a maximum of ±0.4‰ for the driest period in the Holocene.

$\delta^{18}O_P$ records provide quantitative hydro-meteorological information, which is largely independent of the biological response to climate change and thus can be used as a reference in multidisciplinary studies. For Central and Northwest Europe, those $\delta^{18}O_P$ records can be considered as a reasonable proxy for the mean air temperature. A calibration of the $\delta^{18}O$ record from Ammersee for the 200-year long period of instrumental climate observation confirmed the $\delta^{18}O_P$/temperature relationship derived from precipitation measurements (0.58‰/°C) for the last three decades in Europe. Independent qualitative evidence for colder air temperatures during the Younger Dryas could be extracted from water- and isotope-based shallow water temperature reconstruction, which further confirms the temperature-dependence of $\delta^{18}O_P$ in Europe. However, $\delta^{18}O_P$ records provide excellent quantitative paleoclimatic information even for regions where such a temperature-dependence is absent or weak. High quality $\delta^{18}O_P$ records are also very useful to increase the significance of $\delta^{18}O$ records from archives, which are more affected or dominated by secondary isotopic effects. Deep-lake ostracod valves from carefully selected lakes are, at the moment, the best material to obtain quantitative $\delta^{18}O_P$ records for significant time periods from temperate, non-glaciated areas. As such, they merit more consideration as a climate proxy in future research strategies.

Beside $\delta^{18}O_P$-reconstruction, deep-lake ostracod-based oxygen isotope records have been successfully used to unscramble the complex hydrological changes in the Great Lakes Basin, and, in combination with geomorphological lake level reconstruction, to provide information on the water balance development and humidity in hydrologically closed basins. In all cases, deep-lake ostracods provide very precise estimates for the isotopic composition of former lake water, which can be used as a reference to quantify the secondary effects on $\delta^{18}O$ of all other authigenic material produced in the same lake, such as water temperature-dependent fractionation in the case of littoral and sublittoral ostracods.

Perspectives

In future, archived and new core material from Ammersee, Mondsee, Lac d'Annecy, and Lago d'Iseo will be used to close the existing gaps and to provide optimally resolved $\delta^{18}O_P$ records for the entire period since the last glacial maximum. Besides their direct application as proxies for mean annual air temperatures, records of decadal $\delta^{18}O_P$ variations in Europe offer many possibilities for further hydrogeological and paleoclimatological studies. They could serve as 'marker input function' into groundwater and thus allow a better exploitation of stable isotope measurements of fossil groundwater for the last 15,500 years, which, for the moment, is considered as a simple, one-step function from cold, isotopically-depleted to present-day values. They will also facilitate quantification and climatic interpretation of records from less direct continental isotopic archives such as tree rings, shallow-water lake records, and stalagmites. Although lakes with the potential to provide a quantitative $\delta^{18}O_P$ record are characterized by a narrow range of hydrological and geomorphological settings, there are at least ten suitable candidates identified in Europe, for which a limited dataset of water isotopes exists. Those lakes could in future be exploited to provide a network of reasonably well distributed and highly resolved $\delta^{18}O_P$ records. It is likely that at least a similar number of such lakes can be found in the temperate regions of North and South America and within other formerly glaciated mountain ranges around the globe.

References

Alefs, J., Müller, J., & Wunsam, S. (1996). Die Rekonstruktion der epilimnischen Phosphorkonzentrationen im Ammersee seit 1958. *Gwf Wasser Abwasser, 137*(8), 443–448.

Andersen, N., Lauterbach, S., Erlenkeuser, H., Danielopol, D.L., Namiotko, T., Hüls, M., et al., (2017). Evidence for higher-than-average air temperatures after the 8.2 ka event provided by a Central European $\delta^{18}O$ record. *Quaternary Science Reviews, 172*, OP- 96.

Bornemann, A., Pirkenseer, C. M., De Deckker, P., & Speijer, R. P. (2012). Oxygen and carbon isotope fractionation of marine ostracod calcite from the eastern Mediterranean Sea. *Chemical Geology, 310–311*, 114–125.

Chivas, A. R., de Deckker, P., Cali, J. A., Chapman, A., Kiss, E., & Shelley, J. M. G. (1993). Coupled stable-isotope and trace-element measurements of lacustrine carbonates as paleoclimatic indicators. In P. K. Swart, K. C. Lohmann, J. Mckenzie & S. Savin (Eds.), *Climate change in continental isotopic records*. School of Marine and Atmospheric Science.

Chivas, A., De Deckker, P., Wang, S. X., & Cali, J.A. (2002) Oxygen-isotope systematics of the nektic ostracod *Australocypris robusta*. In J. A. Holmes, A. R. Chivas (Eds.), *Ostracoda, applications in quaternary research*. Geophysical Monograph Series (Vol. 131, pp. 301–313). American Geophysical Union (AGU).

Colman, S. M., Keigwin, L. D., & Forester, R. M. (1994). Two episodes of meltwater influx from Lake Agassiz into the Lake Michigan basin and their climatic contrasts. *Geology, 22*, 547–550.

Czymzik, M., Brauer, A., Dulski, P., Plessen, B., Naumann, R., von Grafenstein, U., et al. (2013). Orbital and solar forcing of shifts in Mid- to Late Holocene flood intensity from varved sediments of pre-alpine Lake Ammersee (southern Germany). *Quaternary Science Reviews, 61*, 96–110.

Danielopol, D. L., Ito, E., Wansard, G., Kamiya, T., Cronin, T.M., & Baltanas, A. (2002) Techniques for collection and study of ostracoda. In J. A. Holmes & A. R. Chivas (Eds.), *Ostracoda, Applications in quaternary research*, Geophysical Monograph Series (Vol. 131, pp. 65–97). American Geophysical Union (AGU).

Danis, P. A., von Grafenstein, U., & Masson-Delmotte, V. (2003) Sensitivity of deep lake temperature to past and future climatic changes: A modelling study for Lac d'Annecy, France, and Ammersee, Germany. *Journal of Geophysical Research-Atmospheres, 10*(D19).

Danis, P. A., von Grafenstein, U., Masson-Delmotte, V., Planton, S., Gerdeaux, D., & Moisselin, J. M. (2004) Vulnerability of two European lakes in response to future climatic changes. *Geophysical Research Letters*, 31(21).

Dansgaard, W. (1964). Stable isotopes in precipitation. *Tellus*, 16, 436–468.

De Deckker, P. (2002). Ostracod palaeoecology. In J. A. Holmes & A. R. Chivas (Eds.), *Ostracoda, applications in quaternary research*. Geophysical Monograph Series (Vol. 131, pp. 121–134). American Geophysical Union (AGU).

Decrouy, L., Vennemann, T. W., & Ariztegui, D. (2011) Controls on ostracod valve geochemistry: Part 2. Carbon and oxygen isotope compositions. *Geochimica et Cosmochimica Acta*, 75(22), 7380–7399.

Dettman, D. L., Smith, A. J., Rea, D. K., Moore, T. C., & Lohmann, K. C. (1995). Glacial meltwater in Lake Huron during Early Postglacial time as inferred from single-valve analysis of oxygen isotopes in ostracodes. *Quaternary Research*, 43(3), 297–310.

Devriendt, L. S., McGregor, H. V., & Chivas, A. R. (2017). Ostracod calcite records the 18O/16O ratio of the bicarbonate and carbonate ions in water. *Geochimica et Cosmochimica Acta*, 214, 30–50.

Didié, C., & Bauch, H. A. (2002). *Implications of upper Quaternary stable isotope records of marine ostracods and benthic foraminifers for paleoecological and paleoceanographical Investigations* (pp. 279–299). American Geophysical Union (AGU).

Forester, R. M., Colman, S. M., Reynolds, R. L., & Keigwin, L. D. (1994). Lake Michigan's late Quaternary limnological and climate history from ostracod, oxygen isotope, and magnetic susceptibility. In: D. W. Folger, S. M. Colman, & P. W. Barnes (Eds.), *South. Lake Michigan Coast. Eros. Study* (pp. 93–107). Woods Hole, MA, USA: U. S. Geological Surve.

Fritz, P., Anderson, T. W., & Lewis, C. F. M. (1975). Late Quaternary climatic trends and history of Lake Erie from stable isotope studies. *Science*, 160, 267–269.

Gat, J. R., Lister, G. S., & Frenzel, B. (1995). The 'catchment effect' on the isotopic composition of lake waters; its importance in paleolimnological interpretations. In B. Frenzel, B. Stauffer, & M. M. Weiss (Eds.), *Problems of stable isotopes in tree-rings, lake sediments and peatbogs as climatic evidence for the Holocene*. Issue ESF Proj. 'European Palaeoclim. Man' 10 (pp. 1–16). Stuttgart/New York: Gustav Fischer Verlag.

Gibson, J. J., Edwards, T. W. D., Bursey, G. G., & Prowse, T. D. (1993). Estimating evaporation using stable isotopes: Quantitative results and sensitivity analysis for two catchments in Northern Canada. *Nordic Hydrology*, 24, 79–94.

Hammarlund, D., Edwards, T. W. D., Bjorck, S., Buchardt, B., & Wohlfarth, B. (1999). Climate and environment during the Younger Dryas (GS-1) as reflected by composite stable isotope records of lacustrine carbonates at Torreberga, Southern Sweden. *Journal of Quaternary Science*, 14(1), 17–28.

Hoffmann, G., Werner, M., & Heimann, M. (1998). The water Isotope module of the ECHAM atmospheric general circulation model—A study on time scales from days to several years. *Journal of Geophysical Research*, 103(16), 816–871, 896.

Holmes, J. A., & Chivas, A. R. (2002). Ostracod shell chemistry—overview. in: J.A. Holmes, A.R. Chivas (Eds.), *Ostracoda, applications in Quaternary research*, Geophysical Monograph Series (Vol. 131, pp. 1–4). American Geophysical Union (AGU).

Horne, D. J., Cohen, A., & Martens, K. (2002) Taxonomy, morphology and biology of Quaternary and living Ostracoda. In: J. A. Holmes & A. R. Chivas (Eds.), *Ostracoda, applications in Quaternary research*. Geophysical monograph series (Vol. 131, pp. 5–36). American Geophysical Union (AGU).

Jouzel, J., Russell, G. L., Suozzo, R. J., Koster, R. D., White, J. W. C., & Broecker, W. S. (1987). Simulations of the HDO and $H_2^{18}O$ atmospheric cycles using the NASA GISS general circulation model: The seasonal cycle for present-day conditions. *Journal of Geophysical Research*, 92, 14739–14760.

Keatings, K. W., Heaton, T. H. E., & Holmes, J. A. (2002). Carbon and oxygen isotope fractionation in non-marine ostracods: Results from a 'natural culture' environment. *Geochimica et Cosmochimica Acta*, 66(10), 1701–1711.

Lauterbach, S., Brauer, A., Andersen, N., Danielopol, D. L., Dulski, P., Hüls, M., et al. (2011). Environmental responses to Lateglacial climatic fluctuations recorded in the sediments of pre-Alpine Lake Mondsee (northeastern Alps). *Journal of Quaternary Science*, 26(3), 253–267.

Lewis, C. F. M., & Anderson, T. W. (1992). Stable isotope (O and C) and pollen trends in eastern Lake Erie, evidence for a locally induced climatic reversal of Younger Dryas age in the Great Lakes basin. *Climate Dynamics*, 6, 241–250.

Li, X., & Liu, W. (2010). Oxygen isotope fractionation in the ostracod *Eucypris mareotica*: Results from a culture experiment and implications for paleoclimate reconstruction. *Journal of Paleolimnology*, 43(1), 111–120.

Lister, G. S. (1988). A 15,000-year isotopic record from Lake Zürich of deglaciation and climatic change in Switzerland. *Quaternary Research*, 29(2), 129–141.

Lister, G. S., Kelts, K., Zao, C. K., Yu, J. Q., & Niessen, F. (1991). Lake Qinghai, China; closed-basin lake levels and the oxygen isotope record for Ostracoda since the latest Pleistocene. In A. R. Chivas & P. De Deckker (Eds.), *SLEADS conference*, Arkaroola, South Aust. (pp. 141–162), Aug. 8–16. 62 Refs.

Marco-Barba, J., Ito, E., Carbonell, E., & Mesquita-Joanes, F. (2012). Empirical calibration of shell chemistry of *Cyprideis torosa* (Jones, 1850) (Crustacea: Ostracoda). *Geochimica et Cosmochimica Acta*, 93, 143–163.

Niessen, F., & Kelts, K. (1989). The deglaciation and Holocene sedimentary evolution of southern perialpine Lake Lugano—Implications for Alpine paleoclimate. *Eclogae Geologicae Helveticae*, 82(1), 235–263.

Rea, D. K., Moore, T. C., Jr., Anderson, T. W., Lewis, C. F. M., Dobson, D. M., Dettman, D. L., et al. (1994). Great Lakes paleohydrology: Complex interplay of glacial meltwater, lake levels, and sill depths. *Geology*, 22(12), 1059.

Rozanski, K., Araguás-Araguás, L., & Gonfiantini, R. (1992). Relation between long-term trends of oxygen-18 isotope composition of precipitation and climate. *Science*, 258, 981–985.

Schwalb, A. (2003). Lacustrine ostracods as stable isotope recorders of late-glacial and Holocene environmental dynamics and climate. *Journal of Paleolimnology*, 29(3), 267–351.

Schwalb, A., Lister, G. S., & Kelts, K. (1994). Ostracode carbonate $d^{18}O$- and $d^{13}C$-signatures of hydrological and climatic changes affecting Lake Neuchâtel, Switzerland, since the latest Pleistocene. *Journal of Paleolimnology*, 11, 3–17.

Thompson, L. G., Yao, T., Davis, M. E., Henderson, K. A., Mosley-Thompson, E., Lin, P. N., et al. (1997). Tropical climate instability: The Last Glacial Cycle from a Qinghai-Tibetan Ice Core. *Science*, 276, 1821–1825.

Van der Meeren, T., Ito, E., Verschuren, D., Almendinger, J. E., & Martens, K. (2011). Valve chemistry of *Limnocythere inopinata* (Ostracoda) in a cold arid environment—Implications for paleolimnological interpretation. *Palaeogeography, Palaeoclimatology, Palaeoecology*, 306(3–4), 116–126.

von Grafenstein, U. (2002). Oxygen-isotope studies of ostracods from deep lakes. In J. A. Holmes & A. R. Chivas (Eds.), *Ostracoda, applications in quaternary research*. Geophysical Monograph Series (Vol. 131, pp. 249–265). American Geophysical Union (AGU).

von Grafenstein, U., Belmecheri, S., Eicher, U., van Raden, U.J., Erlenkeuser, H., Andersen, N., et al. (2013). The oxygen and carbon isotopic signatures of biogenic carbonates in Gerzensee, Switzerland, during the rapid warming around 14,685 years BP and the following interstadial. *Palaeogeography, Palaeoclimatology, Palaeoecology, 391*(Part B OP-In Early Rapid Warning, Palaeogeography, Palaeoclimatology, Palaeoecology 1 December 2013 391 Part B: 25–32), 25.

von Grafenstein, U., Eicher, U., Erlenkeuser, H., Ruch, P., Schwander, J., & Ammann, B. (2000). Isotope signature of the Younger Dryas and two minor oscillations at Gerzensee (Switzerland): Palaeoclimatic and palaeolimnologic interpretation based on bulk and biogenic carbonates. *Palaeogeography, Palaeoclimatology, Palaeoecology, 159*(3–4), 215–229.

von Grafenstein, U., Erlenkeuser, H., Kleinmann, A., Müller, J., & Trimborn, P. (1994). High-frequency climatic oscillations as revealed by oxygen-isotope records of benthic organisms (Ammersee, Southern Germany). *Journal of Paleolimnology, 11*, 349–357.

von Grafenstein, U., Erlenkeuser, H., Müller, J., Jouzel, J., & Johnsen, S. (1998). The cold event 8200 years ago documented in oxygen isotope records of precipitation in Europe and Greenland. *Climate Dynamics, 14*(2), 73–81.

von Grafenstein, U., Erlenkeuser, H., Müller, J., & Kleinmann-Eisenmann, A. (1992). Oxygen isotope records of benthic ostracods in bavarian lake sediments—Reconstruction of late and postglacial air temperatures. *Naturwissenschaften, 79*(4), 145–152.

von Grafenstein, U., Erlenkeuser, H., Müller, J., Trimborn, P., & Alefs, J. (1996). A 200 year mid-European air temperature record preserved in lake sediments: An extension of the $d^{18}O_p$—Air temperature relation into the past. *Geochimica et Cosmochimica Acta, 60*(21), 4025–4036.

von Grafenstein, U., Erlernkeuser, H., & Trimborn, P. (1999a). Oxygen and carbon isotopes in modem freshwater ostracod valves: Assessing vital offsets and autecological effects of interest for palaeoclimate studies. *Palaeogeography, Palaeoclimatology, Palaeoecology, 148*(1–3), 133–152.

von Grafenstein, U., Erlenkeuser, H., Brauer, A., Jouzel, J., & Johnsen, S. (1999b). A mid-European decadal isotope-climate record from 15,500 to 5,000 years B.P. *Science, 284*, 1654–1657.

Xia, J., Ito, E., & Engstrom, D. R. (1997). Geochemistry of ostracod calcite: 1. An experimental determination of oxygen isotope fractionation. *Geochimica et Cosmochimica Acta, 61*, 377–382.

Vegetation-Atmosphere Interface: Tree Rings

Joël Guiot and Valérie Daux

A Dendrochonological Approach

In many parts of the world, there is a strong seasonality in the annual distribution of temperatures and rainfall. This seasonality is reflected in the growth of trees, which is the result of the interaction of the tree with its environment, via its leaves (for carbon and water exchanges) and its roots (for nutrients and water). In temperate latitudes, during the winter, the tree is dormant and woody cells are not produced. In spring, when the thermal conditions are met, the tree becomes active and starts to produce large, dispersed woody cells (early wood). Towards late spring and early summer, the cells produced are denser and smaller (late wood), then, at the end of the summer, cells are no longer produced along the trunk and the tree starts to store reserves for the following year. If a trunk is cut through, alternating light and dark bands can be observed, which, combined, constitute an annual growth ring. Comparison between rings shows a high level of variability. This variability is the direct consequence of the climate conditions (temperature, rainfall, sunlight) which prevailed during or before the formation of the cells (Fritts 1976).

These seasonal growths are produced by tissues in the cambium, a layer of cells between the wood and the bark which causes expansion in the diameter of the roots, trunk and branches of the tree. Besides climate, the thickness of the ring also depends on many other parameters including the species, the age of the ring, the availability of nutrients in the soil, the tree's ability to retain water, its exposure etc.

J. Guiot (✉)
European Centre for Research and Teaching in Environmental Geosciences CEREGE, Aix-Marseille University, CNRS, IRD, INRAE, Collège de France, BP 80, 13545 Aix-en-Provence Cedex 04, France
e-mail: guiot@cerege.fr

V. Daux
Laboratoire des Sciences du Climat et de l'Environnement, LSCE/IPSL, CEA-CNRS-UVSQ, Université Paris-Saclay, 91190 Gif-sur-Yvette, France

© Springer Nature Switzerland AG 2021
G. Ramstein et al. (eds.), *Paleoclimatology*, Frontiers in Earth Sciences,
https://doi.org/10.1007/978-3-030-24982-3_16

Douglas (1920) was the first to recognize the potential of series of annual growth rings to provide information about past climates, and he established the fundamentals of what is now known as dendroclimatology. The main difficulty is to distinguish the impact of climate from other factors. The greater the climate stress the tree is subjected to, the easier this is to decode. Generally, there are two types of climate stress: heat stress and water stress. In arid or semi-arid regions, tree growth is limited by the availability of water. Therefore, this will be the main parameter recorded by the tree. Trees in the far north or at high altitudes are constrained by temperature. This makes them very good thermometers. The climate conditions in the months prior to the growing season also affect the ring. These may be the replenishment of groundwater reserves, or carbon reserves ready to be activated for the following year. This further complicates the decoding of climate information.

The dendroclimatic approach (Cook and Kairiukstis 1990; Trouet 2020) involves taking a number of cores from a given forest, at a rate of between two and four cores per tree from ten to twenty trees per stand. A core is a small tube taken from the bark to the middle of the tree, from which the sequence of rings can be read. The trees most likely to provide the best information on the climate parameter to be reconstructed are chosen. For example, when reconstructing a water parameter, trees growing in shallow soils which are unable to store much water will be selected. Each core is then dated by counting the rings from the bark to the heart of the tree. Each ring is supposed to be annual, but sometimes growth stops during the season because of a temporary drought and resumes if it starts to rain. This resulting growth arrest produces what is called a false ring. For this year, there are two rings. In other years, conditions are so unfavorable that the ring appears to be missing. These two types of phenomena will produce errors in the dating of rings. Cross dating, that is, the comparison of series between cores, helps eliminate these errors.

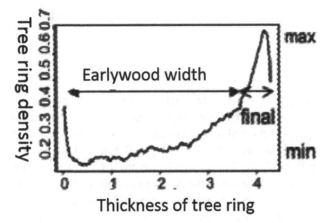

Fig. 16.1 Typical profile of a conifer ring (above) with the corresponding density curve (below) and the main parameters that can be deduced

Each ring is measured and a set of between twenty and forty time series is produced for each stand. As well as the thickness, the density may also be measured using a micro-densitometer (see typical profile in Fig. 16.1). One can thus obtain the thicknesses of the early wood and the summer wood as well as the average density of the early wood and the maximum density. These are the most frequently used parameters (Schweingruber et al. 1978).

The next step is to try to reconstruct the climate signal. Tree-ring series are often affected by trends associated with the age of the tree. Statistical methods are used to eliminate these. Various methods have been developed: polynomial curves, digital filters, exponential decay are removed from the raw data by division. This process is called indexing. The main problem with this is that often a part of the trend linked to climate is also removed. Figure 16.2 shows how one can get an indexed series from an untreated series of rings. The various indexed series from the same stand are averaged to produce a representative series for the stand, called the master timeline, in order to remove intra-site variability. This is done in an attempt to maximize the climate signal. Other indexing methods are increasingly used to try to best preserve climate variations over the long-term (Cook and Kairiukstis 1990).

A single average series (also called master series) of a stand is rarely able to provide a reliable climate reconstruction because the interactions between climate and growth are complex.

A network of master timelines for a given region needs to be established, including different species of trees, in order to better isolate the relevant climate variable. This is then followed by a statistical approach called transfer function (Cook and Kairiukstis 1990). On one side, a matrix of dendrochronological series, necessarily of variable length, is established, and on the other side, weather series for the same region are assembled. Over the time period common to both the meteorological and tree-ring series, a statistical relationship by regression methods can be calibrated.

The following example illustrates how it is possible to reconstruct summer temperatures in Europe from tree-ring series, supplemented by other proxies (Guiot et al. 2010). Among these proxies, we used series of harvest dates from several wine regions in France and Switzerland and the isotopic series for oxygen-18 in Greenland, which is considered to be related to the global climate. All these series are of variable length and resolution. They also represent different characteristics of the climate, but the climate parameter that best explains tree growth and the precociousness of the grape harvest is the temperature between April and September. Therefore, it is this summer temperature that was estimated using a statistical technique of similarities, called the analog method (Guiot et al. 2010).

The graph (Fig. 16.3) shows that the Little Ice Age (the defined span of which varies, but, here, is fixed at 1400–1900, and is marked by advancing glaciers in the Alps) was on average 0.35 °C colder than the 1961–1990 reference period, with a maximum cooling around 1600 (around −0.5 °C). The 1940–2007 period was 0.2 °C warmer than the reference period, and the last decade of the twentieth century topped this with a difference of 0.5 °C. Even taking into account that these averages are calculated over varying periods, the warming of the late twentieth century is significant. This is confirmed by the polynomial curve indicating a warming since the early nineteenth century of 0.6 °C for the century.

Dendro-isotopic Analysis

Tree rings are made up of organic material containing mainly carbon, oxygen and hydrogen. Each one of these elements has several isotopes. The isotopes of an element have chemical properties that are qualitatively identical. However, physical, chemical and biological processes can bring about a fractionation between light isotopes and heavy isotopes of the same element during physical or chemical reactions in which this element is involved. As a result, the isotopic ratios of oxygen ($\delta^{18}O$), carbon ($\delta^{13}C$) and hydrogen (δD) are a source of environmental information.

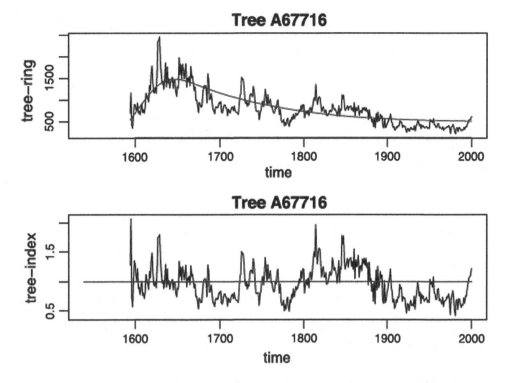

Fig. 16.2 Example of indexing of a tree-ring series: the top graph shows the raw data for the ring thicknesses with the trend line superimposed; the lower curve shows the indexed series, that is to say the raw series divided by the trend

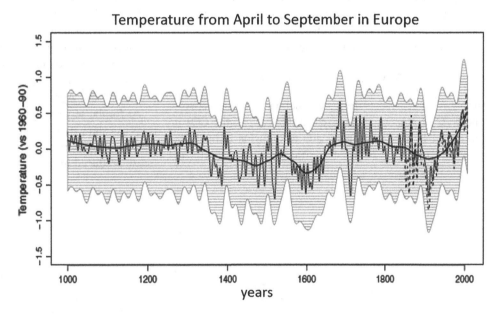

Fig. 16.3 Evolution of the summer temperatures reconstructed for Europe [10° O 60° E-and-30° N 85° N] taken from multiple proxies and smoothed. Shaded: the associated uncertainty (at 95%); dashed line: the corresponding observations (1850–2007); and, bold line: the trend line of the reconstruction (Guiot et al. 2010)

The isotopic composition of oxygen in precipitations is a function of temperature at high and middle latitudes. Water from precipitation infiltrates the soil. In the superficial levels, evaporation can result in oxygen enrichment of the infiltrated water compared to the original precipitation. Trees take most of their water from the superficial soil levels. Therefore, the water they draw can be enriched to some extent in heavy isotopes compared to the precipitation. Nevertheless, there is little or no enrichment in wooded areas where the soils are covered by thick litter that limits evaporation. The isotopic composition of the oxygen in the water of the xylem, through which water and mineral nutrients are transported, is the same as that of the soil water absorbed by the root system. Significant fractionation occurs before the water isotopes are fixed in the wood structure (for a summary of the isotopic composition of wood constituents see Gessler et al. 2014). The critical phase occurs in the leaves where evapotranspiration results in a loss of light isotopes. The extent to which the leaf is enriched with heavy isotopes depends on the difference between the isotopic compositions of the water in the xylem and in the ambient vapor, as well as on the difference between the vapor pressure inside and outside the leaf. As a result of the evaporative fractionations, the isotopic composition of the glucose produced in the leaf is enriched by 27‰ compared to the water in the leaf. During synthesis of cellulose from glucose, about 40% of the oxygen atoms are exchanged with the xylem water. Consequently, the isotopic composition of the cellulose in tree rings reflects that of the source (soil water is more or less equivalent isotopically to precipitations) and the degree of enrichment by evaporation in the leaf. The link between the isotopic compositions of cellulose and rainfall is complex. Nevertheless, a growing number of studies have reported statistically significant correlations between the $\delta^{18}O$ of the cellulose in tree rings (oak, pine, larch, cedar) and that of the rainfall during the growing season, as well as certain other climate parameters (atmospheric temperature, relative humidity, water stress). Figure 16.4 shows examples of these correlations.

The $\delta^{13}C$ of atmospheric carbon is close to −8‰ (relative to the standard Pee Dee Belemnite). That of the leaves and the wood in the trees is in the region of −20‰ to −30‰. The isotopic fractionations which create the differences between the $\delta^{13}C$ of CO_2 in the air and the $\delta^{13}C$ of CO_2 in the plant, occur primarily in the leaf. Farquhar et al. (1982) has expressed this in the following equation:

$$\Delta^{13}C(‰) = a + (b - a)p_i/p_a,$$

where $\Delta^{13}C$ represents the discrimination of carbon between the $\delta^{13}C$ of glucose synthesized in the leaf and the CO_2 in the air, a is the fractionation due to diffusion through the stomata (4.4‰), b is the fractionation caused by carboxylation (27‰), p_i and p_a are the partial pressures of CO_2 in the substomatal cavity and the atmosphere respectively. The partial pressure of CO_2 within the substomatal cavity is conditioned by the opening of the stomata (resulting from a compromise between water loss and uptake of CO_2 from the ambient air). Additional isotopic fractionations, accentuating the depletion in ^{13}C, occur during the synthesis of cellulose and lignin. Plant-air exchanges are determined by the tree's environment (climate, water status of the soil). The outcome is that the $\delta^{13}C$ of the cellulose in tree rings is dependent on atmospheric temperature (Fig. 16.5).

The isotopic composition of hydrogen (δD) is linked to the atmospheric temperature, as is that of oxygen; it is not modified in the transfer from soil to tree, but it is affected by evapotranspiration, which causes isotopic enrichment of the water in the leaf and is subjected to fractionation during the process of photosynthesis. During photosynthesis, enzyme activity causes kinetic fractionations which may differ depending on the position of hydrogen in the glucose molecule (Augusti et al. 2006). During the transformation of glucose to cellulose in the trunk, the catalytic action of the enzymes generates isotopic exchanges between the sugars and water in the xylem which involve about 40% of the hydrogen atoms in the sugars (Waterhouse et al. 2002). As it is difficult to distinguish between the climate and physiological influences on the abundance of deuterium, the use of isotopic ratios of hydrogen in cellulose is complicated (Pendall 2000; Augusti et al. 2008).

As for the series of ring widths, the $\delta^{13}C$ cellulose of tree rings can show trends linked to age. This so-called 'juvenile' effect is characterized by low values of $\delta^{13}C$ in wood cellulose produced in the first decades of the tree (Francey and

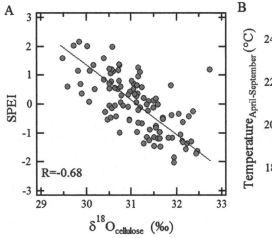
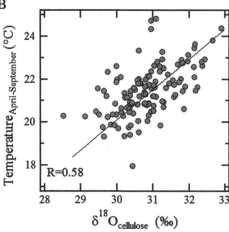
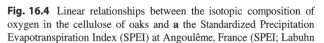

Fig. 16.4 Linear relationships between the isotopic composition of oxygen in the cellulose of oaks and **a** the Standardized Precipitation Evapotranspiration Index (SPEI) at Angoulême, France (SPEI; Labuhn et al. 2016), **b** the average maximum temperatures from April to September at Fontainebleau, France

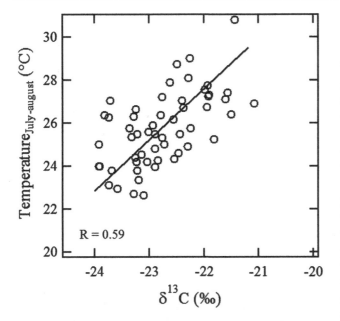

Fig. 16.5 Correlation between the isotopic composition of the carbon in the cellulose of larches in the Névache region (Alps, France) and the average maximum temperature in July–August (Daux et al. 2011)

Isotopic Reconstruction of the Variations in Climate Parameters Over Time

The variations in the isotopic compositions of the cellulose in rings are interpreted in different ways. Indeed, the reconstructed climate parameters are, depending on the case: temperature, relative humidity, sunshine, the amount of summer or winter rainfall, the average isotopic composition of this rainfall. The isotopic fractionations during the manufacture of cellulose are determined, as we have seen, by several factors. The conditions under which the growth takes place will determine which factor is dominant. For example, for trees in areas experiencing drought, stomatal conductance tends to dominate the carbon isotope fractionation processes. In this case, the environmental control factors are the relative humidity of the air and soil moisture; when, on the contrary, trees are growing in conditions where water is not limited, the fractionations are instead conditioned by the rate of photosynthesis which itself depends on the incident radiation and, to a lesser extent, by the temperature. The disparity in the reconstructions mentioned earlier, is a reflection of environmental disparities. The possibility of changes in the dominant factor over time must be contemplated when reconstructing temporal variations in the climate parameters.

Farquhar 1982). In most cases, the isotopic compositions of the oxygen and hydrogen in the cellulose are not affected by the juvenile effect, which makes these parameters particular interesting for the reconstruction of low-frequency climate variations.

Two reconstructions of atmospheric temperature are presented below, one from the $\delta^{13}C$ of the cellulose in *Fitzroya cupressoides* (Patagonia cypress, Argentina) cov-

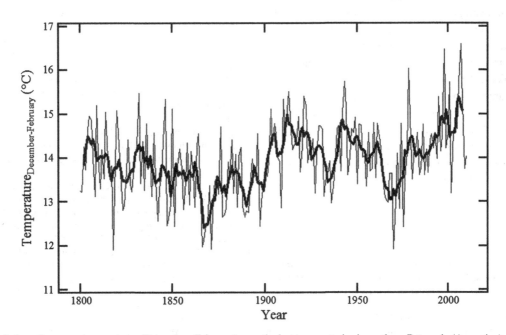

Fig. 16.6 Evolution of summer temperatures (December–February) over the last two centuries in northern Patagonia (Argentina), calculated from the isotopic composition of oxygen in the cellulose of Patagonian cypresses

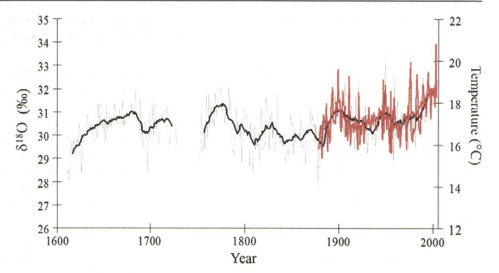

Fig. 16.7 Values for the $\delta^{13}C$ of cellulose taken from the summer wood of forest oaks (*Quercus robur*) and from buildings in Rennes (France) over the period 1610–1996 AD. The values of the reconstructed temperatures are indicated on the scale on the left. Instrumental temperatures, measured by Météo-France, from 1890 to 2003, are shown for comparison (Masson-Delmotte et al. 2005)

ering the period 1800–2011 (Lavergne et al. 2017) (Fig. 16.6) and the other from the $\delta^{18}O$ of the cellulose of *Quercus* (forest oaks and beams from old buildings in the Rennes region covering the last 400 years (Raffali-Delerce et al. 2004; Masson-Delmotte et al. 2005) (Fig. 16.7). The relationship between isotope data and temperature was established through calibrations carried out over the 1931–2011 period for the cypress from Patagonia, and over the 1951–1996 period for the oaks in Brittany. For Lavergne et al. (2018), the evolution of $\delta^{13}C$ in the cellulose of the cypresses implies that the summers (December to February) of the nineteenth century, particularly in the second half, were cool (an average temperature of 13.4 °C) and the summers of the twentieth and twenty-first centuries have higher temperatures (average temperature of 14.2 °C).

The reconstruction produced from the $\delta^{18}O$ of the cellulose of the oaks in Brittany also shows an increase in summer temperatures from the end of the nineteenth century onwards, in line with thermometer data. The similarity in the general shape of the graphs of the oak $\delta^{18}O$ and the instrumental temperatures shows that the dendro-isotopic parameter can be used with a high degree of confidence to reconstruct climate trends on a multi-year scale.

References

Augusti, A., Betson, T. R., & Schleucher, J. (2006). Hydrogen exchange during cellulose Synthesis Distinguishes climatic and biochemical Isotope fractionation in tree rings. *New Phytologist, 172*, 490–499.

Augusti, A., Betson, T. R., & Schleucher, J. (2008). Deriving correlated climate and physiological signals from deuterium isotopomers in tree rings. *Chemical Geology, 252*, 1–8.

Cook, E. R., & Kairiukstis, L. A. (Eds.). (1990). *Methods of dendrochronology: Applications in the environmental sciences*. Boston, MA: International Institute for Applied Systems Analysis, Kluwer Academic Publishers.

Daux, V., Edouard, J. L., Stievenard, M., Mestre, O., Guibal, F., Masson-Delmotte, V., & Thomas, A. (2011). Ring width, and carbon and oxygen Isotopic Composition of the cellulose in *Larix decidua* as climatic proxies: A case study in the French Alps. *Earth and Planetary Science Letters*.

Douglass, A. (1920). Evidence of climatic effects in the annual rings of Trees. *Ecology, 1*(1), 24–32.

Farquhar, G. D., O'Leavy, M. H., & Berry, J. A. (1982). On the relationship between carbon isotope discrimination and intercellular carbon dioxide concentration in leaves. *Australian Journal of Plant Physiology, 9*, 121–137.

Francey, R. J., & Farquhar, G. D. (1982). An explanation of $^{13}C/^{12}C$ variations in Tree rings. *Nature, 297*, 28–31.

Fritts, H. C. (1976). *Tree-rings and climate* (p. 567). New York: Academic Press.

Gessler, A., Ferrio, J. P., Hommel, R., Treydte, K., Werner, R. A., & Monson, R. K. (2014). Stable isotopes in tree rings: towards a mechanistic understanding of isotope fractionation and mixing processes from the leaves to the wood. *Tree Physiology, 34*, 796–818.

Guiot, J., Corona, C., & ESCARSEL Members. (2010). Growing season temperatures in Europe and climate forcings over the last 1400 years. *PLoS-one, 5*(4), e9972. https://doi.org/10.1371/journal.pone.0009972.

Labuhn, I., Daux, V., Girardclos, O., Stievenard, M., Pierre, M., & Masson-Delmotte, V. (2016). French summer droughts since 1326 CE: a reconstruction based on tree ring cellulose $\delta^{18}O$. *Climate of the Past, 12*, 1101–1117.

Lavergne, A., Daux, V., Villalba, R., Pierre, M., Stievenard, M., & Srur, A. M. (2017). Improvement of isotope-based climate reconstructions in Patagonia through a better understanding of climate influences on isotopic fractionation in tree rings. *Earth and Planetary Science Letters, 459*, 372–380.

Lavergne, A., Daux, V., Pierre, M., Stievenard, M., Srur, A. M., & Villalba, R. (2018). Past summer temperatures inferred from dendrochronological records of on the eastern slope of the northern patagonian andes. *Journal of Geophysical Research: Biogeosciences, 123*(1), 32–45.

Masson-Delmotte, V., Raffali-Delerce, G., Danis, P., Yiou, P., Stievenard, M., Guibal, F., et al. (2005). Changes in European precipitation seasonality and in drought frequencies revealed by a four-century-long tree-ring isotopic record from Brittany, Western France. *Climate Dynamics, 24*, 57–69.

Pendall, E. (2000). Influence of precipitation seasonality on Piñon pine cellulose δD values. *Global Change Biology, 6*, 287–301.

Raffali-Delerce, G., Masson-Delmotte, V., Dupouey, J. L., Stievenard, M., Breda, N., & Moisselin, J. M. (2004). Reconstruction of SUMMER droughts using tree-ring cellulose isotopes: A calibration study with living Oaks from Brittany (Western France). *Tellus, 56*, 160–174.

Schweingruber, F. H., Fritts, H. C., Braeker, O. U., Drew, L. G., & Schaer, E. (1978). The X-ray technique as applied to dendroclimatology. *Tree-Ring Bulletin, 38*, 61–91.

Trouet, V. (2020). *Tree Story, the History of the World Written in Rings* (p. 256). Johns Hopkins University Press.

Waterhouse, J. S., Switsur, V. R., Barker, A. C., Carter, A. H. C., & Robertson, I. (2002). Oxygen and hydrogen isotope ratios in tree rings: How well do models predict observed values? *Earth and Planetary Science Letters, 201*, 421–430.

Air-Vegetation Interface: An Example of the Use of Historical Data on Grape Harvests

Valérie Daux

Historical data provide clues, often precise, of climate phenomena experienced by societies. The example below illustrates the scientific approach applied to records of grape harvest dates in order to quantitatively estimate the evolution of temperature in a region.

The vine is a hardy perennial Mediterranean plant, well adapted to hot dry weather conditions. Temperature is a key environmental factor which triggers its various phenological stages. Buds appear when daytime temperatures reach at least 10 °C for an extended period, usually in April. Flowering dates occur about two months later (usually June), when the daily temperatures cumulatively reach a certain threshold. The date of the onset of ripening, veraison (in August), is dependent on the date of flowering and the subsequent temperatures. Finally, the grape reaches maturity about thirty days after veraison, most often in September. Harvesting takes place when the grapes are perfectly ripe. The time between veraison and harvesting is fairly constant; it depends on the temperature, on the phytosanitary status of the vineyard and the amount of rainfall. The time periods given for the different phenological stages are approximate and vary according to grape varieties and regions.

Harvest dates, since they depend mainly on the temperatures during the growth and maturation of the vine, can be regarded as thermal indicators. Variations in harvest dates over the past centuries reflect historical variations in spring and summer temperatures.

Historical Series of Harvest Dates

In the Middle Ages, the official opening dates of harvest were set by informed feudal decision. After the Revolution, this tradition was theoretically abandoned; each owner could commence the harvest at will. However, the use of an official date was maintained in most parishes, for reasons of security and public order. After 1791, it was the mayor, advised by the vineyard owners, who fixed the date of the beginning of the harvest. The actual date of harvesting could take place some days after this date. The official and real dates of harvesting have been recorded in different registers depending on the place and time: these might be in monastic records, chapters of canons, hospital records (such as those from the *Hospices de Beaune, Burgundy, France*), municipal registers or private records (by the vineyards). During the 'Medieval Climate Optimum' (around 10th to thirteenth centuries), vineyards were planted at latitudes as far north as the shores of the Baltic or the south of England. Conversely, due to the cooling of the 'Little Ice Age' (from the 14th to the nineteenth centuries), most northern vineyards collapsed, and the vine growing season was shortened so much that the harvests were difficult even in some southern vineyards.

The series of published harvest dates are mostly composite series constructed from multiple sets of dates from different sites of the same vineyard. Ideally, the composite series should be established from local series of harvesting dates of the same grape variety, to avoid introducing bias related to a phenology which may vary by grape variety. The longest and most comprehensive series published to date is that of the vineyards of Burgundy (Fig. 17.1). This is a composite series constructed from data from multiple sites in the Côtes de Beaune and perhaps also from the Côtes de Dijon. The grapes harvested in that area since the fourteenth century are mostly pinot noir and chardonnay.

Reconstruction of Spring-Summer Temperatures Based on Grape Harvesting Dates

Grape harvesting data from the twentieth century, from different vineyards and varieties, were compared with regional meteorological data (Daux et al. 2007).

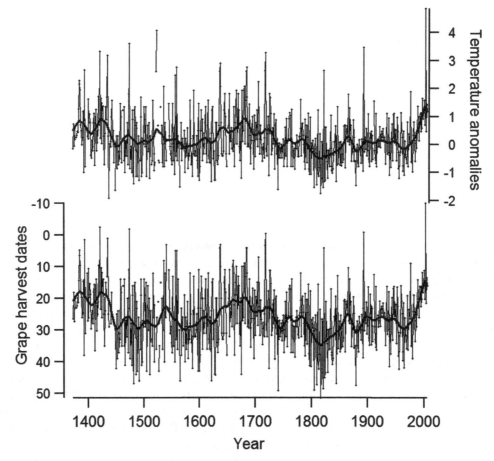

Fig. 17.1 Bottom: Harvest dates in Burgundy since 1370 (Daux et al. 2012). Dates are arbitrarily expressed in days after August 31st. Top: Anomalies in the corresponding temperatures, calculated using a phenological model. The anomalies are calculated compared to the reference period 1960–1989; according to Garcia de Cortazar et al. (2010). The bold lines are 29-year moving averages. The major climate events of the past seven centuries may be described as follows: warming in the 1380s, and from 1415 to 1435, a period that includes the wheat 'blast' that led to the famine of 1420; cooling during the second half of the 'Quattrocento' particularly evident in the famine of 1481, arising from rain and cold conditions; hot spells during the 1520s, 1530s and 1550s; the strong cold surge of the Little Ice Age (LIA) from 1560. A cold seventeenth century'? This was the case from 1570 to 1630 (with a slight warming around 1600–1620), in 1675, during the 1690s and in the following century, from 1709 to 1715; heatwaves in summers of the 1630s, 1660s and 1680s; warming during the eighteenth century, particularly obvious in the years 1704–1707, 1718–1719, in the 1720s and 1730s, the years 1757–1765 and especially from 1778–1781 and during the 1780s; nevertheless, years of cool-cold-wet (1725, 1740, 1770); the Little Ice Age, which never quite ceased, became vigorous again from 1812 to 1860; added to this were the very cold snaps of 1812–1817; heatwave of 1846, affecting grain harvests; the definitive end of the Little Ice Age in 1860; warming during the twentieth century from around 1900 onwards with an acceleration since 1976 and the 1990s. After Le Roy Ladurie et al. (2006)

The harvest date series and temperatures present many similarities, at the annual, decadal and multi-decadal scale (see example in Fig. 17.2). The best correlations (statistically significant with a maximum R) between harvest dates and climate data are those with the average maximal temperatures from April to August. For these correlations, the slopes are around −6 to −10 days/°C for all the vineyards (Fig. 17.3). In other words, regardless of the earliness or lateness of the vines and regardless of the geographic location (including soil, orientation and meteorology), a variation of six to ten days on the date of the harvest reflects a difference of about 1 °C from the average maximum temperatures for the growing season (April in August; see also for example Cook and Wolkovich 2016).

The phenology of woody species can be simulated by models that express the relationship between the maturation process and daytime temperatures. A model is adapted to describe the phenology of a species, or a variety in the case of the vine. Chuine et al. (2004) have thus used recent observations of phenological development of the pinot noir variety, made between 1964 and 2001 in Colmar (France), by INRA, to model the phenology of this variety and to calculate the veraison and harvest dates from temperatures. The inverted model has been used here

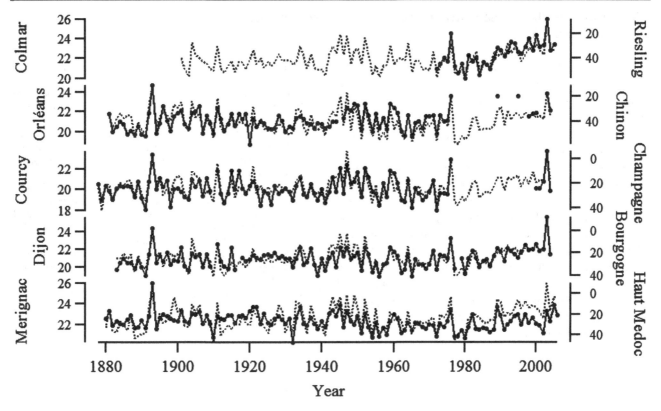

Fig. 17.2 Series of harvest dates (bold lines with symbols, scale on right; in days after August 31) and maximum daytime temperatures for the corresponding region (dotted lines, scales on left) versus time

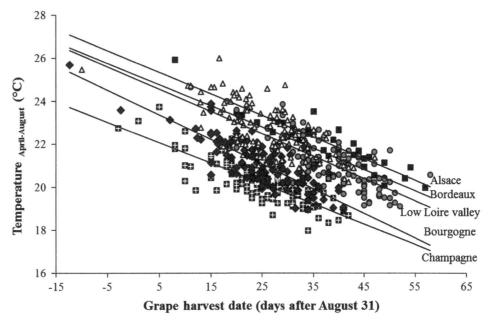

Fig. 17.3 Correlations among harvest dates (in days after August 31) and average monthly daytime temperatures from April to August (in °C) for the vineyards of the Champagne, Burgundy, Alsace, Bordeaux and Loire valley regions

to reconstruct variations in spring-summer temperatures in Burgundy since the fourteenth century from the series of dates shown in Fig. 17.1. This reconstruction highlights that the cooling of the climate in the period 1690–1800, the start of a long cool period that continued until 1970, was by a little more than one degree. The exceptionally hot summer of 2003 also shows up very clearly (Chuine et al. 2004).

Limitations of These Reconstructions: The Anthropic Effects

The decision to harvest is based on a simple principle: the grapes need to be harvested when they are ripe and not too long afterwards so as to avoid exposure to unfavorable weather conditions, such as hail or heavy rain. Judgment of the ideal state of ripeness may have evolved over time depending on the political and economic context, cultivation techniques, oenological practices and consumer tastes. Variations over time in these parameters can create trends and cause breaks in the data series. These variations caused by factors not related to the climate may not necessarily be detectable in a single series. To be as rigorous as possible, a climate reconstruction must be based on a collection of series (ideally of the same variety, from the same wine region), which is tested for disruptions and inter-correlations. Historical information of this type can provide insights and help to identify the proportion of variability attributable to anthropic effects.

References

Chuine, I., Yiou, P., Viovy, N., Seguin, B., Daux, V., & Le Roy Ladurie, E. (2004). Grape ripening as a past climate indicator. *Nature, 432*, 289–290.

Cook, B. I., & Wolkovich, E. M. (2016). Climate change decouples drought from early wine grape harvests in France. *Nature Climate Change, 6*, 715–719.

Daux, V., Garcia de Cortazar-Atauri, I., Yiou, P., Chuine, I., Garnier, E., Le Roy Ladurie, E., et al. (2012). An open-access database of grape harvest dates for climate research: data description and quality assessment. *Clim. Past, 8*, 1403–1418.

Daux, V., Yiou, P., Le Roy Ladurie, E., Mestre, O., Chevet, J.-M., & the OPHELIE team. (2007). Température et dates de vendanges en France', Articles du colloque 'Changement climatique, quels impacts probables sur le vignoble?', Dijon, 28–29 mars 2007, 10p.

Garcia de Cortazar-Atauri, I., Daux, V., Garnier, E., Yiou, P., Viovy, N., Seguin, B., et al. (2010). An assessment of error sources when using grape harvest date for past climate reconstruction. *Holocene, 20*, 599–608.

Le Roy Ladurie, E., Daux, V., & Luterbacher, J. (2006). Le climat de Bourgogne et d'ailleurs. *Histoire, Economie et Sociétés, 3*, 421–436.

Air-Ground Interface: Sediment Tracers in Tropical Lakes

David Williamson

The climate archives found at low latitudes are mostly sedimentary or geomorphological. Their formation and preservation depend fundamentally on favorable geomorphological and climatic conditions. Therefore, they differ according to the environment, the region and the climate and include moraines left behind by mountain glaciers, ancient shorelines of old lakes, dunes, flood levels, archeological sites occupied by animals or humans, continental carbonates (speleothems, travertine, stromatolites) and sedimentary sequences collected in cores from lakes and bogs. Additional archives consist of coastal deposits (e.g. corals) and marine hemipelagic sediments with a high deposition rate, especially near river deltas or coastal upwellings off Mauritania, Benguela and Somalia. The dating and study of aquifers, as well as of tree rings complete the range of climate records currently available and studied, particularly in semi-arid areas.

The physical, chemical and biological controls of the sedimentary records vary greatly according to the local or regional specificity of the areas of deposition. For example, the mixing or overturning of the water column of a lake is generally recorded by biological (diatoms) or organic proxies. It may be activated by an earthquake, outgassing from anoxic hypolimnion or by sediments, but also by the cooling of surface waters, the melting of ice, by wind intensity or by a drop in lakewater level. Hence, the climate or hydrological proxies (e.g. physicochemical properties or isotopic signatures of water and sediments, terrestrial or aquatic vegetation markers, erosional or atmospheric inputs) do not have the same fundamental meaning from one site to another.

To reduce the uncertainties in timing and interpretation, it is essential to undertake a detailed reconstruction of the deposition processes in several sites characteristic of the region and for a given climate. Among the non-destructive physical and chemical methods of measuring the variability of depositional environments, the measurement of magnetic sedimentary properties (low-magnetic field susceptibility, induced and remanent magnetizations) is a particularly appropriate one, because multiple measurements can be performed. The magnetic signature of sediments—and other surface materials—depends on changes in the concentration, grain size and mineralogy of iron minerals (mainly oxides, sulfides, carbonates). Iron mineral assemblages closely depend on the iron biogeochemical cycle, in particular on the weathering of primary minerals, pedogenetic processes, biomass fires, gravity deposition, redox changes, microbial activity, authigenesis and early diagenesis. Magnetic methods thus allow for the reconstruction of the dynamics of surface processes recorded along sedimentary sequences, and the identification of the main modes of environmental and climate variability (Thompson and Oldfield 1986).

Intertropical Hydrological Variability in Africa

Africa is characterized by a wide variety of climates, mainly defined by the amount and distribution of rainfall at the seasonal, annual, and year-to-year level (from less than 100 mm per year to more than 3000 mm per year) and temperature (especially at mid-latitudes and at altitude). From the humid equatorial climate of the Congo basin to the hyperarid or subtropical climates of the Sahel/Sahara and of the Kalahari/Namibia, regional climates respond to the annual and seasonal characteristics of global and tropical atmospheric circulation (in particular the activity of the Intertropical Convergence Zone, ITCZ). The temperatures and surface currents of the neighboring oceans and the regional orographies (e.g. East-African rift, great escarpment of Namibia) determine the regional rainfall patterns. Together with the local geomorphology, these regional climates have a determining influence on the presence, nature and historical continuity of most of the land-based climate archives.

D. Williamson (✉)
World Agroforestry Centre, Research Institute for Development, UMR LOCEAN, Université Pierre et Marie Curie, P.O. Box 30677-00100 Nairobi, Kenya
e-mail: david.williamson@ird.fr

Sahara, Kalahari and Arid Zones: Discontinuous Evidence of Hydrological Inversions

Since the 1950s and 1960s, the first robust evidence of major hydrological changes during the late Pleistocene and Holocene, have been found in the two African subtropical deserts in the form of proof of the existence of gigantic interdune lakes; flora and fauna assemblages; sites demonstrating a sedentary human presence and a positive water balance over centuries or millennia. A key outcome of this research is indications that, during the first half of the Holocene, there was an intensification of the African monsoon in the Sahara and a shift in the tropical summer rains (at the northern limit of the ITCZ) north of 20°N as well as a significant reduction in the Sahara as a biogeographical barrier during this period (Lézine et al. 1990; Petit-Maire and Riser 1981). Very similar evidence of major hydrological changes has been obtained for the Kalahari (Thomas et al. 2003). As shown in Fig. 18.1, these records show the asynchronism of the humidity 'optima' during the Holocene, north (from 11 to 6 ka BP) and south (after 6 ka BP) of the equator, illustrating the forcing of the changes in regional insolation on the monsoon circulation and summer tropical rainfall (from June to September in the Sahara and from December to March in the Kalahari). Furthermore, at the millennial scale, numerous sedimentary discontinuities suggest the occurrence of hydroclimatic thresholds, apparently linked with rapid changes in activity of the ITCZ from one latitude to another, corresponding to rapid climate fluctuations in the high latitudes (Mayewski et al. 2004). However, the difficulty with these archives remains in the discontinuity of the records and the lack of detailed chronological precision on dry periods. The preserved deposits contain the initial stages of wet periods, but the preservation of the signal, in particular for the phases of aridification, is hypothetical.

The depositional gaps and the differential removal of remaining deposits do not allow a continuous hydrological and climate dynamic to be reconstructed, particularly for the glacial period.

The dynamics of the African deserts must then be reconstructed from the proximate and hemipelagic marine field, especially at the mouths of the major rivers (Senegal, Niger, Orange River, Zambezi) and in upwelling zones (Mauritania, Benguela), where sedimentation rates permit a resolution at the millennial scale, coupled with isotopic stratigraphy for the last ice age(s). The data obtained confirm the important role of insolation in the low latitudes in monsoon circulation and the hydrology of the subtropical areas. They also confirm the instability of the edges of the Sahara and the Kalahari at the millennial scale (DeMenocal et al. 2000; Little et al. 1997), already highlighted for the

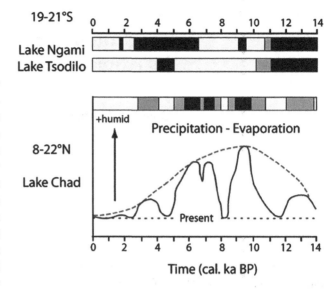

Fig. 18.1 Chronostratigraphy comparing lacustrine sequences from the Kalahari (Thomas et al. 2003) and Lake Chad (Servant and Servant-Vildary 1980). Black represents high water level; gray, the intermediate level; and white the low water level or drying out. Note the phase difference between the water balances of the southern tropical region (Kalahari) and the edge (Chad)

Holocene from Lake Chad (Servant and Servant-Vildary 1980). The coupling of the millennial climate variability in the high latitudes during the glacial period with the humidity of the Sahara is clearly demonstrated from marine sediments, in particular through erosion markers (Adegbie et al. 2003) and ecosystems (Schefuß et al. 2005). However, the information obtained often lacks spatial accuracy, and the critical contribution of transportation processes with regard to the interpretation of markers often remains hypothetical (Pichevin et al. 2005).

(Sub)Equatorial Zone: Changes in the Activity and Position of the ITCZ

The continental archives providing the most continuous recording of climate variability over the last 25 ka are located in sites that have remained relatively humid, located in equatorial regions or at altitude, and where the hydric balance [Precipitation—Evaporation, P-E) remained positive. These are cores taken either from large terminal lake reservoirs found in central Africa (Lake Victoria) or the East African Rift (e.g. Lakes Tanganyika and Malawi), or from small lake basins suitable for the reconstruction of regional conditions. As shown from the complete drying up of Lake Victoria at 18 and 16 ka BP (Stager et al. 2002), the drop by several hundred meters of the levels of lakes Tanganyika and Malawi during the last glacial period, or the rapid rise 18 (over a few decades) of great lakes that are today very shallow or drying out (Rukwa, Victoria, Chad) (Williamson et al. 1993), the whole intertropical zone has recorded major and rapid changes in its water balance, linked mostly to changes in regional temperature. The changes affected the continental surfaces, the soils, the distribution and the biodiversity of terrestrial and aquatic ecosystems. For example, the iron mineralogy of soils was modified, and variations of several orders of magnitude of the concentration in magnetic minerals are evidenced (Williamson et al. 1993).

Owing to unrivaled chronological and spatial resolution, the data obtained from peat bogs or crater lakes, most especially the molecular markers of microbial origin (Coffinet et al. 2014), show that these rapid changes in atmospheric circulation and in the composition of the atmosphere (water, CO_2) were responsible for a fundamental reorganization of ecosystems, particularly with regard to the distribution of mountain ecosystems and most wetland areas.

These records confirm the dominant forcing of insolation and monsoon circulation over tropical hydrology. At the decennial to millennial scale, however, the depositional record clearly indicates the coupling of the tropical climate with the climate in the high latitudes, and with the activity of the thermohaline circulation, particularly during the glacial interval where the Dansgaard-Oeschger variability is identified in numerous hydrological records. New high resolution chronological records are required to identify the mechanisms and the potential phase-lags linked with such changes (Zhang et al. 2018).

Recent Anthropological Influence on Climate Archives: Both a Proof and a Tool to Assess the Impacts of Local and Regional Development

Superimposed on natural climate stress, the human impact on the hydrology and land-based environments of Africa remains poorly documented, and presents a challenge to a better understanding of the interactions between climate, environment and society in developing countries. The first large-scale human impacts on ecosystems (livestock domestication and pastoralism) are recorded as early as ca. 4000 yr B.P., especially during the last 2500 yr with deforestation and biomass burning (Thevenon et al. 2003) being associated with iron metallurgy, the expansion of farming activities and demography: rapid increases in the fluxes of microcharcoal, detrital inputs (especially magnetic) and pollen from ruderal vegetation (near habitations) and pioneer taxa (Zhang et al. 2018). In East Africa, the expansion of trade routes between the coast and the interior at least 1300 years ago increased the abundance of exotic plants, especially cultivated plants from Asia such as banana, taro, rice, and, more recently, plants from the new world such as maize, tobacco, tomatoes. However, it is only since the late nineteenth century and especially after the Second World War, that the archives show radical changes in the environment caused by man (Marchant et al. 2018): the organic components (on land and in water) and inorganic components, dissolved or suspended (notably magnetic), present in the environment are affected (Garcin et al. 2007).

In this context, just as changes in atmospheric composition have been reconstructed from ice cores, the land-based 'pre-industrial' archives recording changes in the climate and the tropical environment are now the local and regional references for environments relatively untouched by man, from the decadal to the millennial scale. Such references are required to establish a frame of reference so as to identify the various controls of climate and society on the structural changes in the landscape and in biodiversity, in vegetal production, in erosional processes, or in stored material (including pollutants), and thus to assess the extent and sustainability of the changes imposed by humans on the Earth's surface.

References

Adegbie, A. T., Schneider, R. R., Röhl, U., & Wefer, G. (2003). Glacial millennial-scale fluctuations in central African precipitation recorded in terrigenous sediment supply and freshwater signals offshore came.

Coffinet, S., Huguet, A., Williamson, D., Fosse, C., & Derenne, S. (2014). Potential of GDGTs as a temperature proxy along an altitudinal transect at Mount Rungwe (Tanzania). *Organic Geochemistry, 68*, 82–89.

DeMenocal, P., Ortiz, J., Guilderson, T., Adkins, J., Sarnthein, M., Baker, L., et al. (2000). Abrupt onset and termination of the African humid period: Rapid climate responses to gradual insolation forcing. *Quaternary Science Reviews, 19*, 347–361.

Garcin, Y., Williamson, D., Bergonzini, L., Radakovitch, O., Vincens, A., Buchet, G., et al. (2007). Solar and anthropogenic imprints on lake masoko (Southern Tanzania) during the last 500 years. *Journal of Paleolimnology, 37*, 475–490.

Lézine, A. M., Casanova, J., & Hillaire-Marcel, C. (1990). Across an early holocene humid phase in Western Sahara: Pollen and isotope stratigraphy. *Geology, 18*, 264–267.

Little, M. G., Schneider, R. R., Kroon, D., Price, B., Summerhayes, C. P., & Segl, M. (1997). Trade wind forcing of upwelling, seasonality, and heinrich events as a response to sub-milankovitch climate variability. *Paleoceanography, 12*, 568–576.

Marchant, R., Richer, S., Boles, O., Capitani, C., Courtney-Mustaphi, C. J., Lane, P., et al. (2018). Drivers and trajectories of land cover change in East Africa: Human and environmental interactions from 6000 years ago to present. *Earth-Science Reviews, 178*, 322–378.

Mayewski, P. A., Rohling, E. E., Stager, J. C., Karlén, W., Maasch, K. A., Meeker, L. D., et al. (2004). Holocene climate variability. *Quaternary Research, 62*, 243–255.

Petit-Maire, N., & Riser, J. (1981). Holocene lake deposits and palaeoenvironments in central Sahara, Northeastern Mali. *Palaeogeography Palaeoclimatology Palaeoecology, 35*, 45–61.

Pichevin, L., Cremer, M., Giraudeau, J., & Bertrand, P. (2005). A 190 Ky record of lithogenic grain-size on the namibian slope: Forging a tight link between past wind-strength and coastal upwelling dynamics. *Marine Geology, 218*, 81–96.

Servant, M., & Servant-Vildary, S. (1980). L'Environnement quaternaire du bassin du Tchad. The Sahara and the Nile. In M. A. J. Williams, & H. Faure (Ed.), (pp. 133–162). Leiden: Balkema.

Schefuß, E., Schouten, S., & et Schneider, R. R. (2005). Climatic controls on central African hydrology during the Past 20,000 Years. *Nature, 437*, 1003–1006.

Stager, J. C., Mayewski, P. A., & Meeker, L. D. (2002). Cooling cycles, heinrich event 1, and the desiccation of lake victoria. *Paleogeography, Paleoclimatology, Paleoecology, 183*, 169–178.

Thevenon, F., Williamson, D., Vincens, A., Taieb, M., Merdaci, O., Decobert, M., et al. (2003). A late holocene charcoal record from Lake Masoko, SW Tanzania: Climatic and anthropologic implications. *The Holocene, 15*, 785–792.

Thomas, D. S. G., Brook, G., Shaw, P., Bateman, M., Appleton, C., Nash, D., et al. (2003). Late pleistocene wetting and drying in the NW Kalahari: an integrated study from the Tsodilo Hills, Botswana. *Quaternary International, 104*, 53–67.

Thompson, R., & Oldfield, F. (1986). *Environmental magnetism*. London: Allen and Unwin.

Williamson, D., Taieb, M., Damnati, B., Icole, M., & Thouveny, N. (1993). Equatorial extension of the younger dryas event: Rock-magnetic evidence from lake Magadi. *Global and Planetary Change, 7*, 235–242.

Williamson, D., Jelinowska, A., Kissel, C., Tucholka, P., Gibert, E., Gasse, F., et al. (1998). Mineral-magnetic proxies of erosion/oxidation cycles in tropical maar-lake sediments (Lake Tritrivakely, Madagascar): Paleoenvironmental implications. *Earth and Planetary Science Letters, 155*, 205–219.

Zhang, Y. C., Chiessi, C. M., Mulitza, S., Sawakuchi, A. O., Haggi, C., Zabel, M., et al. (2018). Different precipitation patterns across tropical South America during Heinrich and Dansgaard-Oeschger stadials. *Quaternary Science Reviews, 177*, 1–9.

Air-water Interface: Tropical Lake Diatoms and Isotope Hydrology Modeling

Florence Sylvestre, Françoise Gasse, Françoise Vimeux, and Benjamin Quesada

Regardless of the timescale, a distinctive characteristic of the tropics is a high variability in the water cycle, in particular in the *P-E* water balance [precipitation minus (evaporation + evapotranspiration)]. This variability has more impact on life in the tropics than the variability in temperature which only has a moderate influence compared to the high and middle latitudes. The time frame, speed and magnitude of hydrological events need to be established so that their causes and mechanisms can be understood. These events are recorded in terrestrial sediments (lacustrine deposits, deposits in caves), in fossil waters in deep aquifers, and imprinted in the landscape by geomorphological features (dried-up river channels, ancient lake shorelines).

Lacustrine archives are particularly interesting because they contain the imprint of hydrological variations recorded in lakes. They preserve a wide range of complementary indicators (or proxies), providing access to different environmental and climate parameters if they can be calibrated.

Lakes are today numerous in the humid tropics, and evidence of ancient lakes (i.e. paleolakes) are plentiful in certain regions that are now arid. They react to changes in climate with often dramatic fluctuations, accompanied by profound changes in chemistry, biology, water-mass dynamics and water-atmosphere heat transfers. Indeed, one way which is often forgotten that lakes can have a significant interaction on the continental hydrological cycle is through the source of water vapor they represent (Vallet-Coulomb et al. 2008). In this chapter, we will examine specifically the example of a hydro-isotopic reconstruction established from sedimentary records from a paleolake located in the tropical Andes (Fig. 19.1) and its impact on the local hydrological cycle during the last glacial-interglacial transition (18–12 ka) (Quesada et al. 2015). Over this period, this paleolake, called 'Tauca' (from the name of a village where the eponymous outcrops were discovered, Servant and Fontes 1978) underwent a transgression phase due to an increase in rainfall in the tropical Andes (Sylvestre et al. 1999; Blard et al. 2011), before disappearing abruptly around 14.3 ka. Here, we propose a reconstruction of the isotopic composition of the lake water and we explore how its disappearance might have disturbed the local atmospheric water cycle. In particular, we explore the possible link between a sudden evaporation of the Tauca paleolake and an abrupt event found in the isotopic composition (oxygen 18) of the glacier covering Mount Sajama overlooking the former lake (Thompson et al. 1998).

F. Gasse—Deceased.

F. Sylvestre (✉) · F. Gasse · B. Quesada
Aix-Marseille Université, CNRS, IRD, Collège de France, INRAE, CEREGE, Europôle Méditerranéen de l'Arbois, 13545 Aix-en-Provence cedex 4, France
e-mail: sylvestre@cerege.fr

F. Vimeux
Institut de Recherche pour le Développement (IRD), Laboratoire HydroSciences Montpellier (HSM), UMR 5569 (CNRS, IRD, UM), 34090 Montpellier, France

F. Vimeux · B. Quesada
Institut Pierre Simon Laplace (IPSL), Laboratoire des Sciences du Climat et de l'Environnement, LSCE/IPSL, CEA-CNRS-UVSQ, Université Paris-Saclay, 91190 Gif-sur-Yvette, France

Site Selection and Collection of Samples

A fundamental criterion for the selection of study sites is the sensitivity of the lake system (the lake and its watershed) to variations in *P-E*. The change in volume of a lake during the time interval Δt is represented by the equation:

$$\Delta V_L / \Delta t = S_L (P_L - E_L) + A - D. \quad (19.1)$$

where V_L is the volume of the lake (m^3) and S_L is the surface area (m^2). V_L and S_L are functions of the level of the lake (H_L). P_L and E_L are, respectively, the rates of precipitation

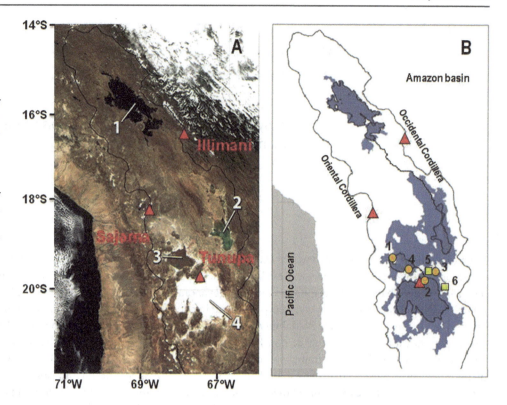

Fig. 19.1 a The Bolivian Altiplano (MODIS image modified from http://www.nssl.noaa.gov/projects/pacs/web/ALTIPLANO/) with 1. Lake Titicaca, 2. Lake Poopó, 3. The Salar de Coipasa, 4. The Salar de Uyuni. The red triangles indicate the position of the ice cores taken from the glaciers of Illimani, Mount Sajama and Tunupa. b Diagram of the Bolivian Altiplano indicating the contemporary lakes, the outline of the watershed (black line) and the extent of the Tauca paleolake shown in blue. The study sites are indicated by orange dots and yellow squares

and evaporation at the surface of the lake (m), A (m^3) is the sum of the surface and subterranean inflows. A is dependent on the P-E balance of the watershed ($P_B - E_B$) and its surface S_B. D (m^3) is the sum of the losses from the surface and from infiltration at the bottom of the lake. Equation (19.1) shows that closed lakes ($D = 0$), common in semi-arid or arid areas, are the most sensitive because they cumulate variations in the P-E of the lake and of its watershed. Thus, they amplify the responses to climate change.

The Bolivian Altiplano is located in the central tropical Andes between 15°S and 22°S latitude and 65°W and 69°W longitude at an average altitude of 3800 m. Today, it is occupied by a chain of four lakes of decreasing altitude and increasing salinity from north to south: Lake Titicaca, a deep lake of fresh water at 3810 m; Lake Poopó, a very shallow, hypersaline lake at 3686 m; and the Coipasa and Uyuni lakes at 3657 m and 3653 m which are now covered by two large salt crusts of 2,500 km^2 and 11,000 km^2 respectively, the remaining legacy following evaporation of the Tauca paleolake (Fig. 19.1a). This hydrological system is endorheic, and Lake Titicaca can flow into Lake Poopó, which itself, in very rainy seasons, can feed the Salars of Coipasa and Uyuni (Fig. 19.1a). The Bolivian Altiplano receives most of its precipitation during the rainy season which occurs from November to March. Today, precipitation comes mainly from the Atlantic, transported to the Altiplano by the easterly winds (Garreaud et al. 2003).

In a paleohydrological study of lacustrine archives, the maximum morphological, geological, meteorological, hydrological and hydrochemical information on the lake watershed is collected by satellite imagery, aerial photographs, field observations, and instrumental data. The spatial distribution of exposed lake deposits and former perched shorelines (beach deposits, notches eroded in the cliffs by waves, etc.) is measured by GPS and localized using a digital terrain model (DTM). Contemporary samples of water and silts are collected as references for the interpretation of past tracers. Sediments are taken at the outcrop and by coring. The sites are chosen according to the bathymetry and the continuity of the identified sediments. On the Bolivian Altiplano, the samples were selected from outcrops according to their altitude around the lake, making it possible to quantify the depth of the lake during its evolution through time (Fig. 19.1b) (Sylvestre et al. 1999).

Reconstruction of Paleohydrological Conditions

The study commences with a lithological description of the samples, a prerequisite to understanding the organization of the sedimentary deposits, before selecting the levels to establish their absolute chronology using radiometric methods such as radiocarbon. In stratigraphy, we find the strata with remarkable sedimentary facies that allow the different profiles to be correlated. Indicators of a small sliver of water or of a drought located in the profile (mollusk shells, coarse sand beds, or paleosols) as well as former perched shorelines provide direct data on lake level fluctuations. The mineralogical composition of the sediments, their grain size, the geochemical properties of the various fractions (mineral or organic, detrital or authigenic), and the assemblages of fossilized organisms are indirect indicators of past environmental conditions.

In the outcrops of the Bolivian Altiplano, we focused our investigations on the diatoms. These are microscopic unicellular algae (3–100 μm), which are very sensitive to variations in environmental parameters, such as the depth or salinity of water, and each cell produces an easily fossilized bivalve siliceous outer skeleton. This external skeleton, called a frustule, consists of an inner layer of silica tetrahedra (Si_2O, nH_2O) bonded to each other, and of a hydrated outer layer made of organic material, mainly proteins and polysaccharides. Diatom valves are extracted from sediments by physicochemical treatment. In optical microscopy, the species are identified by the structure of their valves: their absolute (number of valves/g) and relative (percentage) abundance can then be estimated. Knowing the self-ecology of each species identified makes it possible to characterize the aquatic environment in which they developed.

The presence of oxygen in the inner layer of the frustule allows the isotopic composition in oxygen 18 to be measured. Preliminary dehydration is necessary because of the hydrated nature of the silica so that only the isotopes in the oxygen of the internal structure are measured. After extraction of oxygen and purification in a preparation line, the O_2 molecule is analyzed by mass spectrometry (Crespin et al. 2008, Crespin et al. 2010).

Quantification of the Oxygen Isotopic Composition of Lakes

In lacustrine environments, the isotopic oxygen composition of diatoms ($\delta^{18}O_{diatoms}$) varies with the temperature and isotopic composition of the lake water ($\delta^{18}O_{lake\ water}$). Diatomaceous silica precipitates in isotopic equilibrium with the lake water, and calibration studies have established the thermo-dependent relationship $[([\delta^{18}O_{diatoms} - \delta^{18}O_{lake\ water}]$ (‰ vs. VSMOW) = a*T lake water (°C) + b)] to express the variation in the isotopic composition of diatoms as a function of temperature (Brandriss et al. 1998; Moschen et al. 2005; Crespin et al. 2010; Dodd and Sharp 2010; Dodd et al. 2012; Alexandre et al. 2012). These relationships between the oxygen isotopic composition of the diatoms and the water during formation have been experimentally established for lacustrine diatoms under contemporary environmental conditions, taking the seasonal to multi-annual variations into account (Fig. 19.2).

These relationships all show a low dependence on temperature, of around −0.20‰/°C. On the other hand, the fractionation factors are very different, which may be due either to sampling problems or to different methodological approaches when carrying out these calibrations (Crespin et al. 2010; Alexandre et al. 2012). Nevertheless, applied in different contexts, this method proved relevant to the reconstruction of the isotopic composition of lakes in the past (Leng and Barker 2006). In addition, the application of these different calibrations enables sensitivity tests to be conducted depending on the potential temperature range applicable during the period considered.

So that the thermo-dependence of the isotopic fractionation between the diatoms and the formation water is taken into account, the temperature of the lake water in which they develop must also be known. Diatoms are photosynthetic organisms which grow in the epilimnion and reflect the surface temperatures of the lake. For current periods, field observations are essential; for past periods, assumptions based on simulated atmospheric temperature data from climate models and data obtained from other proxies in the same region are used.

Hydro-Isotopic Modeling and Paleoclimatic Interpretation

A quantified estimate of the hydrological balance of lakes or of the P-E is obtained by hydrological modeling. The more or less sophisticated models used all rely on the balance equation for lake water (Eq. 1). The first step is to derive the contemporary water balance from the available instrumental data and the relationships between H_L, V_L and S_L constructed from a digital terrain model (DTM). Estimating evaporation, a function of temperature, solar radiation, humidity, wind and vegetation cover, is always complex. It involves the use of different hydrological and climatic concepts and methods. It is better constrained by connecting the water balance to the salt balance (Vallet-Coulomb et al. 2001) with a watershed runoff model (Legesse et al. 2004) or with an energy balance model (Kutzbach 1980). Application of the model to the past is usually done for a time interval where the lake is considered to have been in equilibrium ($\Delta V_L = 0$). We know S_B, H_L (from which we calculate V_L and S_L), the solar radiation provided by astronomical calculations, paleotemperatures and paleo-vegetation on the watershed can be deduced from data e.g. from palynology.

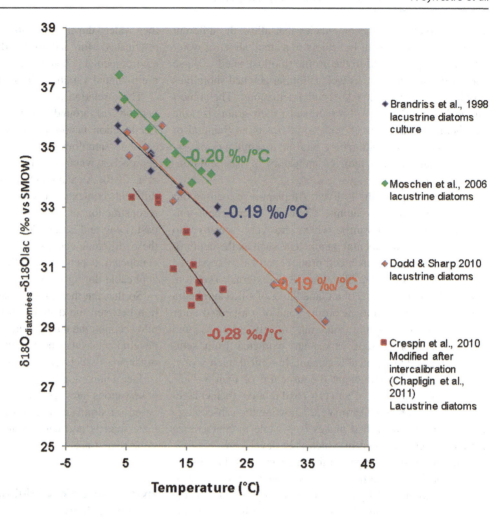

Fig. 19.2 Calibration of the thermo-dependent relationship between $\delta^{18}O_{diatoms}$ and $\delta^{18}O_{lake\ water}$ based on different calibrations of freshwater diatoms Modified from Crespin et al. (2010), Alexandre et al. (2012)

On the Bolivian Altiplano, paleohydrological data (Fig. 19.3) show that the Bolivian Altiplano experienced a first lacustrine transgression at 18,500 BP, interrupted between 18,100 BP and 15,800 BP by a phase of stagnation of the lake level, followed by a subsequent phase of maximum extension up to 15,000 BP years. After this date, this lake dried up. A second lacustrine transgression of lesser magnitude, named 'Coipasa', took place between 12,500 BP and 11,900 BP.

The reconstruction of the isotopic composition of the lake produces a very interesting result showing a spectacular decrease in its isotopic composition during the transgression phases (increase of the lake level) in response to rainfall inputs, followed by phases of isotopic enrichment during the phases when its level stabilizes (Fig. 19.3).

Simple hydro-isotopic modeling has shown that the increase in precipitation during the establishment of the lake is quantitatively consistent with the decrease recorded in the reconstruction of $\delta^{18}O_{lake\ water}$ (Quesada et al. 2015). It also partly explains the increase in the isotopic composition of the lake during its stability phases. This isotopic enrichment of lake waters is caused by interplay between lake evaporation and precipitation processes, both of which alter the isotopic composition of the lake differently. This confirms the establishment of the Tauca paleolake, probably caused by a massive influx of precipitation (significant decrease in $\delta^{18}O$ of the water of the paleolake) from the tropical Atlantic in the Altiplano watershed. The abrupt disappearance of the paleolake is contemporaneous with an isotopic excursion of +7‰ recorded in the ice of Mount Sajama (Thompson et al. 1998). Calculation of the isotopic composition of the vapor flux produced by evaporation of the paleolake, with its volume and surface taken into account, shows that this isotopic excursion could have been caused by an evaporation of 5–60% of the total volume of the lake. If this hypothesis is proven correct, it shows that the Tauca paleolake has certainly influenced the local, or even regional, hydrological cycle.

In conclusion, this example shows the potential impact of a lake on the local hydrological conditions, especially during

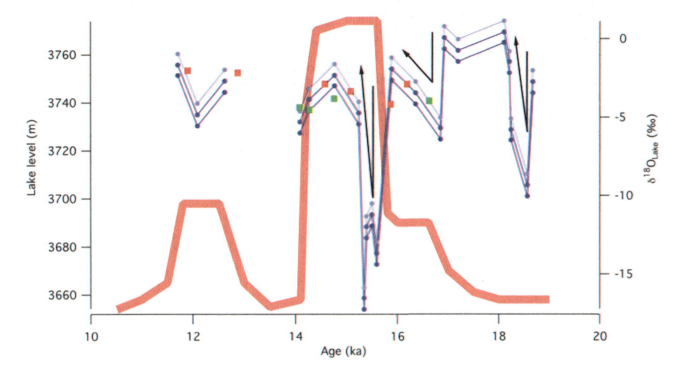

Fig. 19.3 Reconstruction of lake levels (m) and the isotopic oxygen composition $\delta^{18}O_{lake\ water}$ (‰ vs. V-SMOW) from (1) the isotopic oxygen composition of diatoms (blue lines) using the thermo-dependent relationship established by Dodd and Sharp (2010) for T = 5 °C, T = 7.5 °C and T = 10 °C, and (2) the isotopic oxygen composition of ostracods (green squares) and carbonates (red squares; Blard et al. 2011) for T = 7.5 °C

strong fluctuations in its level. Sites where several climate archives co-exist, allowing the large-scale climate signal to be differentiated from local conditions are relatively rare. The example discussed here shows the benefits to be gained from the co-location of a glacier archive and a lacustrine archive.

References

Alexandre, A., Crespin, J., Sylvestre, F., Sonzogni, C., & Hilbert, D. W. (2012). The oxygen isotopic composition of phytolith assemblages from tropical rainforest soil tops (Queensland, Australia): Validation of a new paleoenvironmental tool. *Climate of the Past, 8*, 307–324.

Blard, P.-H., Sylvestre, F., Tripati, A. K., Claude, C., Causse, C., Coudrain, A., et al. (2011). Lake highstands on the altiplano (Tropical Andes) contemporaneous with Heinrich 1 and the Younger Dryas: New insights from 14C, U-Th dating and δ18O of carbonates. *Quaternary Sciences Review, 30*, 3973–3989.

Brandriss, M. E., O'Neil, J. R., Edlund, M. B., & Stoermer, E. F. (1998). Oxygen isotope fractionation between diatomaceous silica and water. *Geochimica Cosmochimca Acta, 62*, 1119–1125.

Crespin, J., Alexandre, A., Sylvestre, F., Sonzogni, C., Paillès, C., Garreta, V. (2008). IR-laser-extraction technique applied to oxygen isotopes analysis of small biogenic silica samples. *Analytical Chemistry, 80*, 2372–2378.

Crespin, J., Sylvestre, F., Alexandre, A., Sonzogni, C., Paillès, C., & Perga, M.-E. (2010). Re-examination of the temperature-dependent relationship between $\delta^{18}O_{diatoms}$ and $\delta^{18}O_{lake\ water}$ and implications for paleoclimate inferences. *Journal of Paleolimnology, 44*, 547–557.

Dodd, J. P., & Sharp, Z. D. (2010). A laser fluorination method for oxygen isotope analysis of biogenic silica and a new oxygen isotope calibration of modern diatoms in freshwater environments. *Geochimica Cosmochimca Acta, 74*, 1381–1390.

Dodd, J. P., Sharp, Z. D., Fawcett, P. J., Brearley A. J., & McCubbin F. M. (2012). Rapid post-mortem maturation of diatom silica oxygen isotope values. Geochemistry, Geophysics, Geosystems, *13*. https://doi.org/10.1029/2011gc004019.

Garreaud, R. D., Vuille, M., & Clement, A. C. (2003). The climate of the Altiplano: Observed current conditions and mechanisms of past changes. *Palaeogeography, Palaeoclimatology, Palaeoecology, 194,* 5–22.

Kutzbach, J. E. (1980). Estimate of past climate at paleolake Chad, North Africa, based on a hydrological and energy balance model. *Quaternary Research, 14,* 210–223.

Legesse, D., Vallet-Coulomb, C., & Gasse, F. (2004). Analysis of the hydrological response of a tropical terminal lake, lake Abiyata (Main Ethiopian Rift Valley) to changes in climate and human activities. *Journal of Hydrological Processes, 18,* 487–504.

Leng, M. J., & Barker, P. A. (2006). A review of the oxygen isotope composition of lacustrine diatom silica for palaeoclimate reconstruction. *Earth-Science Review, 75,* 5–27.

Moschen, R., Lücke, A., & Schleser, G. (2005). Sensitivity of biogenic silica oxygen isotopes to changes in surface water temperature and palaeoclimatology. *Geophysical Research Letters, 32,* L07708. https://doi.org/10.1029/2004GL022167.

Quesada, B., Sylvestre, F., Vimeux, F., Black, J., Paillès, C., Sonzogni, C. et al. (2015). Impact of Bolivian paleolake evaporation on the $\delta^{18}O$ of the Andean glaciers during the last deglaciation (18.5–11.7 ka): Diatom-inferred $\delta^{18}O$ values and hydro-isotopic modeling. *Quaternary Science Review, 120,* 93–106.

Servant, M., & Fontes, J. C. (1978). Les lacs quaternaires des hauts plateaux des Andes boliviennes: Premières interprétations paléoclimatiques. *Cahiers ORSTOM. Série Géologie, 10,* 9–23.

Sylvestre, F., Servant, M., Servant-Vildary, S., Causse, C., Fournier, M., & Ybert, J.-P. (1999). Lake-level chronology on the Southern Bolivian Altiplano (18–23°S) during late-glacial time and the early holocene. *Quaternary Research, 51,* 54–66.

Thompson, L. G., Davis, M. E., Mosley-Thompson, E., & Sowers, T. A. (1998). A 25,000-Year tropical climate history from bolivian ice cores. *Science, 282,* 1858–1864.

Vallet-Coulomb, C., Legesse, D., Gasse, F., Travi, Y., & Chernet, T. (2001). Lake evaporation estimates in tropical Africa (Lake Ziway, Ethiopia). *Journal of Hydrology, 245,* 1–18.

Vallet-Coulomb, C., Gasse, F., & Sonzogni, C. (2008). Seasonal evolution of the isotopic composition of atmospheric water vapour above a tropical lake: Deuterium excess and implication for water recycling. *Geochimica Cosmochimica Acta, 72,* 4661–4674.

Air-Ice Interface: Tropical Glaciers

Françoise Vimeux

Just as the polar caps are excellent archives of information on past climates, so too are the high-altitude tropical glaciers. Indeed, the prevailing temperature and humidity conditions there usually ensure very good preservation of chemical and isotopic tracers. These archives have therefore been used for the past 25 years to study the variability of the tropical climate over past centuries and millennia. They are mostly found in the Andes in South America between 0° and 20° S (Vimeux 2009; Vimeux et al. 2009), although drilling has been conducted on Kilimanjaro (Thompson et al. 2002) and the southern Himalayas (Thompson et al. 2000, 2006). The focus of this chapter is on climate information taken from the Andean glaciers.

The rapid dynamic of these glaciers, the high level of snow accumulation per year (0.5–1 m) and reduced ice thickness (100–150 m) mean that it is not possible to access climate periods as old as those possible in polar cores. The oldest cores date back to the last glacial maximum, about 20 000 years ago, with the Sajama core (Bolivia, 6542 m, Western Cordillera, 18° 06′ S, 68° 53′ W) dating back to ~ 25,000 years (Thompson et al. 1998). On the other hand, the tropical ice allows our climate to be studied with a very good temporal resolution, at the seasonal level for recent centuries.

Paleoclimate Markers

Several types of quantifiable climate variables can be extracted from tropical glaciers: the net accumulation, the temperature in the borehole and regional precipitation. We propose to review these and to present the main results in terms of climate variability. Although complementary, we will not discuss here the qualitative results from the chemical analysis of ice, which essentially informs us about changes in the environmental, atmospheric transport processes and air pollution.

Over recent centuries, the seasonal cycles of the chemical elements and stable isotopes in precipitation allow us to date the annual layers with a relatively good degree of accuracy (±5 to 10 years around 1900). It is thus possible to calculate the annual net accumulation. To correct for the effects of snow compaction in the depths, an ice flow model is applied, or alternatively, the thinning of the layers is corrected by directly observing the relationship between the annual thickness and the depth, along the core. The latter method cannot however be used to discuss climate trends which are, in principle, corrected using the same method. The question that then arises is: what does the net accumulation represent given that it is actually the combination of the total accumulation (controlled by the precipitation and wind) minus removal (sublimation, erosion by wind), and that the latter processes can have different seasonalities? If one considers sites where accumulation is low (0.31 m of water/year) with a high level of sublimation throughout a long dry season, such as at Cerro Tapado (Chile, arid diagonal, 5550 m, 30° 08′ S, 69° 55′ W), it is difficult to use this parameter as a marker of the amount of deposited precipitation. This is less the case for sites such as the one at Illimani (Bolivia, eastern cordillera, 6350 m, 16° 37′ S, 67° 46′ W) where the rainfall season is longer, the annual snow accumulation heavier (0.58 m of water/year) and the sublimation is concentrated over a shorter time of the year which is different to the accumulation season (Ginot et al. 2006). Only the end of the rainy season which represents about 10% of the annual accumulation may be truncated in the records. In that case, we can we assume that large variations in the net accumulation are representative of the amount of precipitation. The accumulation estimates made on the Quelccaya core (Peru, 5670 m, 13° 56′ S, 70° 50′ W, the only core with dating by

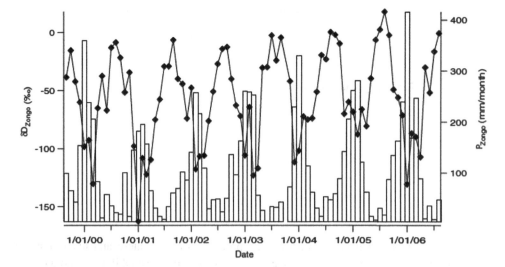

Fig. 20.1 Example of relationship between the isotopic composition of deuterium (‰) in rainfall collected in the Zongo Valley at multiple sites (Bolivia, 16° S, a valley linking the Andean peaks to the Amazon) (connected points) and the quantity of precipitation (mm/month) (bars), on a monthly scale over several years. During the rainy season, the isotopic composition is strongly depleted of heavy isotopes, while during the dry season, it is enriched. The correlation with local precipitation explains only 50% of the isotopic signal. The remainder of the variance can be explained by precipitation at the regional level. This figure is adapted from Vimeux et al. 2005

seasons over the past 1500 years) show sequences of high (the seventeenth and eighteenth centuries and in the twentieth century) and low (the nineteenth century) accumulations (Thompson et al. 2006).

As in polar ice, it is possible to measure the temperature in the borehole. This method does not allow past temperature changes to be reconstructed with a high temporal resolution, but it does offer the possibility of measuring slow fluctuations. In high altitude glaciers, this profile depends on the energy balance at the surface and on the flow of geothermal heat in the depths. The proximity of the ice to the bedrock does not allow these profiles to be applied to the second half of the cores although the surface profiles can be interpreted. In the cores where these profiles are available, an increase in temperature is observed for recent decades (Vimeux et al. 2009), reaching 1.1 °C in the twentieth century.

The measurement of the isotopic composition ($\delta^{18}O$ and δD) of tropical ice provides information on rainfall patterns caused by the air mass along its trajectory. The linear relationship between the isotopic composition of snow and surface air temperature, well established for the polar regions, does not hold true in the tropics. This relationship is mainly due to the fact that, in the middle and high latitudes, the amount of precipitation formed and air temperature are closely linked as per the Clausius-Clapeyron law. This is not the case in the tropics, where the majority of precipitation is convective and where the water cycle is complex (recycling of water vapor from the surface). The coupling of observations, through rainfall collection network systems, and modeling of atmospheric cycling of stable isotopes of water with a hierarchy of models (general circulation atmospheric model, mesoscale model correctly representing the topography and one dimension convection model) has shown that at the seasonal and interannual scale, the isotopic composition of Andean snow is strongly related to precipitation upstream from the drilling sites along the trajectories, in the Amazon and over the tropical Atlantic regions where the most intense convection phenomena are located (Vimeux et al. 2005; Vuille and Werner 2005; Vimeux et al. 2011) (Fig. 20.1). It was shown that the relationship between water isotopes in the Andes and precipitation is strongly dependent on convection conditions (re-evaporation of water droplets and recycling of the resulting vapor in the convective column) (Risi et al. 2008).

Some Important Results from the Interpretation of Andean Isotopic Records

Recent studies have sought to link the changes in precipitation in tropical South America to larger scale processes over the last century. Most of the interannual variability in rainfall for this region is linked to variations in intensity and geographical extension of the ascending and convective branch of the Hadley-Walker cell, affecting the South

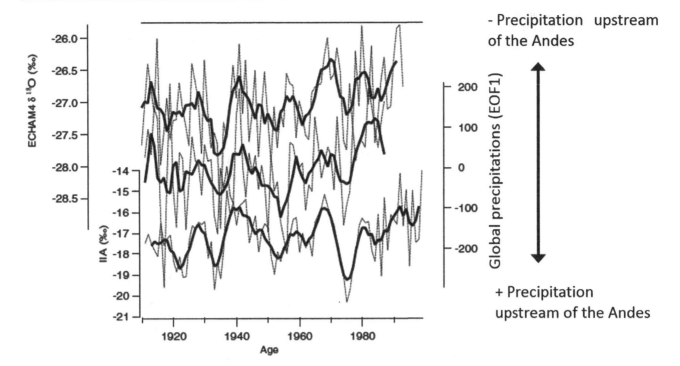

Fig. 20.2 The Andean Isotopic Index (AII) is built from the isotopic composition of four Andean ice cores having similar interannual variations and a sufficiently precise dating over the last century (in Bolivia: Illimani (16° S, 6300 m) and Sajama (18° S, 6542 m) in Peru: Huascaran (9° S, 6048 m) and Quelccaya (14° S, 5670 m)). It is compared here to: a) the first component of an analysis of the principal components of global precipitation (EOF1) which reflects the first mode of interannual climate variation that is ENSO; b) the isotopic composition of the oxygen-18 in Amazonian water vapor, simulated by the atmospheric model ECHAM-4. The lines in bold represent a moving average over 5 years. This figure is adapted from Hoffman et al. (2003)

American monsoon pattern. This cell is strongly disrupted by anomalies in the temperature of surface waters in the tropical Pacific.

We therefore think that ENSO events are likely to mark the isotopic composition of Andean ice (Bradley et al. 2003; Hoffmann et al. 2003). The isotopic composition recorded in several Andean cores over the last century effectively shows a common decadal signal, called Isotopic Andean Index strongly linked to rainfall variations in the Amazon, caused by the ENSO phenomenon (Fig. 20.2). However, variations in temperature of the surface waters of the tropical Atlantic also have a strong impact on the South American monsoon system (and so on regional precipitation), and it becomes difficult to untangle the different causes in the isotopic signal (Hoffmann 2003).

The Illimani and Quelccaya ice cores with their precise dating back to about 1700 show a significant loss of about 1.5‰ of $\delta^{18}O$ between the late seventeenth and early nineteenth century. This period also corresponds to the maximum spread of glacial moraines in Bolivia, Peru and Ecuador, dated by lichenometry (Jomelli et al. 2009). This comparison between the spread of glaciers and the $\delta^{18}O$ of ice suggests that during the Little Ice Age, the tropical Andes were wetter (depletion of isotopes and higher accumulation on the glaciers to increase the mass at their base) and colder (maximum spread of the glaciers).

At the glacial-interglacial scale, isotope profiles obtained on three Bolivian and Peruvian cores (Illimani; Huascaran, Peru, 6050 m, 9° 06′ S, 77° 30′ W and Sajama) show a common isotope signal in terms of variability and amplitude (Fig. 20.3). This signal, very similar to that recorded by the isotopic composition in the polar cores, highlights highly depleted values during the last glacial maximum and a gradual enrichment during deglaciation, with a return to depleted conditions before reaching an optimum around 11,000 years. This structure is similar to that described in polar ice, where known climate periods are recorded (glacial maximum, Younger Dryas and the Holocene optimum), although the interpretation of the isotopes in Andean ice (moisture) is different from that of polar ice (temperature). The translation of this glacial-interglacial variation in terms of humidity shows that 20,000 years ago, the air masses sustained a more major drain along their trajectories, corresponding to a precipitation increase of about 10% (Vimeux et al. 2005).

Fig. 20.3 Isotopic composition in deuterium (‰) over the last meters of the Illimani core and isotopic composition of oxygen (‰) of air trapped in the bubbles

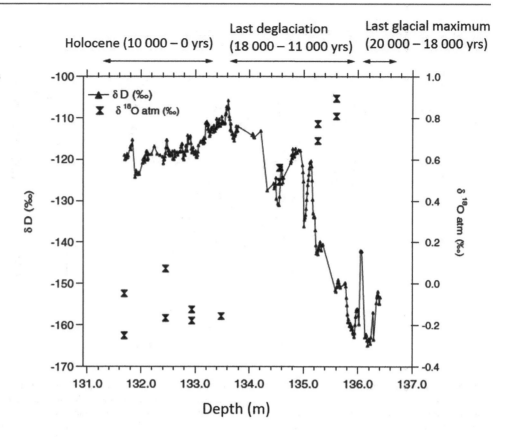

A study using coupled ocean-atmosphere climate models for this glacial period shows that rainfall was indeed more intense 20,000 years ago in Northeast Brazil and the tropical South Atlantic, the cradle of Andean precipitation (Khodri 2009).

References

Bradley, R. S., Vuille, M. Hardy, D., & Thompson, L. G. (2003). Low latitude ice cores record Pacific sea surface temperatures. *Geophysical Research Letters, 30*. https://doi.org/10.1029/2002gl016546.

Gilbert, A., Wagnon, P., Vincent, C., Ginot, P., & Funk, M. (2010). Atmospheric warming at a high-elevation tropical site revealed by englacial temperatures at Illimani, Bolivia (6340 m above sea level, 16° S, 67° W). *Journal Geophysical Research, 115*, D10109. https://doi.org/10.1029/2009JD012961.

Ginot, P., Kull, C., Schotterer, U., Schwikowski, M., & Gäggeler, H. W. (2006). Glacier mass balance reconstruction by sublimation induced enrichment of chemical species on Cerro Tapado (Chilean Andes). *Climate of the Past, 2*, 21–30.

Hoffmann, G., Ramirez, E., Taupin, J. D., Francou, B., Ribstein, P., Delmas, R., Dürr, H., Gallaire, R., Simoes, J., Schotterer, U., Stievenard, M., & Werner, M. (2003). Coherent isotope history of Andean ice cores over the last century. Geophysical Research Letters, 30, 1179–1184.

Hoffmann, G. (2003). Taking the pulse of the tropical water cycle. *Science, 301*, 776–777.

Jomelli, V., Favier, V., Rabatel, A., Brunstein, D., Hoffmann, G., & Francou, B. (2009). Fluctuations of glaciers in the tropical andes over the last millennium and palaeoclimatic implications: A review. *Palaeogeography, Palaeoclimatology, Palaeoecology, 281*, 269–282. https://doi.org/10.1016/j.palaeo.2008.10.033.

Khodri, M. (2009). Sensitivity of South American tropical climate to last glacial maximum boundary conditions: Focus on teleconnections with tropics and extratropics. In F. Vimeux, F. Sylvestre, & M. Khodri (Eds.), *Developments in paleoenvironmental research series (DPER)* (Vol. 14, XVII, 418 p. 106 illus., 61 in color). Springer. ISBN 978-90-481-2671-2.

Risi, C., Bony, S., & Vimeux, F. (2008). Influence of convective processes on the isotopic composition ($\delta^{18}O$ and δD) of precipitation and water vapor in the tropics, part 2: Physical interpretation of the amount effect. *Journal of Geophysical Research, 113*, D19306. https://doi.org/10.1029/2008JD009943.

Thompson, L. G., Davis, M. E., Mosley-Thompson, E., Sowers, T.A., Henderson, K. A, Zagorodnov, V. S., Lin, P.-N., Mikhalenko, V. N., Campen, R. K., Bolzan, J. F., Cole-Dai, J., & Francou, B. (1998). A 25,000-Year tropical climate history from bolivian ice cores. *Science, 282*, 1858–1864.

Thompson, L. G., Yao, T., Mosley-Thompson, E., Davis, M. E, Henderson, K. A., & Lin, P. -N. (2000). A high-resolution millennial record of the South Asian monsoon from himalayan ice cores. *Science, 289*, 1916–1919.

Thompson, L. G., Mosley Thompson, E., Davis, M. E., Henderson, K. A., Brecher, H. H., Zagorodnov, V. S., et al. (2002). Kilimanjaro ice core records: Evidence of Holocene climate change in tropical Africa. *Science, 298*, 589–593.

Thompson, L. G., Mosley Thompson, E., Brecher, H., Davis,M., Leon, B., Les, D., Lin, P.-N., Mashiotta, T., & Mountain, K. (2006). Abrupt tropical climate change: Past and present. *Proceedings of the National Academy of Sciences, 103*, 10536–10543.

Vimeux, F., Gallaire, R., Bony, S., Hoffmann, G., Chiang, J., & Fuertes, R. (2005). What are the climate controls on isotopic composition (δD) of precipitation in Zongo Valley (Bolivia)? Implications for the Illimani ice core interpretation. *Earth and Planetary Science Letters, 240,* 205–220.

Vimeux, F., Ginot, P., Schwikowski, M., Vuille, M., Hoffmann, G., Thompson, L. G., et al. (2009). Climate variability during the last 1000 years inferred from Andean ice cores: A review of methodology and recent results. *Palaeogeography, Palaeoclimatology, Palaeoecology, 281,* 229–241. https://doi.org/10.1016/j.palaeo.2008.03.054.

Vimeux, F. (2009). Similarities and discrepancies between polar and andean ice cores over the last deglaciation in past climate variability from the last glacial maximum to the holocene in South America and surrounding regions. In F. Vimeux, F. Sylvestre, M. et Khodri (Eds.), *Developments in paleoenvironmental research series* (DPER) (Vol. 14, XVII, 418 p., 106 illus., 61 in color.). Springer. ISBN 978-90-481-2671-2.

Vimeux, F., Tremoy, G., Risi, C., & Gallaire, R. (2011). A strong control of the South American see-saw on the intra-seasonal variability of the isotopic composition of precipitation in the Bolivian Andes. *Earth and Planetary Sciences Letters, 307,* 47–58.

Vuille, M., & Werner, M. (2005). Stable isotopes in precipitation recording South American summer monsoon and ENSO variability: Observations and model results. *Climate Dynamics, 25.* https://doi.org/10.1007/s00382-005-0049-9.

Climate and the Evolution of the Ocean: The Paleoceanographic Data

Thibaut Caley, Natalia Vázquez Riveiros, Laurent Labeyrie, Elsa Cortijo, and Jean-Claude Duplessy

Introduction: The Development of Tools and Concepts

The idea of reconstructing the history of oceans and climates in the past using marine sediment cores arrived quite late after the beginnings of oceanography. It was initiated in the twentieth century, well after the first attempts to measure variations in seawater temperature down the water column, which date back to the eighteenth century with the great circumnavigation expeditions. Land geologists were the first to propose paleoceanographic reconstructions from exposed marine series, limiting the collected information to former coastal waters. The first reconstructions of past seawater temperatures were made possible by the piston corer developed by the Swedish oceanographer Kullenberg, capable of collecting continuous sedimentary deposits without layer disruption. Geologists were therefore able to collect uninterrupted sedimentary series, sometimes over 20 m in length, for laboratory analysis, and thus study long records of the environmental conditions from the time the sediments were deposited. With this type of corer, the Swedish expedition of 1947–1948 collected over 300 different cores from various deep ocean basins that became the basis of the first studies on the geological history of the oceans. In parallel, during the 1950s, Maurice Ewing, the founder of the Lamont-Doherty Geological Observatory (USA), and one of the developers of seismic sediment mapping of the ocean floor, initiated the first systematic collection of marine sediment cores. The first descriptions of the main sedimentary systems, changes in fossil faunas and the timeframe for the first biostratigraphic age scales were proposed based on these cores. One of the main results of these studies was the continuous reconstruction of the alternating warm and cold phases that took place during the Pleistocene.

Unquestionably, the honor for the initiation of quantitative paleoceanography belongs to Cesare Emiliani. After the discovery of isotopic fractionation and the development of an accurate method to measure isotopic ratios, Harold Urey and his group in Chicago refined the use of the isotopic ratio $^{18}O/^{16}O$ in fossil carbonates as a paleothermometer. They realized that the $^{18}O/^{16}O$ ratio of foraminiferal shells and other carbonates depended on two variables: the temperature and the $^{18}O/^{16}O$ ratio of the water where the carbonate was formed. Changes in water temperature are reflected in variations in the isotopic fractionation between the carbonate and the water during the formation of the shell: for water with a given isotopic composition, the higher the temperature, the lower the $^{18}O/^{16}O$ ratio (Epstein et al. 1951, 1953). Emiliani (1955) applied this tool to foraminifera shells sampled along a sediment core from the Caribbean Sea to propose the first reconstruction of the variations in sea surface temperature (SST) over the past 400 ka (Fig. 21.1). He also established the major methodological guidelines for this type of study: use of continuous records, precise dating, and interpretation of parameters quantitatively linked to key variables of the climate system.

To extract a temperature signal from the $^{18}O/^{16}O$ ratio of planktonic foraminifera, Emiliani had to constrain the changes in the isotopic composition of the ocean water in which the foraminifera developed. The $^{18}O/^{16}O$ ratio of

---

Thibaut Caley, Natalia Vázquez Riveiros and Laurent Labeyrie—These authors have contributed equally to this chapter.

T. Caley (✉)
EPOC, UMR 5805, CNRS, University of Bordeaux, Pessac, France
e-mail: thibaut.caley@u-bordeaux.fr

N. V. Riveiros · L. Labeyrie · E. Cortijo · J.-C. Duplessy
Laboratoire des Sciences du Climat et de l'Environnement, LSCE/IPSL, CEA-CNRS-UVSQ, Université Paris-Saclay, 91190 Gif-sur-Yvette, France

N. V. Riveiros
Institut Français de Recherche pour l'Exploitation de la Mer (Ifremer), Unité de Geosciences Marines, Pointe du Diable, 29280 Plouzané, France

L. Labeyrie
LGO Université Bretagne Sud, 56000 Vannes, France

Fig. 21.1 a Cesare Emiliani, founder of isotopic marine paleoclimatology, in the early 1950s at the University of Chicago. (Photo from the archives of the Rosenstiel School of Marine and Atmospheric Science, University of Miami). **b** First attempt to evaluate surface water temperature changes in the Caribbean Sea, Emiliani (1955). Subsequent studies have shown that the timescale was underestimated by about 25% (the last interglacial, called Sangamon in American literature, is dated at about 125,000 years and not 100,000 years), and that the amplitude of temperature variations, calculated from a simple model (see below), was overestimated

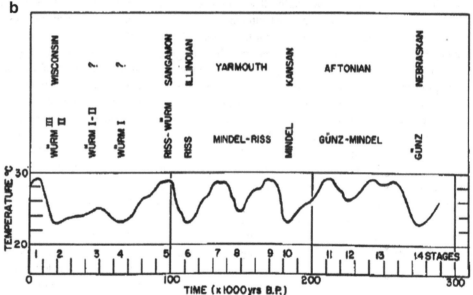

seawater is affected by evaporation and precipitation: the vapor phase is depleted in the heavy ^{18}O isotopes relative to the liquid phase. Conversely, when the water vapor in clouds condenses, the precipitation is richer in ^{18}O than the vapor. Thus, the transport of air masses from low to high latitudes is accompanied by a large-scale isotopic distillation process in the water vapor that results in the gradual decline of the $^{18}O/^{16}O$ ratio in precipitation. For this reason, the $^{18}O/^{16}O$ ratio of snowfall feeding the high-latitude ice caps is depleted by more than 30‰ compared to that of the tropical ocean.

The growth and melting of ice caps, which involves considerable volumes of water (several million cubic kilometers), directly affect the salinity and the average $^{18}O/^{16}O$ ratio of the ocean, and therefore that of the foraminifera that develop there. Regional climate changes are also accompanied by local variations in evaporation and precipitation, which induce further regional variations in the salinity of surface seawater and its $^{18}O/^{16}O$ ratio.

From the data available at the time, Emiliani (1955) estimated that the development of large ice sheets covering Canada (the Laurentide ice sheet) and northern Europe (the

Fennoscandian ice sheet) at the maximum of the glaciation caused an enrichment in ^{18}O of the global ocean of +0.4‰. We now know that the enrichment in ^{18}O in the glacial ocean was in fact close to +1.0‰ (Schrag et al. 2002). Despite these inaccuracies, the work of Emiliani was the first to demonstrate from paleoclimate observations that glacial—interglacial periods indeed oscillated with cyclicities predicted by the Milankovitch theory several decades earlier (see Chap. 28). Emiliani also proposed the 'Marine Isotopic Stage' (MIS) nomenclature, now universally adopted, to characterize the alternation of warm and cold Pleistocene phases, with odd numbers for interglacial periods and even numbers for glacial ones (1 for the Holocene, 2 for the last glacial period, and so on). He also discovered that the last interglacial, or MIS 5 (Fig. 21.1), was interrupted by two colder periods, which led him to divide it into three warm subperiods (designated 5a, 5c and 5e from the most recent to the oldest) and two cold ones (5b and 5d). The term '5e' is still frequently used, as it has been incorporated into the European continental reconstructions as the equivalent to the Eemian warm period. The isotopic stratigraphy formalism has since been generalized, with subdivisions either numbered as decimals between alternating warmer (e.g., 5.1 for 5a, 5.3 for 5c and 5.5 for 5e) and colder (5.2 and 5.4) periods (Pisias et al. 1984) or as letters (Railsback et al. 2015).

With the assumption that past variations in foraminiferal $^{18}O/^{16}O$ ratios in cores from different ocean basins had to be approximately synchronous across global climate changes, Emiliani paved the way for a global marine isotopic stratigraphy. The demonstration that the volume of ice caps was indeed the dominant component of the isotopic signal recorded in marine cores reinforced the stratigraphic value of the marine isotopic stage age scale, which became a major reference tool for past climate change studies. The routine use of drilling ships as part of the International Ocean Drilling Program has allowed the recovery of sediment cores that cover the last tens of millions of years, extending the isotopic sequences not only to the Quaternary (Fig. 21.2), but as far back as the Paleocene, 60 Ma ago.

By 1970–1980, the paleoclimate community had arrived at the conclusion that variations in the oxygen isotopic composition provided a remarkable stratigraphic tool to establish long-term correlations. However, new tools still needed to be developed to precisely reconstruct past SST, as well as variations in other oceanic features such as salinity, or the direction and intensity of deep-water currents.

This chapter will focus on the development of classic and new paleoceanographic tracers over the last decades. We will mainly, but not exclusively, focus on tracers that are based on foraminifera, since these abundant microfossils have been extensively used because of their ubiquity in the oceans and their great preservation potential. Their faunal associations and isotopic composition have been widely used as paleoclimate indicators for decades, and the advent of relatively new indicators implies an undiminished interest. Other indicators such as corals will be briefly discussed in association with some of the tracers.

Sea Surface Temperature

The surface temperature of the ocean is an essential climate parameter that governs heat exchange with the atmosphere. SST also modulates the solubility of gases, oxygen and CO_2 in particular, and their exchange rates with the atmosphere. The amplitude of the spatial variability of SST is well known: it ranges between −1.96 °C, freezing point for seawater at 35 psu, and 30–35 °C, maximum temperature recorded for the open ocean. However, its temporal variability is more difficult to constrain because it not only varies on a daily basis, but also seasonally and annually. In situ measurement sensors are precise to ±0.001 °C at a given location, water depth and time. Satellite data provide global coverage and allow long-term monitoring of the evolution of SST, but their accuracy is, at best, close to 0.1 °C, and surface values are averaged over tens or hundreds of square kilometers. Paleoceanographers cannot aim to reconstruct SST variations with this level of precision. Nonetheless, given the magnitude of changes in the past, relevant information may be acquired when SST changes are estimated to the nearest degree. Paleoceanographers also aim to estimate, whenever possible, the amplitude of the seasonal cycle, and the temperature distribution of the upper water column.

For these studies, two major types of paleotemperature indicators are used: (i) changes in the distribution of fossil planktonic flora or fauna (foraminifera, diatoms, dinoflagellates, radiolarians), and (ii) geochemical tracers produced by these organisms or recorded in their fossil skeletons.

The Distribution of Marine Fauna and Flora

The distribution of the various groups that make up the marine planktonic ecosystem was extensively studied during the major exploration campaigns that marked the nineteenth and early twentieth centuries. Foraminifera, single-celled protozoans that secrete a calcareous shell, were the most generally recorded group. They are very diverse and inhabit all the oceans, from the coldest to the warmest. However, each individual species has a limited tolerance to environmental changes, particularly temperature, which allowed biologists at the end of the nineteenth century to highlight the zonal distribution of many species. They also established existing relationships between climate and the abundance of certain

Fig. 21.2 Variations in the isotopic composition of the oxygen in benthic foraminifera from 70 million years ago (Zachos et al. 2008) with a detail of the last 1.8 Ma showing periodicities predicted by the astronomical theory of paleoclimate (Lisiecki and Raymo 2005). Presence of Northern Hemisphere and Antarctic ice sheets are from Zachos et al. (2008). Geological epochs (Pl. = Pleistocene, Plio = Pliocene) are indicated at the bottom of the top panel

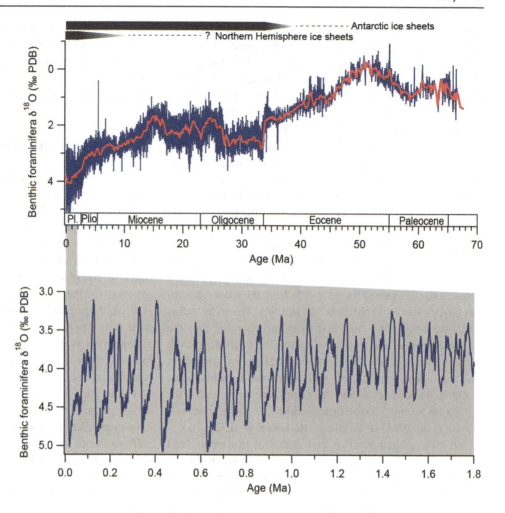

species. These observations, along with the statistical method of 'transfer functions' developed by Imbrie and Kipp, served as a basis for the quantitative reconstruction of surface temperatures from fossil fauna found in marine sediments. Transfer functions allow the estimation of seawater temperature during both the cold and warm seasons, when adequate data is available. The basic principle for the reconstruction of SST changes in the past is to assume that the ecological requirements of the species present nowadays have not changed since the past period under consideration. The specific abundance of planktonic foraminifera samples from recent core tops may be expressed as a matrix (x species relative distributions within each of n sampling stations) from which the vectors corresponding to the main factors that describe the specific faunal (or floral) variance can be extracted. The method proposed by Imbrie and Kipp (1971) provides the best correlation between changes in these factors and the associated changes within the modern environmental parameters, systematically selected for the n stations (such as summer and winter temperatures, or other parameters, provided they are statistically independent). A similar factor analysis, when applied to fossil assemblages down sediment cores, allows the paleotemperature to be estimated using the ecological equations calculated from the modern core tops. This work led to the great success of the CLIMAP group (Climate Long-range Investigation, Mapping And Prediction), which reconstructed the first global map of summer and winter SST distribution during the Last Glacial Maximum (LGM) period, about 21 ka ago (Fig. 21.3).

One of the main problems with this paleo-reconstruction method is that it is based on the a priori stable statistical correlation between changes in the specific factors defined by the species distributions and the arbitrarily chosen environmental parameter, temperature, in this case. However, other environmental factors, such as the availability of food supply, may also be involved and change the sensitivity of foraminifera to temperature from one region to another.

There is also another potential problem with the transfer function approach: it will provide accurate estimates only if modern conditions are good analogs of the past hydrological conditions. This is not always necessarily the case. For example, the fossil fauna of the eastern Mediterranean Sea during the LGM period has no modern analogs. Indeed, this basin experienced hydrological and climate conditions very

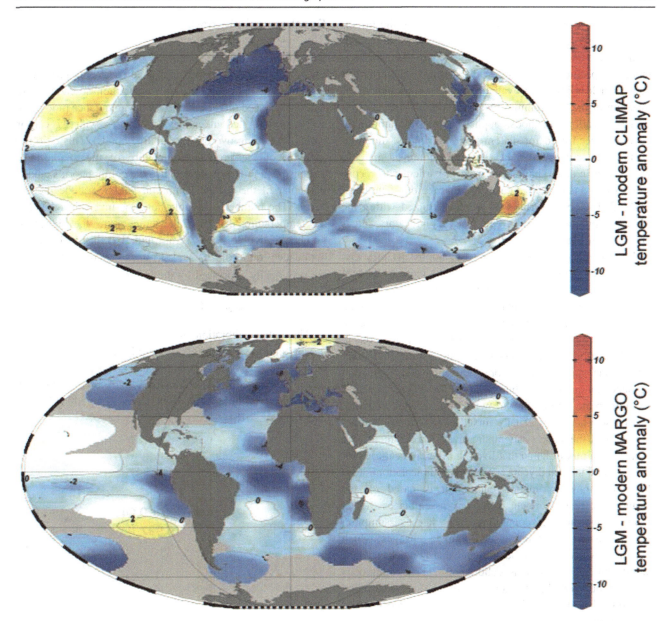

Fig. 21.3 Reconstructions of temperature anomalies (°C) between the Last Glacial Maximum (LGM) and modern surface waters as obtained by the CLIMAP (1981) (August temperatures, top figure) and MARGO (2009) programs (July–August–September temperatures, bottom figure). While the two reconstructions have many features in common, the cores studied under the MARGO program revealed the presence of larger longitudinal gradients in SST in all ocean basins than the estimates of CLIMAP (1981)

different from the modern ones, due to the development of large ice sheets over Northern Europe and to a very different hydrological cycle. In fact, the absence of direct analogs becomes the rule rather than the exception for the distant past: most fossil species from periods earlier than the Quaternary are not found in recently deposited sediments.

This is why the method used by Imbrie and Kipp (1971) was gradually replaced by the *best analogs* method. Its principle is simple: to compare the changes in fossil assemblages to modern references, without a preconceived idea of the origin of the observed changes. The closest analogs are defined using a mathematical distance calculation. The method is based on the assumption that the closer a fossil assemblage is to one or more modern references, the more similar their optimal growth conditions are (temperature and nutrient supply, in particular). The distance to the best analogs and the dispersion of associated environmental conditions provide an estimate of the reconstruction uncertainties (Waelbroeck et al. 1998).

In parallel with these statistical developments, recent advances in the field of artificial intelligence and neural networks have helped to improve paleoceanographic reconstructions without fundamentally changing the principle. Still based on the comparison between fossil fauna and modern reference fauna, they do not require the establishment of specific mathematical relationships, but establish their own learning from available databases through minimization of the uncertainties (Malmgren et al. 2001).

These various methods, applied here to foraminifera, have also been used for diatom flora, coccoliths, dinoflagellate cysts and radiolarians. Diatoms, dominant in cold waters rich in silica, have been used in particular to reconstruct variations in sea ice cover in polar regions. The latest LGM SST global distribution, as reconstructed within the MARGO program, used several types of indicators and is presented in Fig. 21.3.

Statistical methods have many limitations: (i) they can only be used if the faunal assemblages are close to modern assemblages; (ii) the genetic diversity of species among different ocean regions may induce variations in their faunal responses to temperature and the corresponding statistical links; (iii) temperature reconstructions based on variations in the abundance of fossil fauna or flora assume that other factors, such as productivity for example, have no significant influence on the relative abundances of the different species; (iv) due to the activity of burrowing animals (bioturbation), marine sediments are usually mixed over several centimeters, so that the same stratigraphic level of a sediment core represents a mix of fauna that lived in different centuries (or even several thousand years apart if the sedimentation rate is low); (v) transfer function calibration is based on the assumption that sediment core top assemblages reflect modern hydrological conditions. This latest assumption ignores in particular ocean and climate changes which occurred over the last millennia, and that could have been significant enough to bias calibrations. These limitations have encouraged the development of new reconstruction methods based on either biological or geochemical mechanisms.

The biological approach is still in its infancy, and it is derived from ecological studies of the requirements of the different species in the modern ocean. A first approach directly calibrates the proxies (foraminifera or others) from controlled laboratory cultures with varying physiological and geochemical constraints that duplicate those observed in the marine environment. A more theoretical approach complements these calibrations by modeling the growth conditions within the natural environment, using the experimentally calibrated variables. Using such methods, it may be possible to obtain a reliable reconstruction of the hydrology corresponding to the specific habitat of the different species of planktonic foraminifera (e.g., Lombard et al. 2009).

Geochemical Methods

Organic Tracers

The organic geochemistry of marine sediments provides a different set of tracers. The most common so-called 'biomarker' is based on the changes in the abundance ratio of di- and tri-unsaturated alkenones (molecules with 37 carbon atoms containing two or three double bonds). This ratio is a function of the growth temperature of the synthesizing organisms, a group of algae called coccolithophorids, and in particular of the species *Emiliania huxleyi* for the modern ocean. The number of double bonds is inversely related to the temperature: the lower the temperature, the higher the number of double bonds (Prahl and Wakeham 1987). The abundance ratio of di- and tri-unsaturated alkenones is conventionally expressed by the index $U^{k'}_{37}$:

$$U^{k'}_{37} = [C_{37:2}]/[C_{37:2} + C_{37:3}]$$

The initial calibration of the $U^{k'}_{37}$ index is based on *E. huxleyi* cultures in controlled conditions (Prahl and Wakeham 1987) (Fig. 21.4), and has been subsequently verified using samples collected from ocean water or from the sediment surface (Müller et al. 1998; Conte et al. 2006; Tierney and Tingley 2018).

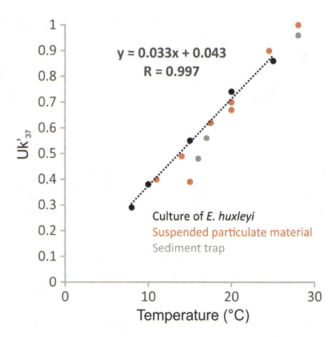

Fig. 21.4 Relationship between the unsaturation index $U^{k'}_{37}$ and SST. The line represents the temperature calibration curve based on cultures of *E. huxleyi* grown under laboratory conditions (Prahl and Wakeham 1987). Natural particulate samples collected are indicated (Prahl and Wakeham 1987)

However, like all paleoclimate indicators, the $U_{37}^{k'}$ index also presents various biases that limit its use in specific oceanographic contexts. A few of the most significant ones are as follows:

- The temporal evolution of the alkenone-producing species. *E. huxleyi*, which is currently the main producer of alkenones, was not present prior to MIS 8 (Thierstein et al. 1977). Alkenones are also produced by other species such as *Gephyrocapsa oceanica*, but with different temperature—$U_{37}^{k'}$ index relationships. Care must therefore be taken when applying the $U_{37}^{k'}$ method to ancient sediments.
- The tiny coccolithophorids are easily transported by sea currents, so these algae can travel long distances between their place of production and place of sedimentation. A significant portion of the residual input to sediments may thus originate from remote areas with very different hydrological conditions. This problem is particularly significant in areas of low productivity or in frontal zones that separate two distinctly different water masses (Sicre et al. 2005).
- Reconstructions could be biased toward a specific season (Rosell-Melé and Prahl 2013) and a degree of nonlinearity may exist in the relation between alkenones and SST at the higher and lower ends of the temperature range (Conte et al. 2006).

Another organic tracer to reconstruct past SST is based on the quantification of the average number of cyclopentane rings found in glycerol dialkyl glycerol tetraethers (GDGTs) of archaea membrane lipids. An index, called TEX$_{86}$, was deduced after analyzing the GDGTs distribution in marine surface sediments in comparison to annual mean SSTs (Schouten et al. 2002).

Recently, a number of different TEX$_{86}$ calibrations have been developed (Kim et al. 2010; Tierney and Tingley 2014; Ho and Laepple 2016), in response to possible differences in membrane adaptation of the resident archaea communities at different temperatures, and to the differences found between the TEX$_{86}$ ratio and other SST reconstruction proxies. A number of pre- and post-depositional processes can influence the TEX$_{86}$ ratio. For some processes, this influence can be constrained. For example, the BIT index is used to track the amount of terrestrial GDGT input, using a ratio of branched versus isoprenoid GDGTs (Weijers et al. 2006; Schouten et al. 2013). Nonetheless, the scientific understanding of TEX$_{86}$ remains imperfect, particularly since the effects of environmental factors such as salinity, nutrient concentrations, and water column structure may modulate the TEX$_{86}$–SST relationship (Tierney and Tingley 2014).

Chemical Tracers

The chemical composition of the carbonate from foraminiferal tests and coral skeletons may also provide paleotemperature or paleoenvironment indicators. For example, the concentration of magnesium incorporated in the calcium carbonate of foraminifera is an empirical function of the temperature at which that foraminifer crystallized its test. On time scales where the Mg/Ca of the oceans has remained constant, the sensitivity of Mg/Ca to temperature has been determined using either a culture-based, sediment trap or core top calibrations (see for example, Lea et al. 1999; Elderfield and Ganssen 2000; Anand et al. 2003; Mashiotta et al. 1999) (Fig. 21.5), and it takes the form:

$$\mathrm{Mg/Ca} = B \exp(A * T)$$

where A and B are the exponential and pre-exponential constants, respectively, and T is the temperature in °C.

Magnesium replaces calcium more easily at high than low temperatures, so the Mg/Ca ratio from carbonates increases with temperature at the time of calcite formation. Thermodynamic considerations suggested an exponential temperature dependence of Mg uptake into calcite (Rosenthal et al. 1997).

However, the growth temperature is not the only factor to be considered. Seawater salinity and alkalinity have also been shown to significantly alter the Mg/Ca ratio in

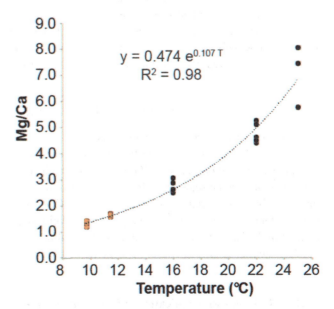

Fig. 21.5 Mg-temperature calibration results from culturing experiments with one species of planktonic foraminifera (*Globigerina bulloides*) (black dots) and core top samples (orange dots). Mg/Ca results are plotted versus calibration temperature (black dots) or World Ocean Atlas mean annual SST (orange dots) (modified from Mashiotta et al. 1999)

foraminiferal tests (Elderfield et al. 2006; Mathien-Blard and Bassinot 2009; Hönisch et al. 2013). Another issue is the observed offsets in both culture and field studies between Mg/Ca ratios among individual species, that indicate the need for single-species calibrations. In addition, the geographical extension of genotypes must be assessed when choosing to develop calibrations (Vázquez Riveiros et al. 2016). Recent studies try to simultaneously assess the relationship between foraminiferal Mg/Ca, and temperature, salinity, and the carbonate system using statistical approaches (Khider et al. 2015; Gray et al. 2018).

Mg/Ca in foraminifera has been cited here as a main example of a geochemical temperature tracer. Other ratios, such as Sr/Ca or Li/Mg in corals, are also used as tracers of temperature (Corrège 2006; Montagna et al. 2014).

Isotopic Tracers

As presented in this chapter's introduction, the first isotopic approach developed was the relationship between temperature, the isotopic composition of seawater and the isotopic composition of the biocarbonate that developed within that water. Traditionally, isotopic compositions are expressed using the notation δ, which is the relative difference (expressed in parts per thousand) between the isotope ratio R of the sample and that of a reference standard:

$$\delta = [(R_{sample}/R_{st}) - 1] \times 1000$$

The $\delta^{18}O$ of the water is denoted δ_w, that of a carbonate, δ_c, and the relationship between temperature T, δ_w and δ_c is known as the 'paleotemperature equation'. This relationship was experimentally determined by Urey's group in the 1950s and improved by Shackleton (1974) in the form below:

$$T = 16.9 - 4.38 \times (\delta_c - \delta_w) + 0.10 \times (\delta_c - \delta_w)^2 \quad (21.1)$$

In this empirical formula, δ_c represents the $\delta^{18}O$ of the CO_2 extracted from the carbonate through dissolution with phosphoric acid, and δ_w is the $\delta^{18}O$ of the CO_2 obtained by equilibration with the seawater to be analyzed. δ_c and δ_w are measured by mass spectrometry using the same CO_2 laboratory standard. Other $\delta^{18}O$—temperature relationships defined in the last decades (Bemis et al. 1998; Mulitza et al. 2003; Marchitto et al. 2014) use the same terminology.

Box 1. Practical Application of the Paleotemperature Formula

Nowadays, isotopic geochemistry laboratories have adopted the convention of expressing the δ_c isotopic compositions against the PDB (Pee-Dee Belemnite) international standard and δ_w against the SMOW (Standard Mean Ocean Water) international standard. These standards are distributed by international agencies for laboratory calibration. To properly apply the paleotemperature formula, which presumes that all isotopic compositions are expressed relative to the same standard, it is necessary to compare the isotopic ratio of the CO_2 extracted from the PDB standard by controlled phosphoric acid attack with the isotopic ratio of the CO_2 isotopically equilibrated with the SMOW standard. The latter is lower in ^{18}O content by 0.27‰ than the CO_2 extracted from PDB, so that for every water sample:

$$\delta_w(\text{vs.PDB} - CO_2) = \delta_w(\text{vs.SMOW} - CO_2) - 0.27. \quad (21.2)$$

If, as in the paleotemperature formula, PDB is used as the standard for carbonates and SMOW as the standard for waters, then Shackleton's equation becomes:

$$T = 16.9 - 4.38 \times (\delta_c - \delta_w + 0.27) + 0.10 \times (\delta_c - \delta_w + 0.27)^2.$$

One major disadvantage of the paleotemperature formula is that temperatures can only be determined if the isotopic composition of the water is known, which is almost never the case for geological samples.

A more recent isotopic method, still under development, is expected to overcome this constraint (Ghosh et al. 2006; Schauble et al. 2006). The crystal lattice of a carbonate consists of CO_3^{2-} groups and of cations (Ca^{2+}, for example). Among the CO_3^{2-} ions in a sample, the heavy isotopes ^{13}C and ^{18}O do not spread out randomly. Their relative abundance will depend on the isotopic equilibrium reaction:

$$^{13}C^{16}O_3^{(2-)} + {}^{12}C^{18}O^{16}O_2^{(2-)} \Leftrightarrow {}^{13}C^{18}O^{16}O_2^{(2-)} + {}^{12}C^{16}O_3^{(2-)}$$

so that the distribution of these four isotopic species depends on their own binding energy, itself a function of temperature.

The abundance of the various isotopic species is assessed by dissolving the carbonate with phosphoric acid and measuring the abundance of $^{13}C^{18}O^{16}O$ molecules (with a mass of 47) in the extracted CO_2, and comparing this to the abundances of other isotopic species with masses 45 and 46. The 'stochastic' state is taken as a reference and is defined by a random distribution of the isotopes of C and O within the molecules.

The thermodynamic variable, denoted as $\Delta 47$, which describes the state of the carbon dioxide and from which we deduce a paleotemperature, is defined by the relationship:

$$\Delta 47 = [(R47/R47*-1) - (R46/R46*-1) \\ - (R45/R45*-1)] \times 1000$$

where $R45$, $R46$ and $R47$ are the abundance ratios of masses 45/44, 46/44 and 47/44, respectively, in the CO_2, and where $R\,45*$, $R\,46*$ and $R\,47*$ denote these ratios for a gas with the same overall composition but in 'stochastic' state (Ghosh et al. 2006).

The main advantage of this method (called the $\Delta 47$ method) is that the thermodynamic-measured value reflects an internal balance of the crystal lattice and requires no knowledge of the composition of the original water. This method is based on properties obeying thermodynamical principles, so it can be applied to a variety of environments without changes (Fig. 21.6).

However, as it is the case for the paleotemperature formula, the calibration performed by Ghosh et al. (2006) only applies to carbonates precipitated at thermodynamic equilibrium. Therefore, testing for the possible existence of effects of parasitic isotopic fractionation of kinetic, biological or diagenetic origin should be carried out (Eiler et al. 2014; Saenger et al. 2012).

The temperature sensitivity of the $\Delta 47$ proxy is low (~0.003‰/°C) (Kele et al. 2015). It requires high measurement precision, which is commonly achieved by increasing counting times and/or the number of replicates analyzed per sample, a challenge for foraminifer-based reconstructions that use low carbonate samples. Recent studies have focused on the development of precise standardized calibrations that are applicable to paleoceanographic studies (Peral et al. 2018).

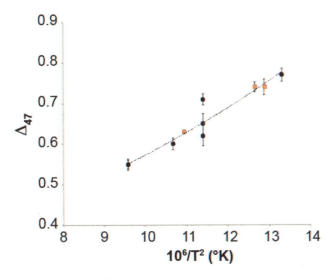

Fig. 21.6 Δ_{47} of CO_2 extracted from calcites grown from aqueous solution and of deep-sea corals (in orange) and surface corals (in black), plotted against $10^6/T^2$, where T is the known growth temperature in Kelvin (modified from Ghosh et al. 2006)

Sea Surface Salinity

While the temperature of seawater varies over a range of more than 30 °C in the open ocean, salinity changes much less (between 33 and 38 g of salt per liter (psu)). Salinity is highest in tropical areas, where evaporation exceeds precipitation (Fig. 21.7), and it decreases where precipitation dominates, in the equatorial belt and at high latitudes. As the hydrological cycle is greatly affected by the glaciations, significant variations in the salinity of the ocean during the Quaternary are to be expected.

Temperature and salinity jointly determine the density of seawater, the driver of deep ocean circulation. Dense waters sink at high latitudes and are progressively redistributed by deep currents through the various basins of the world's oceans. The reconstruction of the distribution of surface water salinity in the past would thus contribute to the understanding of why and how ocean circulation changed when climate conditions were different from today. It will also provide modelers with quantitative estimates to use as a forcing of numerical climate models.

Estimating the surface water salinity distribution of past oceans is difficult, partly due to the close correlation of temperature and salinity. Because of this correlation, changes in plankton distribution and transfer functions do not differentiate between changes in SST and changes in salinity. Moreover, the dominant signal recorded by most indicators is often temperature.

Nonetheless, reconstructions of past sea surface salinity (SSS) using transfer functions of dinoflagellate or diatom assemblages have been proposed in specific marine environments (DeSève 1999; De Vernal et al. 2001) with an accuracy of ±1.8 psu for the present day (De Vernal et al. 2001). However, these methods are difficult to extrapolate unambiguously to a global scale due to non-analogue situations in the past.

The most common method presently used to reconstruct past SSS is the calibration of salinity against stable oxygen isotope ratios measured on foraminifera (Duplessy et al. 1991; Malaizé and Caley 2009). Geochemical methods based on the analysis of trace metals have recently been developed; we will briefly discuss these two approaches in the sections that follow.

Isotopic Methods

Paleosalinities from Stable Oxygen Isotopes ($\delta^{18}O$)

In the open ocean, the isotopic composition of seawater is closely correlated to salinity (Figs. 21.7 and 21.8): the vapor pressure of $H_2^{18}O$ being lower than that of $H_2^{16}O$, isotopic ratio of vapor in the atmosphere is systematically lower than

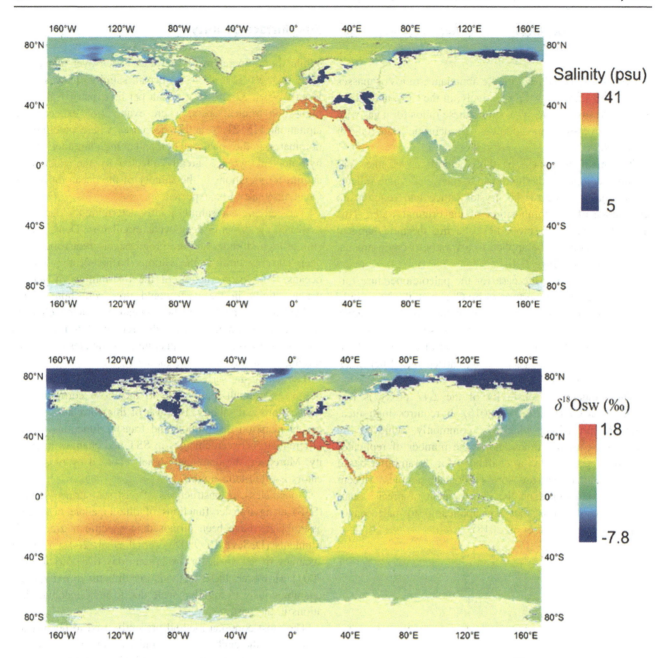

Fig. 21.7 Surface salinity of the modern oceans (WOA 2013: Zweng et al. 2013) and gridded data set of surface $\delta^{18}O$ of seawater (δ_w) (Legrande and Schmidt 2006)

in the condensed phase (ocean and rain). The more evaporation exceeds precipitation, the higher the surface water isotopic ratio. At local to regional scale, there is a linear relationship between salinity and δ_w (Fig. 21.8). This is the relationship which allows the estimation of past seawater salinities.

The major uncertainty in the 'paleotemperature formula' now becomes an advantage. δ_w can be determined when δ_c is measured if the foraminiferal habitat temperature T is estimated independently (by Mg/Ca ratios analysis of the same shells for example).

In the past, δ_w has been seen to vary globally. When the ice caps grew on land, they trapped snow poor in ^{18}O, and so the δ_w of the ocean increased. Thus, we observe a simultaneous drop in sea level and an increase in δ_w. Conversely, when the ice caps melt, sea level rises and δ_w decreases. The most recent studies (see Chap. 24) estimate that sea level dropped by about 120 m during the LGM, and that the average isotopic composition of the ocean was then at +1.0‰ SMOW (while the current value is 0‰ SMOW by definition). Various approaches aim to reconstruct changes in sea level linked with changes in continental ice volume.

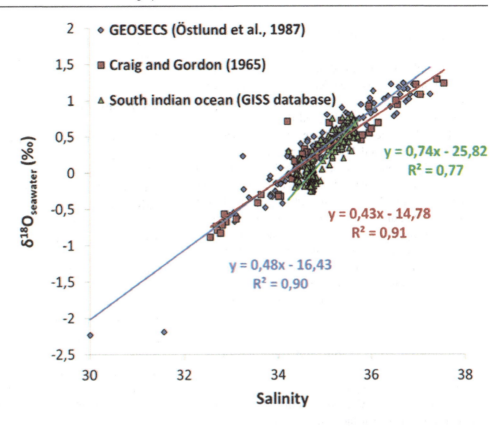

Fig. 21.8 The relationship between surface δ_w and salinity for the global ocean as measured within the international GEOSECS program (Ostlund 1987), by Craig and Gordon (1965), and for the South Indian Ocean with the GISS database (Schmidt et al. 1999)

It is then feasible to estimate the change in mean ocean δ_w related to variations in continental ice volume and sea level.

Locally, changes in the hydrological cycle (evaporation, precipitation, water mass movements, melting events) can cause additional variations in δ_w, of both climatic and hydrological origin.

Reconstructing the evolution of the isotopic composition of seawater in the past is in itself an interesting task because δ_w tracks changes in the hydrologic cycle. However, we can try to qualitatively interpret a record of paleo δ_w in terms of paleosalinities (see Box 2).

Box 2. Practical Calculation of Paleosalinities

Estimating paleosalinity changes along a sediment core requires the following steps:

- Measure δ_c and T at each level so as to derive a recording of paleo δ_w over the time period covered by the core, using the paleotemperature equation (Eq. 21.1).
- Estimate global δ_w changes $\left(\delta_w^{ice}\right)$ related to continental ice volume variations over the study period using known records of changes in sea level. A drop of 120 m in sea level is accompanied by an increase in δ_w^{ice} of +1.0‰. Given an average depth of the modern ocean of ~ 3900 m and an average salinity of 34.7 psu, and since the amount of salt in the ocean remains constant, a drop in sea level of 120 m is also accompanied by an increase in salinity. The average ocean salinity becomes:

$$(34.7 \times 3900)/3780 = 35.8\,\text{psu}.$$

Salinity has thus increased by about 1.1 psu and δ_w^{ice} by +1.0‰. Coupled models of the Northern Hemisphere ice sheets allow the ice-sheet contribution to the variability in oxygen isotope composition and sea level changes to be determined (Bintanja et al. 2005).

- Estimate the variation in local isotopic composition δ_w^{local} due to hydrological changes by subtracting δ_w^{ice} changes from the δ_w value reconstructed for each core level. The corresponding change in salinity can be estimated from current observations (Fig. 21.7).

The statistical error associated with this approach is high and rarely permits meaningful quantitative salinity reconstructions because of the associated large uncertainties (Rohling and Bigg 1998; Schmidt 1999; Rohling 2000; Legrande and Schmidt 2011; Caley and Roche 2015). The structural/analytical error is in the range of 0.8–1.8 psu (Schmidt 1999). In addition, the spatial and temporal evolution of the slope of the δ_w-salinity relationship, tested in isotope-enabled numerical climate models, can lead to very large errors (Legrande and Schmidt 2011), up to 25 psu in certain regions for the LGM (Caley and Roche 2015).

To reduce these very large errors on past SSS reconstructions, model-derived temporal slopes of the δ_w-salinity relationship can be used directly in the calculation. This approach has been tested with success for the LGM on a marine sediment record located in Gulf of Guinea and influenced by West African monsoon hydrology (Caley and Roche 2015). However, allowing model-derived regional δ_w-salinity relationships to vary through time can lead to significant uncertainties related to the shortcomings of the models, so complementary approaches should also be developed.

Paleosalinities from Stable Hydrogen Isotopes (δ^2H)

Another method uses hydrogen isotope changes to reconstruct paleosalinities. Culture experiments have found a constant offset between the hydrogen isotopic composition of water and the hydrogen isotopic composition of alkenones synthesized in that water (Paul 2002; Englebrecht and Sachs 2005). Schouten et al. (2006) demonstrated that this offset was dependent on salinity via biological fractionation processes. Reconstructing salinity by using the biological fractionation factor that is linked to it requires information on the past hydrogen isotope ratio of seawater (δ^2H_w).

Isotope-enabled climate model results indicate a rather stable dependence between δ^2H and surface δ_w in the past (Caley and Roche 2015). As δ_w can be reconstructed (see Sect. "Chemical Methods") this suggest that δ^2H_w can also be obtained. An estimation of paleosalinities based on δ^2H measurements in alkenones might therefore be possible if the slope and the intercept of the regression between the biological fractionation factor and salinity can be sufficiently constrained. The impact of species composition and growth phase on the use of alkenone δ^2H to reconstruct paleosalinity currently requires further investigations (Wolhowe et al. 2009; Chivall et al. 2014; M'Boule et al. 2014).

Pairing information from water isotopes, $\delta^{18}O$ and δ^2H (isotopologues), could yield better estimates for paleosalinity (Rohling 2007; Leduc et al. 2013). Numerical modeling experiments for the Holocene and the LGM periods have demonstrated that this combination of water isotopologues may indeed allow for a better estimation of paleosalinity variability (Legrande and Schmidt 2011; Caley and Roche 2015). Nonetheless, ecological biases introduced by combining proxies based on two different organisms (foraminifera are zooplankton and coccoliths are phytoplankton) could emerge, together with differences in dissolution and bioturbation in a sediment core.

Chemical Methods

Calibrations established using the modern Ba/Ca-salinity relationship (Carroll et al. 1993; Weldeab et al. 2007) (Fig. 21.9) have suggested that the Ba/Ca ratio of foraminiferal $CaCO_3$ can be used as a proxy for river runoff. This approach is limited to coastal regions affected by river runoff (i.e. prone to relatively large salinity changes) and assumes that (1) the Ba/Ca ratio in planktonic foraminifera shells is dominated by the Ba/Ca concentration of seawater (Hönisch et al. 2011) and not by other factors and (2) that the present-day calibration is applicable to the past.

Another recent study has established the potential of the Na/Ca ratio of foraminiferal calcite as a quantitative proxy for past salinities. In culture experiments, Wit et al. (2013) studied sodium incorporation in the benthic foraminifera *Ammonia tepida* at a range of salinities and suggested that foraminiferal Na/Ca could serve as a robust and independent proxy for salinity. More recently, the field study of Mezger et al. (2016) on planktonic foraminifera also suggested that salinity controls foraminiferal Na/Ca. Incorporation of Na in foraminiferal calcite could therefore constitute a potential proxy for salinity, although species-specific calibrations are still required and more research on the effect of temperature is needed.

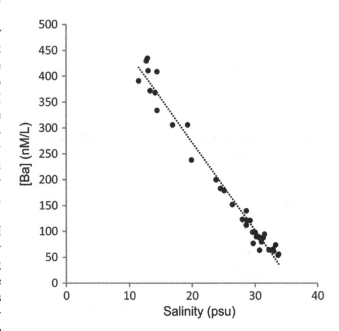

Fig. 21.9 Relationship between Ba in seawater and salinity in the Bay of Bengal (Carroll et al. 1993)

Reconstruction of the Hydrology of the Deep Ocean

Main Features of Modern Circulation

On a rotating planet like the Earth, surface ocean circulation is governed by the winds and the position of the continents that define the shape of the ocean basins. Deep circulation, on the other hand, is governed by the small variations in the density of the water masses. Since density depends on temperature and salinity, the term 'thermohaline circulation' is used. Reconstructing past changes in temperature, salinity and density in the intermediate and deep ocean is thus needed to understand the temporal variations of the circulation and distribution of water masses; it also provides a benchmark for simulations provided by general circulation models.

As we saw in Chap. 1, deep waters are currently formed in winter in highly localized areas of the high latitudes: the Norwegian Sea and the Labrador Sea in the northern hemisphere and the Weddell Sea and the Ross Sea near Antarctica. During winter, surface waters here become denser as they cool down but also because the formation of sea ice is accompanied by a release of salt. When surface waters become as dense as deep waters, large-scale convection movements are initiated and the waters sink into the abyssal depths. Once at depth, very small changes in density of the various deepwater masses govern their circulation through the different basins. Surface waters sinking in the Norwegian Sea cross the sills separating it from the Atlantic Ocean to form North Atlantic Deep Water (often referred to by the acronym NADW). This water mass then follows the American coast to join the Southern Ocean and is caught up in the Antarctic divergence, a large upwelling zone, where it mixes with the surface waters of southern high latitudes. Here, these very cold surface waters increase their density through winter sea ice formation, and sink along the Antarctic continental shelf forming the densest waters in the world. This water mass is called Antarctic Bottom Water (or AABW), and it lines the bottom of all ocean basins. At present, abyssal waters are not formed from the surface waters of the Indian and Pacific Oceans. The Antarctic waters that rush into these basins create the Pacific and Indian Deep Waters (PDW and IDW, respectively) by mixing with the waters of the main thermocline, and then return to the Southern Ocean at around 3 km depth. We can devise a simplified view of the global ocean circulation, where the Norwegian Sea is the main source of deep waters, and the Southern Ocean acts as a recirculation pump returning to the depths the surface waters surrounding Antarctica that have received upwelled NADW via the Antarctic divergence. This circulation pattern is critically dependent on the climate of the high latitudes of the Northern Hemisphere (North Atlantic Ocean, Norwegian and Labrador Seas) and of the Southern Ocean in the Southern Hemisphere.

Reconstructing the Temperature and Salinity of Deep Waters

The past evolution of the deep ocean has been the subject of extensive research. However, the reconstruction of the basic properties of bottom waters has been hindered by the lack of transfer functions linking the abundance of benthic species to the temperature of seawater in the vicinity of the sediment. In many environments, these abundances are essentially governed by the availability of food and by the proportion of dissolved oxygen. However, paleoceanographers have attempted to apply other approaches to the reconstruction of bottom water temperature.

Searching for a Reference Zone with Constant Temperature

As early as 1967, Shackleton (1967) suggested that the $\delta^{18}O$ of benthic foraminifera must closely resemble the $\delta^{18}O$ of deep waters, because these waters are formed close to freezing conditions and their temperature cannot drop much further during a glacial period. Labeyrie et al. (1987) further elaborated on this concept by analyzing the $\delta^{18}O$ of benthic foraminifera from a core in the Norwegian Sea where temperature, well below 0 °C in the deep basins, is constrained by exchanges with the ice. By comparing the isotopic record from this core with others from the Pacific and Indian Oceans, the authors were able to demonstrate, contrary to the assumptions of Shackleton, that deep water temperature in the major ocean basins did change significantly at the beginning of the last glaciation, with a cooling in all deep oceans to a temperature close to the freezing point (~ -1 °C) during the LGM. This result has been confirmed since by other tracers such as the Mg/Ca ratio of benthic foraminifera (see Sect. "Estimating the Temperature Independently of the Paleotemperature Formula"). Unfortunately, the isotopic benthic record of the Norwegian Sea is far from continuous and this method could not be successfully applied to reconstruct the evolution of the abyssal water temperature over the whole of the Quaternary.

It should be emphasized, however, that very low deep-water temperatures during glacial periods are to be expected. Indeed, under current conditions, NADW is formed from very cold water (close to the freezing point) that overflows from the Norwegian Sea through the sills located between Scotland, the Faroe Islands, Iceland and Greenland. However, the water that crosses these shallow sills (with a depth of less than one kilometer) mixes with the much warmer waters of the North Atlantic, so that the newly formed NADW is characterized by a temperature of +3 °C and a salinity of 34.95 psu. Its density remains high, close to but slightly less than that of AABW (temperature of −1 to 0 °C, salinity of 34.6 psu). It is therefore AABW that lines the great ocean depths, and, in the Atlantic Ocean, is topped by NADW. During the last glaciation, the sinking of very cold water directly into the North Atlantic (and not into the Norwegian Sea anymore) explains why the deep waters of the world are all found to be at temperatures close to the freezing point.

Estimating the Temperature Independently of the Paleotemperature Formula

The simplest way to determine the temperature of the water close to the sediment is to use the concentration of trace metals (Mg/Ca) contained in the carbonate shells of benthic foraminifera. This independent estimate of bottom water temperature allows the calculation of the δ_w of deep water using the paleotemperature formula.

However, the temperature dependence of Mg incorporation in benthic foraminiferal tests is species-specific, and may depend on different hydrological factors such as salinity or carbonate ion saturation (Elderfield et al. 2006). In addition, the expected bottom water temperature variations during glacial-interglacial cycles are small compared to surface temperature changes, implying relatively small Mg/Ca variations. The recently discovered Δ47 method is only starting to be applied to this problem (Peral et al. 2018), although more precise measurements in benthic foraminifera are needed to confirm the utility of this technique.

Searching for the Geochemical Signature of Ancient Waters in Pore Waters

Adkins et al. (2002) found that in long cores extracted by drilling ships, pore water trapped within the sediments shows measurable differences in salinity and in δ_w. These differences increase initially with core depth, then reach a maximum after which they decrease slowly. They interpreted this maximum as the signature of highly saline water from the LGM that had diffused into the sedimentary column. Using a simple diffusion model, the authors estimated the values for the salinity and the δ_w of the bottom waters 20 ka ago. Using this estimate of the δ_w of bottom waters, and combining it with $\delta^{18}O$ measurements on the calcite of benthic foraminifera, the paleotemperature formula confirmed that the deep waters of the glacial ocean were actually at a temperature near freezing point and hypersaline. The sediment cores that have been measured for δ_w are too few to give a complete picture of the ocean during the last glaciation. However, they show a significant disparity in the salinities from one basin to another, with the Southern Ocean being the saltiest, in contrast with the modern situation.

Reconstructing Changes in Water Mass Distribution

Lynch-Stieglitz et al. (1999) and (2014) showed that an approximate direct relationship could be established between the $\delta^{18}O$ of benthic foraminifera and the density of seawater, within the temperature range where the temperature-salinity-density relationship is roughly linear (T greater than 2 °C). In this way, the authors were able to study the geostrophic deformations of the deep thermocline in the Straits of Florida and propose estimates of the changes in the meridian flow linked to the Atlantic thermohaline circulation between the LGM and the present.

Reconstructions using benthic foraminiferal $\delta^{18}O$ have also shown marked changes in the distribution of deep and intermediate water masses during the LGM compared to the present day. In the Atlantic Ocean, the temperature gradient currently observed at the base of NADW at around 3000 m was to be found at around 2000 m, and was much more pronounced than the one that currently separates NADW and AABW (Labeyrie et al. 1992). In the Indian Ocean, a strong gradient separated two water masses with distinctly different characteristics at a depth of around 2000 m (Kallel et al. 1988). More recent research even suggests the presence of a third deep water mass in the deepest North Atlantic at the LGM, formed by brine rejection and not by heat loss to the atmosphere (Keigwin and Swift 2017).

Reconstructing the Circulation of Deep Waters

Searching for Lines of Current from the $\delta^{13}C$ of Benthic Foraminifera

An original approach, independent of temperature and salinity, tries to characterize the main features of deep water circulation without trying to a priori understand the underlying physical mechanisms governing it. It is based on the carbon cycle and its tracer, the $^{13}C/^{12}C$ ratio usually denoted as $\delta^{13}C$. At the ocean surface, waters easily exchange their gas content with the atmosphere; they contain carbon dioxide and are rich in dissolved oxygen. During photosynthesis, phytoplankton preferentially absorbs $^{12}CO_2$ over $^{13}CO_2$. The organic material thus produced has a $\delta^{13}C$ close to −20‰, while the $\delta^{13}C$ of dissolved CO_2 in surface waters varies between +1 and +2‰. This surface organic matter forms the base of the ocean's food chain, and eventually falls to the depths carried in fecal pellets of zooplankton and higher animals. In the water column, settling organic matter undergoes a slow remineralization, which consumes any dissolved oxygen that may remain and produces CO_2

depleted in ^{13}C. Consumption of dissolved oxygen and production of CO_2, accompanied by a decrease in δ^{13}C, take place in the deep waters (Fig. 21.10). The δ^{13}C of dissolved CO_2 in deep waters is thus lower than that of the surface waters.

The remineralization of organic matter is a slow process. This is why deep waters are characterized by a high δ^{13}C in regions close to their formation area (this is the case for the North Atlantic Ocean). Gradually, as they move away and circulate at depth, without the opportunity to exchange with the atmosphere, they become increasingly deprived of dissolved oxygen, while their δ^{13}C decreases through the mechanism described above. To give an order of magnitude, we can consider that the δ^{13}C of deep water decreases by about 1‰ per thousand years.

It is therefore the waters of the deep basins of the Pacific and Indian Oceans, at the end of the circulation scheme described in Sect. "Main Features of Modern Circulation", which have the lowest δ^{13}C. The evolution of δ^{13}C in the deep ocean can thus be used as a tracer to characterize the lines of current and the exchanges between the various deep water masses. Epibenthic foraminifera, such as the species *Cibicides wuellerstorfi*, reflect this evolution of the water in which they grew, and variations in their δ^{13}C in cores extracted from different ocean basins are used to reconstruct changes in ocean circulation through time (Duplessy et al. 1984; Schmittner et al. 2017). For example, this proxy has been used to reconstruct ocean circulation during the LGM, when well ventilated (with a high δ^{13}C) waters of the Atlantic Ocean formed Glacial North Atlantic Intermediate Waters (GNAIW), at a shallower depth than today's NADW, while deep waters (AABW) were even more poorly ventilated than today (Fig. 21.11). The understanding of this variability in the thermohaline circulation, which is a major regulatory mechanism of climate, is the subject of substantial research, both analytical and in modeling.

Using Trace Elements Measured in Benthic Foraminifera

In the modern ocean, geochemists have showed that cadmium (Cd) is included in organic matter, so that its cycle follows that of phosphate. The concentration of dissolved Cd in ocean waters shows therefore very similar variations to that of dissolved phosphate, a nutrient with a well-known cycle. It is assimilated by phytoplankton to ensure growth, so, as for all organic matter formed by photosynthesis, it falls into the water column with organic debris and is gradually released in the deep waters as bacteria oxidize it. Consequently, in the deep waters of the ocean, the consumption of dissolved oxygen and production of carbon dioxide (depleted in ^{13}C as we have seen) occur in parallel with increases in phosphate and cadmium.

The Cd ion has a charge and an ionic radius similar to that of Ca. It is therefore easily incorporated in trace amounts into the carbonate shells of benthic foraminifera, so that their Cd/Ca ratio reflects the concentration of Cd in the seawater

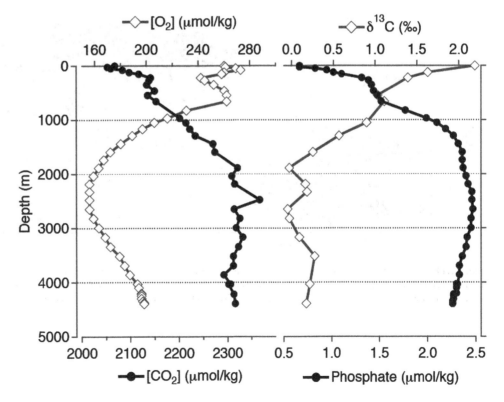

Fig. 21.10 Variations in the concentration of dissolved oxygen, the concentration of total dissolved CO_2, the $^{13}C/^{12}C$ ratio of the total dissolved CO_2 and phosphate with respect to depth at GEOSECS station 322 in the Pacific Ocean (43.0 °S/129.9 ° W). The oxygen minimum at depth indicates consumption by marine bacteria. The resulting CO_2 production is characterized by a maximum of dissolved inorganic carbon concentrations and by a minimum of δ^{13}C (since carbon in organic matter is depleted by about 20‰ relative to dissolved inorganic carbon)

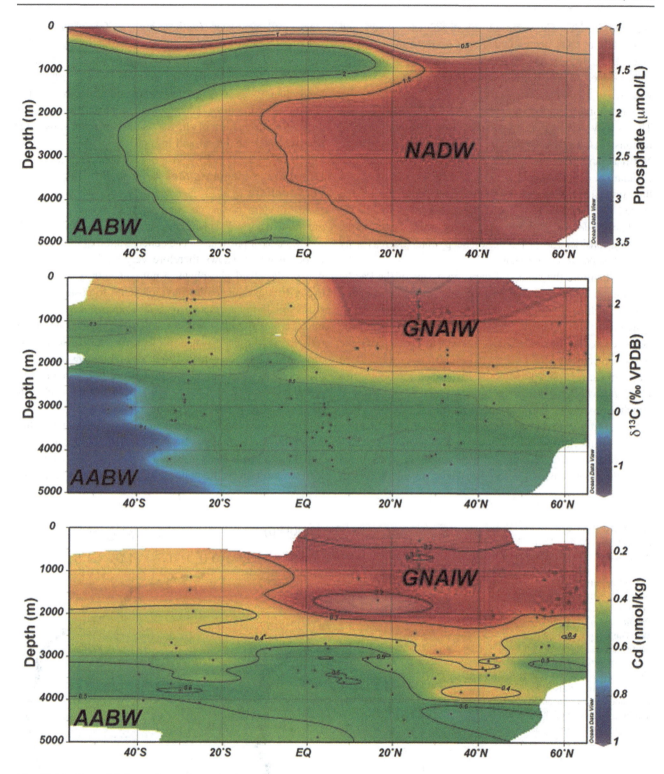

Fig. 21.11 (Top) Modern distribution of dissolved phosphate (μmol/liter) in the western Atlantic; (middle) reconstructed benthic $\delta^{13}C$ in the western and central Atlantic during the LGM; (bottom) estimates of the Cd concentration (nmol/kg) during the LGM based on the ratio of Cd/Ca in the shells of benthic foraminifera (figure modified from Lynch-Stieglitz et al. 2007)

in which the foraminifera developed. The Cd/Ca ratio is therefore a tracer for nutrient concentration of the water masses, present and past (Marchitto and Broecker 2006; Fig. 21.11).

Cd and ^{13}C have similar geochemical behaviors, with the essential difference that surface waters can exchange their dissolved carbon dioxide with the atmosphere, while Cd is not involved in ocean-atmosphere exchanges. In general, there is an excellent anti-correlation between the variations in the Cd/Ca ratio and the ^{13}C/^{12}C of benthic foraminifera measured in sediment cores. One notable exception is the Southern Ocean, where the benthic foraminifera that lived during the last ice age have very negative δ^{13}C, while their concentrations in Cd are very similar to those of recent sediments from the Holocene. Despite inter-specific differences that may have affected δ^{13}C reconstructions based on benthic foraminifera (Gottschalk et al. 2016), or diagenetic and metabolic effects influencing the incorporation of trace metals into biogenic calcite (Marchitto and Broecker 2006; McCorkle et al. 1995), this discrepancy between the two indicators remains still to be explained.

Reconstructing the Dynamics of Water Masses

The tracers we have described so far allow the reconstruction of specific physical or chemical characteristics of water masses, but they do not convey information on their dynamics. In this section, we do not discuss in detail tracers related to particle transport at the ocean floor (particle size distribution, magnetic grain size, sortable silt). However, we will discuss two unstable radioactive tracers in the ocean: the concentration in ^{14}C of benthic foraminifera and the excess ^{231}Pa/^{230}Th ratio in sediments.

When surface waters exchange carbon dioxide with the atmosphere, they absorb ^{14}C. Currently, the ^{14}C concentration in the surface ocean is 95% of that of the atmosphere. When surface waters sink, they bring with them the dissolved carbon dioxide as well as the ^{14}C they contain. Once they reach the bottom of the ocean, these waters are isolated from the atmosphere, and ^{14}C decreases due to its own radioactive decay, with its period of 5720 years. The oldest waters in the northern Pacific and Indian Oceans have an apparent age of around 800 years (see Chap. 4). The planktonic foraminifera (that live in surface waters) and the benthic foraminifera (that live on the ocean floor) incorporate the ^{14}C of the waters around them into their shells. By comparing, at the same level of sediment, the ^{14}C ages of planktonic and benthic foraminifera, we can estimate the apparent age of the deep waters over the last ~40,000 years. This apparently simple method presents in fact many difficulties. Firstly, despite recent analytical developments, benthic foraminifera are not always sufficiently abundant, and it may be difficult to obtain the amount of carbonate required for analysis from a normal-size sample, even using the most sensitive technique, accelerator mass spectrometry. Second, the ^{14}C ages of foraminifera, once in the sediment, are very sensitive to bioturbation: the abundance of one species shows considerable variation over time, and shells that are found at one level may have been displaced by the activity of burrowing animals and therefore come from a significantly different age level than the selected one. Finally, the atmospheric ^{14}C concentration has undergone large scale changes, so much so that the difference in ^{14}C age between planktonic and benthic foraminifera does not directly reflect the residence time of the waters at depth.

Another approach to reconstruct the dynamics of deep-water masses makes use of the geochemical behavior of the decay chain of uranium in seawater. The isotopic composition of dissolved uranium is constant throughout the ocean. Two of the isotopes of uranium, ^{235}U and ^{234}U, decay producing ^{231}Pa and ^{230}Th respectively, with an output ratio that is constant and equal to 0.093. ^{231}Pa and ^{230}Th are very reactive to particles sinking in the water column: they are adsorbed on their surface and settle as sediment along with them. However, ^{231}Pa is less reactive than ^{230}Th, so that the residence time in seawater of dissolved ^{231}Pa is close to 200 years, while that of dissolved ^{230}Th is only thirty years (Yu et al. 1996). The residence time of ^{231}Pa is close to that of NADW in the Atlantic. Because of this, a fraction of the dissolved ^{231}Pa is advected out of the North Atlantic Ocean by NADW (about 50% in the modern ocean), while ^{230}Th is unaffected and settles completely with the particles. The net loss of ^{231}Pa in the water column at depths greater than or equal to the level at which NADW flows, leads to a deficit of ^{231}Pa in the sediments and therefore in the ^{231}Pa/^{230}Th ratios below the production ratio (0.093). If the circulation of NADW becomes slower, less ^{231}Pa is advected out of the basin, and the ^{231}Pa/^{230}Th ratio of the particles settling to the sediment increases to values closer to the production ratio.

It should be noted that sediments also contain ^{231}Pa and ^{230}Th, present as daughter isotopes of the uranium in clays that have reached secular equilibrium with their relevant parents. The addition of ^{231}Pa and ^{230}Th from settling particles thus produces an excess of these two radioisotopes in the sediment. Measuring the ratio of excess ^{231}Pa/^{230}Th in the sediments therefore allows the variations in the circulation of deep waters from the North Atlantic towards the Southern Ocean to be traced, and so to detect the variability associated with major changes in climate (McManus et al. 2004; Gherardi et al. 2009; Lippold et al. 2016). This technique has been used with success to reconstruct the 'strength' of the thermohaline circulation back to ~140 ky ago (Guihou et al. 2010, 2011; Böhm et al. 2015).

Major Fields of Paleoceanography

A prerequisite to reconstruct ocean circulation in the past is that marine sediments have not been buried in the mantle at subduction zones. Due to the renewal of the ocean floor by plate tectonics, the oldest sediments date back to the Triassic, about 200 million years ago. These very old sediments are rare, and found only in the Pacific Ocean. In practice, we can hope to obtain global reconstructions for the whole Tertiary era, but they become increasingly scarce as we go back to the Secondary era. Here, we will restrict ourselves to the analysis of some of the major aspects of the evolution of climate and paleoceanography over the last 25 million years, which correspond to the progressive establishment of glacial conditions in the high latitudes of both hemispheres.

From 'The Greenhouse Effect Era' to the 'Ice Ages'

The Quaternary period, which covers the last 2.6 million years of the history of our planet, is characterized by persistent major ice sheets in the high latitudes of both hemispheres. The volume of these caps and their geographical expansion fluctuate over timescales of 10^4 to 10^5 years, in response to changes in insolation controlled by the orbital parameters of the Earth (see Chap. 28).

The origin of these major glacial phases is discussed in several chapters that present the point of view of geophysicists, geochemists and modelers (see Chaps. 22, 26 and 27). At time scales exceeding a million years, the major causes of the development of glaciations are feedbacks related to plate tectonics. Among these feedbacks, the following may be highlighted:

- the varying shapes and positions of landmasses and ocean basins,
- the existence of passages between basins,
- the location and altitude of mountain ranges, both in the ocean and on land, affecting oceanic and atmospheric circulation and heat transfer,
- volcanism and erosion, and their impact on atmospheric chemistry and on the pCO_2 of the atmosphere.

The record describing the evolution of the $\delta^{18}O$ of benthic foraminifera over time shown in Fig. 21.2 is the result of a compilation of analyses made in more than forty ocean-drilling sites. This global compilation registers both the changes in δ_w and temperature, in accordance with the paleotemperature formula (Eq. 21.1). It shows that the current climate is the result of a long decline that started at the end of the climate optimum of the early Eocene (52–50 Ma ago). This decline is characterized by an increase in $\delta^{18}O$ over time that reflects the drop in ocean temperatures at higher latitudes (where deep waters are formed) in a world without major glaciers, followed by the growth of ice caps. The $\delta^{18}O$ of benthic foraminifera therefore dropped from an average of $\sim 0.2‰$ during the climate optimum of the early Eocene, to an average of $\sim 4‰$ at the end of the Quaternary.

This slow climate decline is generally attributed to two main factors: a gradual reduction in atmospheric CO_2 concentration, and the growing thermal isolation of the Antarctic continent due to the widening of the ocean passages that surround it.

Relatively abrupt incidents (that is, events happening over a much shorter time scale than the general trend), of large amplitude, are superimposed onto this slow drift, indicating that other climate drivers such as thresholds or rapid feedbacks are also involved. The first of these changes occurred at the very end of the Paleogene, at the boundary of the Eocene/Oligocene, about 33.5 Ma ago. It resulted in a rapid increase in benthic foraminiferal $\delta^{18}O$ from +1.6 to +2.8‰ over only 100–200 ka (Fig. 21.2). This change in $\delta^{18}O$ is attributed to the development of the Antarctic ice cap. The northward drift of the Australian continent allowed the opening of the Strait of Tasmania and the establishment of the Antarctic Circumpolar Current (Zachos et al. 2001). The isolation of this large land mass resulted in a drop in temperatures and the establishment of a permanent ice cap on East Antarctica, with an ice volume that may have reached about 50% of its current size. After a period of ten million years when the climate changed little, two warming phases, one at the end of the Oligocene and the other during the Middle Miocene, caused a significant reduction of the Antarctic ice sheet. The increase in temperature and the decrease in ice volume were reflected in a 1.2‰ decrease in the benthic $\delta^{18}O$ signal. The causes of these climatic changes are not clear yet.

The final phase of the climatic decline leading to the major glaciations of the Quaternary began from the Middle Miocene, between 14.2 and 12.2 Ma, and was marked by an increase in benthic foraminiferal $\delta^{18}O$ of 1.0‰ over two million years. Paleotemperature reconstructions based on Mg/Ca ratios in benthic foraminifera suggest that deep water temperature varied little, and that the growth of the Antarctic ice cap was the main contributor to the benthic $\delta^{18}O$ change (around $\sim 0.8‰$). If these estimates were correct, the ice cap located on the eastern part of Antarctica would have reached 85% of its current volume. As for the West Antarctic ice cap, it seems to have only developed from 6 Ma on, as evidenced by the first coarse sediment deposits in the Weddell Sea, coming from melting icebergs emitted at the edge of this cap. These sediments transported by drifting ice result from land erosion caused by the friction of glaciers, and are often referred to by their acronym IRD (*ice rafted debris*).

Although the development of a permanent ice cap on Antarctica started early, during the early Neogene, the development of perennial continental ice caps in the high latitudes of the northern hemisphere did not occur until the end of the Neogene. The first deposits of IRD in the Norwegian Sea, proof of the early development of an ice cap (although perhaps not a permanent one) on Greenland are not observed before 5.5 Ma (Jansen and Sjoholm 1991). The rapid increase in $\delta^{18}O$ from $\sim$3.2 Ma onwards may be interpreted as the beginning of permanent glaciation in high latitudes of the northern hemisphere. This glaciation intensified rapidly around 2.1–2.6 Ma, as evidenced by the massive IRD deposits in the Norwegian Sea.

Many studies have focused on the hypothesis of the 'closure of the Panama isthmus' as a potential trigger for the development of northern hemisphere ice sheets. The mechanism would involve warm intertropical waters no longer being able to cross from the Atlantic to the Pacific. They would therefore deviate into the North Atlantic, increasing oceanic evaporation in this basin and thus snow accumulation in the high latitudes. However, recent studies suggest that the closure of the Panama isthmus could have occurred during the Miocene, well before the intensification of glaciations (Montes et al. 2015). The recent work of Rohling et al. (2014) also observes a large temporal offset during the onset of the Plio-Pleistocene ice ages, between a marked cooling step at 2.73 My ago and the first major glaciation starting 2.15 My ago. Other theories indicate that a decrease in atmospheric CO_2 may have been responsible for a cooling, an increase of deep water formation in the North Atlantic and a change of circulation that together induced the start of the glaciations.

The 'Middle Pleistocene Transition' and the Establishment of 100-Ka Cycles

The trend towards the climatic decline (seen as an increase in benthic $\delta^{18}O$) discussed in Sect. "From 'The Greenhouse Effect Era' to the 'Ice Ages'" continued over the last two million years, as is shown in detail in Fig. 21.2. Superimposed on this trend are quasi-periodic oscillations. They reflect the alternating glacial periods—corresponding to a cooling of deep waters and an increase in ice volume at high latitudes—and interglacial periods, with warming and relative melting of the ice caps. It should be emphasized here that the use of the terms 'interglacial' and 'glacial' does not imply a *total* melting of ice sheets. During interglacial periods, ice sheets do not disappear, even if they are greatly reduced in the northern hemisphere. For example, during the LGM, the ice sheets in the northern hemisphere covered a large portion of North America and Europe. During the Holocene, the interglacial period we currently live in, these caps were largely diminished; the meltwater derived from them has caused sea level to rise by 120 m since the LGM. However, an ice cap of 2.8 million km^3 continues to exist on Greenland that would cause a further sea level rise of about 7 meters, if it were to completely melt.

The amplitude of the glacial-interglacial oscillations increased sharply between 1.2 and 0.6 Ma (Fig. 21.2). During this period, called 'the Middle Pleistocene Transition' (Clark et al. 2006; McClymont et al. 2013), a threshold response to longer-term atmospheric CO_2 decline has been proposed (Raymo et al. 1997). However, recent atmospheric partial pressure CO_2 reconstructions have failed to show this long-term decrease during the Pleistocene (Hönisch et al. 2009). The gradual increase in glacial-interglacial amplitude is mainly due to increasingly high values of $\delta^{18}O$ during glacial periods. The few available reconstructions of deep-water temperature during this period indicate near-freezing temperatures at every glacial maximum instead of a gradual cooling (Elderfield et al. 2012), which suggests that an increase in Antarctic ice volume would be responsible for the rapid and steep increase in seawater $\delta^{18}O$ at 0.9 Ma.

This change in amplitude of glacial-interglacial oscillations is accompanied by a disruption in the frequency content of the global $\delta^{18}O$ signal. While benthic foraminifera $\delta^{18}O$ oscillations show mainly a cycle of $\sim$41 ka over most of the Neogene and early Quaternary, the last 600,000 years are dominated by oscillations with a cyclicity of $\sim$100 ka (Fig. 21.2). Some authors have agreed on the progressive nature of the 'Middle Pleistocene Transition', with the amplification of the 100 ka cycles occurring over hundreds of thousands of years. However, in some ocean regions, the records fail to demonstrate this progressive nature. This is the case, for example, in the equatorial Atlantic, where the dynamics of the thermocline, reconstructed from micropaleontological tracers, suddenly change its variability around $\sim$930 ka. The mechanisms responsible for this transition are still unclear, although it appears that an important role can be attributed to the enormous Laurentide ice sheet, which may have favored the frequency of 100 ka through its inertia (Clark and Pollard 1998) (see Chap. 28).

The Last Glacial Maximum (LGM)

The LGM has long been, and still remains, a major area of interest in paleoclimatology, in particular because it presents another extreme on the climate spectrum on which Earth System models can be validated (Kageyama et al. 2018). Early studies defined this period as the time encompassing the last great cold maximum (as recorded by micropaleontology and pollen) as well as the maximum spread of the ice sheets (marked by the position of moraines on land masses). Radiocarbon dating has placed the maximum at around 16-20 ka ^{14}C (equivalent to 18–23 ka in calendar age). This period was the first to be the subject of a global paleoclimate study, thanks to the CLIMAP group. The isotopic maximum in δ^{18}O of planktonic and benthic foraminifera, interpreted as reflecting the cumulative effects of the cold and ice volume maxima, was used as a stratigraphic marker, and summer and winter sea surface temperatures were determined from micropaleontological transfer functions. This established the CLIMAP maps (CLIMAP 1981) (Fig. 21.3) that served as boundary conditions for the first comprehensive paleoclimate modeling experiments. The CLIMAP results have had a profound impact. For the first time, the magnitude of the temperature change between an ice age (LGM) and an interglacial period (modern times) could be quantified: the average global temperature dropped by 6 °C. However, this cooling was far from uniform: it exceeded 10 °C at high northern latitudes, while it was only a few degrees in the intertropical region. This intense cooling was associated with the development of large ice sheets over the landmasses of the northern hemisphere, which, with about 50 million cubic kilometers of ice, was the most glaciated hemisphere. In addition, analyses of Antarctic ice showed that atmospheric CO_2 concentration was about 100 ppmv below pre-industrial values (Petit et al. 1999).

It quickly became necessary to expand these early studies. Continental tracers in tropical regions, such as pollen series or concentration of noble gases in aquifers (which are dependent on the temperature of the rains feeding these aquifers), indicated a cooling of 3–6 °C during the LGM. In the nearby ocean, sea surface temperature alkenone reconstructions indicated a cooling of only about 2 °C, and micropaleontological transfer functions showed little or no change. Detailed studies were therefore conducted in later decades to explain the observed differences. It appeared that many cores taken from the tropical Pacific Ocean and used for the CLIMAP reconstruction had very low sedimentation rates, so that bioturbation caused the contrasts in fauna over time to disappear. In addition, the fauna from warm waters exhibited variability that did not solely respond to temperature changes, with the result that micropaleontological transfer functions became insensitive at the temperatures above 25 °C common in tropical regions. At the same time, high-resolution studies started to indicate strong climatic variability between 17 ka and 25 ka. In the North Atlantic, for example, the LGM does not correspond to the coldest conditions, which are instead associated with two periods framing the LGM: Heinrich Stadial (HS) 2 at around 24–22 ka and HS1 at around 19–17 ka.

It therefore became essential to reconsider the reconstruction of the surface ocean during the LGM. The latest and most comprehensive synthesis was carried out in the MARGO program (MARGO Project Members 2009), which focused on the period 19–23 ka corresponding to the LGM sensu stricto. This period corresponds to the maximum expansion of ice sheets, as opposed to the coldest conditions of Heinrich Stadials. The LGM MARGO reconstructions (Fig. 21.3) agree relatively well with CLIMAP, but they also revealed some important differences:

- the northern seas were ice-free during the summer;
- latitudinal and longitudinal thermal gradients were strong; the mid-latitudes of the North Atlantic Ocean experienced the strongest cooling (~ -10 °C);
- the decrease in temperature was generally larger on the eastern side than on the western side of the oceans; this was particularly marked along the African margin, especially in coastal upwelling zones of Namibia and South Africa;
- the cooling of tropical waters was close to 2 °C, although some localized waters of the Pacific and Indian Oceans experienced moderated warming;
- in the Southern Ocean, a cooling of 2–6 °C marked a northward displacement of the polar front.

Other studies have focused on the deep ocean during the LGM. We have mentioned several of them in the description of the various methodological techniques that have been developed over the last forty years. Significant differences between the LGM and the present day include:

- the downwelling of surface waters in the North Atlantic happened in open ocean, leading to the formation of very cold deep water that found its density equilibrium at 2000 m depth;
- the very cold and dense bottom waters formed in the southern hemisphere spread throughout the deep ocean, occupying a much larger volume than today;
- the boundary between deep and bottom waters was characterized by a much stronger gradient of physical (T, S, density) and geochemical (δ^{18}O, δ^{13}C) properties than today;
- the ventilation and renewal rates of deep waters are still poorly constrained because of conflicting information from different tracers with a complex geochemical behavior (^{14}C, ^{231}Pa/^{230}Th); this uncertainty is also reflected in the simulations from general circulation models of the ocean and coupled ocean-atmosphere models. Most proxies do indicate, however, lower ventilation of the deep ocean and a resulting large accumulation of carbon dioxide in the deeper waters.

The Last Deglaciation

Several decades ago, continental paleoclimatologists described the warming by steps that occurred during the last disappearance of the large northern ice caps, a period extending from 20 to 8 ka. This 'last deglaciation' was also identified by paleoceanographers in marine sediment cores with high sedimentation rates (Fig. 21.12). The terminology for this succession of warming and relative cooling comes directly from the first descriptions made in the continental records based on pollen assemblages: Older Dryas, Middle Dryas, Bølling-Allerød and Younger Dryas all take their names from plant pollens (in the case of Dryas, it is associated with the reappearance of the cold flower *Dryas octopetala*), or from locations from which the samples were taken (the proglacial lake of Bølling and the city of Allerød in Denmark).

The drivers and feedbacks that led to this specific sequence of events are still being actively studied. The start of the deglaciation is linked to the evolution of the astronomical parameters, with a strong increase in summer insolation in the northern hemisphere between 20 and 10 ka (Milankovitch's theory), and aided by pulses of increases in atmospheric CO_2 likely released from the CO_2-rich deep waters of the Southern Ocean (Marcott et al. 2014). However, the mechanisms that explain the phase differences between the two hemispheres (Fig. 21.12) during the deglaciation are still unclear. A distinct warming trend appeared in Antarctica around 19 ka, while simultaneously, the northern hemisphere, after a brief warming trend, cooled

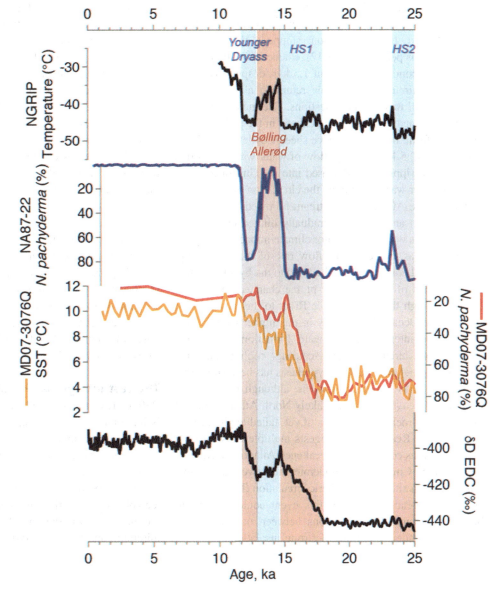

Fig. 21.12 Records of the last deglaciation in the high latitudes of both hemispheres. (Top) Variations in $\delta^{18}O$ in the ice from the NorthGRIP site (Greenland), a proxy for the local variations in atmospheric temperature (Rasmussen et al. 2014); (middle-top) variations in the percentage of the cold species *Neogloboquadrina pachyderma* in core NA87-22 from the North Atlantic (55 °30 ′N, 14 °42 ′W, 2161 m deep) (Waelbroeck et al. 2001; Vázquez Riveiros et al. 2013); (middle-bottom) variations in the isotopic composition of hydrogen δD of ice from Dome C (Antarctica), a proxy for local variations in atmospheric temperature (EPICA 2004); (bottom) variations in the percentage of the cold species *Neogloboquadrina pachyderma* and SST estimated by the Mg/Ca method in core MD07-3076Q from the Southern Ocean (44 ° 09 ′S, 14 °13 ′W, 3770 m deep) (modified from Vázquez Riveiros et al. 2010)

and presented armadas of icebergs linked to the HS1 event (see Chap. 29). Following this event, around 14 ka, northern hemisphere warming was strongly amplified with a culmination during the Bølling-Allerød, while temperatures in the southern hemisphere stabilized and even dropped in Antarctica during the so-called Antarctic Cold Reversal (Fig. 21.12). During this early phase of the deglaciation, the North-South antiphase is similar to what is observed during abrupt events of the last ice age, with the exception of the general deglacial warming trend. Recent studies point to CO_2 as a key mechanism of global warming during the last deglaciation. An anti-phased hemispheric temperature response to ocean circulation changes, superimposed on globally in-phase warming driven by increased CO_2 concentrations, is an explanation for much of the temperature change at the end of the most recent ice age (Barker et al. 2009; Shakun et al. 2012).

However, at the end of the Bølling-Allerød warm event, at about 12.5 ka, the ice caps stopped melting, the sea level stabilized and the deglaciation stopped: this was the Younger Dryas period (Fig. 21.12), characterized by a return to very cold conditions for about 1.5 kyr, despite insolation reaching maximum values. This return to almost ice age conditions still raises many questions. The most commonly accepted explanation is a sudden change in the path taken by meltwater from the Laurentide ice sheet (Leydet et al. 2018). Until about 12.5 ka, this huge flow of water was transported by the Mississippi River. Released into the Gulf of Mexico, the fresh water was drawn in by the circulation of the surface currents of the Atlantic (Gulf Stream followed by the North Atlantic Drift) and was very gradually diluted by the salty tropical waters without any major climate impact. During the Younger Dryas, however, the flow rate of the Mississippi River dropped considerably, which led to the hypothesis that the watershed of the meltwater plume changed and flowed instead through the St. Lawrence River to the northwest of the Atlantic Ocean. The salinity in this higher latitude area was reduced, interrupting deep water formation and thus the thermohaline circulation, and causing cooling and the growth of some glaciers. This hypothesis has been supported by simple ocean circulation models, although marine sediment cores recovered from the likely North Atlantic zone of evacuation of meltwater have, as of yet, failed to yield traces of this event. Recent study suggests multiple causes of the Younger Dryas cold period: a weakened Atlantic Meridional Overturning Circulation, moderate negative radioactive forcing and an altered atmospheric circulation (Renssen et al. 2015). The detailed study of this event could help us to better understand the interactions between ocean, ice and atmosphere under conditions of strong insolation.

Interglacial Periods, the Holocene and the Last Two Millennia

In order to explain the succession of glacial and interglacials periods over the last million years, Milankovitch developed the astronomical theory of paleoclimate. Since then, conceptual models have been able to describe the general trends, as well as the dominant periodicities centered around 100, 40 and 20 kyr fairly accurately (see Chap. 28). Although changes in ice cap volume during glacial periods and the time constants of their response to changes in insolation are relatively well understood, the same cannot be said for the evolution of climate during interglacial periods. In particular, the mechanisms causing the differences in duration, in temperature of the atmosphere and ocean, and in ocean circulation are not well understood, even though differences in forcing are precisely calculated (Past InterGlacialS Working Group of PAGES 2016). This lack of understanding is derived in part from the small ocean temperature differences between past interglacial periods and the present day, with temperature changes that remain close to the error of temperature reconstructions with the usual tracers (Sect. "Sea Surface Temperature"). A further complication arises because the internal mechanisms in the climate system must be investigated through its various components (atmosphere, ocean, continent), which involves the construction of time scales common to the various archives used to reconstruct each of them, and makes the study of the interglacial periods prior to the Holocene particularly difficult.

In this section, we limit ourselves to the analysis of the last two interglacial periods: the Last Interglacial (also called the Eemian), about 125 ka ago, and the Holocene, period in which we now live. Eemian and Holocene, the terms used in this chapter, are names borrowed from palynologists to identify these two interglacial periods. A short subsection will finally be devoted to results recently obtained for the last two millennia, which has the advantage of presenting a wide range of continental and marine records that can, in some cases, be compared with recorded meteorological data.

The Last Interglacial Period

Before presenting our understanding of this period of time, it is important to define what an interglacial is. It may in fact be defined in a number of ways depending on whether one considers, for example, variability in flora, ocean circulation, atmospheric temperature or ocean temperature (Past InterGlacialS Working Group of PAGES 2016). If we take ice volume as a marker, an interglacial period sensu stricto is the time interval during which the ice volume is at its minimum and remains constant for several millennia.

Strictly speaking, the interglacial comparable to the period we live in, and defined by an ice volume minimum is called the Last Interglacial, and runs from about 129–116 ka (Govin et al. 2015; Dutton et al. 2015). From 115 ka, the midpoint of the transition marking the entry into MIS 5d, ice volume had already increased significantly, so much so that sea level dropped by as much as −40 m at the height of MIS 5d at about 110 ka.

During the Last Interglacial, the insolation forcing was characterized by a relatively high eccentricity, the combination of a strong inclination and a perihelion close to the summer solstice. This orbital configuration triggered an increase in summer insolation in the northern hemisphere of more than 30 W/m^2 compared to the present day. Despite these differences in forcing, the general evolution of the Last Interglacial climate is to a first degree quite similar to that of the Holocene: high temperatures at higher northern latitudes until about 123 ka (in line with higher insolation and higher elevation of the sun on the horizon), followed by a gradual cooling linked to the decline in boreal summer insolation in parallel with the progressive growth of glacial conditions (Cortijo et al. 1999). However, the Last Interglacial temperature peak was reached at about 126 ka in the North Atlantic against 129 ka in the southern high latitudes. This hemispheric asynchrony is related to the disruption of the Atlantic overturning circulation due to freshwater discharges into the North Atlantic (in response to ice sheet melting) that led to the persistence of cold conditions in the northern high latitudes and the early warming of southern high latitudes during the early phase of the Last Interglacial (Capron et al. 2014, 2017) (Fig. 21.13).

Nevertheless, despite greenhouse gas concentrations that were similar to pre-industrial times, the larger increase in summer insolation in the northern hemisphere with respect to the current situation did have an impact on the climate of the Last Interglacial optimum. Surface water temperatures were 1–2 °C warmer in the North Atlantic, the Nordic Seas and the Southern Ocean than during the Holocene (Capron et al. 2014, 2017; Hoffman et al. 2017). Such warmer high latitudes during the Last Interglacial had a double impact:

– the warming by about 0.4 °C of the temperatures of the deep Atlantic waters, which was then carried into Antarctic circumpolar deep waters (Duplessy et al. 2007);
– the partial melting of Greenland and West Antarctica (Dutton et al. 2015).

These two combined actions brought about a rise in sea level of 6–9 m (Dutton et al. 2009) compared to current levels.

The Last Interglacial is a good case study to test our mechanistic understanding of the effect of warmer-than-present polar climate on sensitive components of the Earth system (e.g. ice sheets, sea level). It has recently sparked

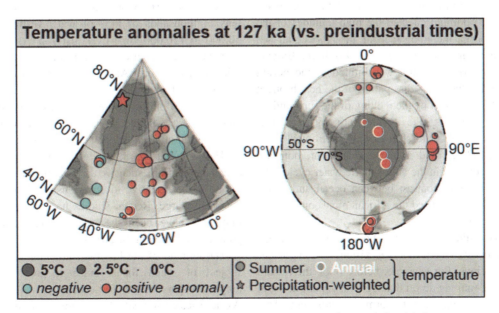

Fig. 21.13 Temperature anomalies at 127 ka compared to preindustrial times (1870–1899 CE) in the northern and southern high latitudes (modified after Capron et al. 2017). Negative (positive) temperature anomalies are shown in blue (red). The bigger the dot, the stronger the temperature anomaly. Most records indicate warmer conditions at 127 ka compared to preindustrial, in response to the high boreal summer insolation. The few cold anomalies suggest remnants of freshwater discharge into the North Atlantic, Nordic Seas and Labrador Seas

interest in the community, as shown by multiple paleo-data compilations and model-data comparison exercises (Otto-Bliesner et al. 2017).

The Holocene

The Holocene period started about 11 ka ago. The last major ice sheets had not completely disappeared, but major changes had occurred since the early deglaciation, both in terms of sea level and continental and oceanic temperatures. At first look, the climate over these past 11 ka seems stable, but this apparent stability hides very pronounced regional variations in the hydrological cycle, in the circulation of surface waters (especially during the final stage of melting of the residual ice caps), and in the general circulation of the Mediterranean Sea, a basin surrounded by land and with limited connections to the open ocean and thus very strongly affected by changes in rainfall intensity over its watershed.

The Holocene is a period of major movement and development of populations. However, for the most part of this period, human activities still had a negligible impact on the global environment, so the study of climate changes over recent millennia provide a benchmark against which disturbances caused by industrial and agricultural activities can be detected. The reconstruction of Holocene climate changes is facilitated by the precise chronology offered by carbon-14 analysis.

The forcing of summer insolation at 65°N at the beginning of the Holocene reached more than 390 W/m^2 and caused a global warming that would last until about 6 ka. The temperature optimum affected the high latitudes of the North Atlantic basin, including Iceland, the Norwegian Sea and the Scandinavian coast (Koç et al. 1993). In the Barents Sea, the temperature maximum was limited to the period from 7.9 to 6.9 ka due to the dissipation of the heat brought by the North Atlantic Drift by the melting of the surrounding ice. At lower latitudes, the temperature increase was accompanied by a northward shift of the Intertropical Convergence Zone (ITCZ) and a major change in monsoon dynamics, and therefore in the atmospheric water cycle. The increase in the thermal contrast between ocean and continent, for example, accentuated the African monsoon as far as the center of the continent.

The study of sediment cores from both the Mediterranean Sea and African lakes indicates the existence of major climate reorganizations. For example, before 6 ka, the Sahara was not the wide-ranging desert that it is today, but grassland dotted with lakes conducive to farming settlement. This period is called the African Humid Period (AHP). Around 6 ka, this wet period ended and conditions degraded at a rate that is still debated (Collins et al. 2017; Shanahan et al. 2016; Tierney and deMenocal (2013). The tropical vegetation of canopy forests along the rivers declined, and the Sahelian vegetation in turn disappeared about 2.7 ka ago to make way for the desert conditions present today. This major change could be related to the gradual decrease in insolation over the past 10 ka aided by the albedo feedback induced by the gradual disappearance of vegetation. Alternatively, a rapid termination of the AHP could have been triggered by northern-latitude cooling combined with biogeophysical feedbacks (Collins et al. 2017).

During this wet period, the Mediterranean Sea received more fresh water, especially in the eastern basin (Kallel et al. 1997). The sinking of well-ventilated, shallow water masses in winter became impossible in the Levantine basin, and bottom waters there became completely anoxic, leading to the disappearance of benthic fauna below 800 m depth. A layer of black sediment rich in organic matter, called a sapropel, marks this event (Rossignol-Strick et al. 1982; Rohling et al. 2015). Although ventilation of the eastern waters of the Mediterranean resumed at 6 ka, the deep fauna of this basin, whose colonization rate is slow, is still very poor.

In addition to these long-term reorganizations, the Holocene also recorded an abrupt event of short duration 8.2 ka ago. Without reaching the amplitude of the rapid and sudden climate changes of the last ice age, this event still left a significant imprint on northern hemisphere temperatures. Like its glacial counterparts, the '8.2 ka event' is associated with a freshwater discharge, in this case due to the rupture of a proglacial reservoir, Lake Agassiz, formed by the retreat of the Laurentide ice sheet (Barber et al. 1999; Wiersma and Renssen 2006; Hoffman et al. 2012). The sudden release of tens of thousands of km^3 of water (estimates vary from 50,000 to 120,000 km^3) over just 1–5 yrs had strong consequences, such as a reduction in the SST (about 1 °C) and salinity of the North Atlantic, a reduction of 2–6 °C in the atmospheric temperature above Greenland, a decrease in the temperature of air and water in the lakes of western Europe, and a decrease in the intensity of ocean circulation for a period of about 100 yrs after the freshwater discharge.

The study of the 8.2 event has shown that interglacial ocean circulation, such as the one of the early Holocene, may also be sensitive to an intense, although brief, freshwater discharge. Recent studies have pointed out that this may also have been the case during earlier interglacial periods (Galaasen et al. 2014).

The climate of the last two millennia has also been the subject of much focus, since it provides a relatively long-term perspective for recent observations from the World Meteorological Organization (WMO) network (restricted to the last 150 years) and from satellites dedicated to the observation of the Earth (limited to a few decades). The reconstructions of air temperature in the northern hemisphere, used as projections for the whole planet, have primarily been based on continental data (Mann et al. 1998).

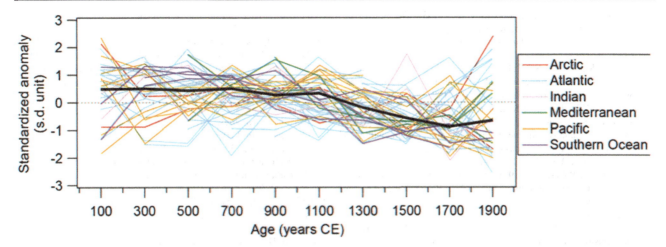

Fig. 21.14 Standardized SST anomalies over the last 2000 years (modified after McGregor et al. 2015). Thin colored lines represent individual SST reconstructions from different ocean basins, which have been averaged into 200-year long bins (e.g. 1–200 CE). The thick black line is the area-weighted median SST value

Reconstructing and understanding changes in the ocean over the last two millennia is particularly difficult, since oceanographic observations are only available for the last century at most, and for paleoceanographers this period is recorded in the uppermost portion of the sediment which is often poorly consolidated or lost.

Recently, much effort has been put on the gathering of the best time-series of the last two millennia as part of the Past Global Changes (PAGES) 2 k network (PAGES 2k Consortium 2013, 2017). Continental-scale temperature reconstructions provide evidence of twentieth century warming over all reconstructed regions except Antarctica (PAGES 2k Consortium 2013). A global SST compilation shows a 1800-year long cooling of the surface ocean over the pre-industrial past 2000 years (Fig. 21.14), and that the cooling from 801 to 1800 CE was likely caused by volcanic eruptions (McGregor et al. 2015). A more recent synthesis of paleoclimate records since 1500 CE has identified that sustained industrial-era warming of the tropical oceans first developed during the mid-nineteenth century and was nearly synchronous with Northern Hemisphere continental warming (Abram et al. 2016).

If we look at the deep ocean, the relative strength of the meridional overturning circulation has also been recently assessed for the last 1.6 ka (Thornalley et al. 2018). The authors suggest that, while it was relatively stable between 400 and 1850 CE, it has declined in strength by ∼15% at the beginning of the industrial era. In addition, the comparison of SST patterns in the North Atlantic with model simulations points to an additional weakening over the last 150 yrs (Caesar et al. 2018).

Compilation studies like the ones presented above highlight the need for paleoclimate reconstructions, which can be compared to instrumental records that at present are too short to comprehensively assess anthropogenic climate change.

References

Abram, N. J., McGregor, H. V., Tierney, J. E., Evans, M. N., McKay, N. P., Kaufman, D. S., & The Pages 2 k Consortium. (2016). Early onset of industrial-era warming across the oceans and continents. *Nature, 536*, 411. https://doi.org/10.1038/nature19082.

Adkins, J. F., McIntyre, K., & Schrag, D. P. (2002). The salinity, temperature and $\delta^{18}O$ of the glacial deep ocean. *Science, 298*, 1769–1773.

Anand, P., Elderfield, H., & Conte, M. H. (2003). Calibration of Mg/Ca thermometry in planktonic foraminifera from a sediment trap time series. *Paleoceanography, 18*(2).

Barker, S., Diz, P., Vautravers, M. J., Pike, J., Knorr, G., Hall, I. R., et al. (2009). Interhemispheric Atlantic seesaw response during the last deglaciation. *Nature, 457*(7233), 1097.

Barber, D. C., Dyke, A., Hillaire-Marcel, C., Jennings, A. E., Andrews, J. T., Kerwin, M. W., et al. (1999). Forcing of the cold event of 8200 years ago by catastrophic drainage of Laurentide lakes. *Nature, 400*, 344. https://doi.org/10.1038/22504.

Bemis, B. E., Spero, H. J., Bijma, J., & Lea, D. W. (1998). Reevaluation of the oxygen isotopic composition of planktonic foraminifera: Experimental results and revised paleotemperature equations. *Paleoceanography, 13*, 150–160.

Bintanja, R., van de Wal, R. S., & Oerlemans, J. (2005). Modelled atmospheric temperatures and global sea levels over the past million years. *Nature, 437*, 125–128.

Böhm, E., Lippold, J., Gutjahr, M., Frank, M., Blaser, P., Antz, B., et al. (2015). Strong and deep Atlantic meridional overturning circulation during the last glacial cycle. *Nature, 517*, 73–76.

Caley, T., & Roche, D. M. (2015). Modeling water isotopologues during the last glacial: Implications for quantitative paleosalinity reconstruction. *Paleoceanography and Paleoclimatology, 30*(6), 739–750.

Capron, E., Govin, A., Stone, E. J., Masson-Delmotte, V., Mulitza, S., Otto-Bliesner, B., et al. (2014). Temporal and spatial structure of multi-millennial temperature changes at high latitudes during the Last Interglacial. *Quaternary Science Reviews, 103*, 116–133.

Capron, E., Govin, A., Feng, R., Otto-Bliesner, B., & Wolff, E. W. (2017). Critical evaluation of climate syntheses to benchmark CMIP6/PMIP4 127 ka Last Interglacial simulations in the high latitude regions. *Quaternary Science Reviews, 168*, 137–150.

Carroll, J., Brown, E. T., & Moore, W. S. (1993). The role of the Ganges-Brahmaputra mixing zone in supplying barium and ^{226}Ra to the Bay of Bengal. *Geochimica et Cosmochimica Acta, 57*, 2981–2990.

Caesar, L., Rahmstorf, S., Robinson, A., Feulner, G., & Saba, V. (2018). Observed fingerprint of a weakening Atlantic Ocean overturning circulation. *Nature, 556*, 191–198.

Chivall, D., M'Boule, D., Sinke-Schoen, D., Sinninghe Damsté, J. S., Schouten, S., & van der Meer, M. T. (2014). The effects of growth phase and salinity on the hydrogen isotopic composition of alkenones produced by coastal haptophyte algae. *Geochimica et Cosmochimica Acta, 140*, 381–390.

Clark, P. U., & Pollard, D. (1998). Origin of the middle Pleistocene transition by ice sheet erosion of regolith. *Paleoceanography and Paleoclimatology, 13*(1), 1–9.

Clark, P. U., Archer, D., Pollard, D., Blum, J. D., Rial, J. A., Brovkin, V., et al. (2006). The middle Pleistocene transition: Characteristics, mechanisms, and implications for long-term changes in atmospheric pCO$_2$. *Quaternary Science Reviews, 25*, 3150–3184.

CLIMAP. (1981). Seasonal reconstructions of the Earth's surface at the last glacial maximum. *Geological Society of America, Map and Chart Series, 36*.

Collins, J. A., Prange, M., Caley, T., Gimeno, L., Beckmann, B., Mulitza, S., et al. (2017). Rapid termination of the African Humid Period triggered by northern high-latitude cooling. *Nature Communications, 8*, 1372.

Conte, M. H., Sicre, M. A., Rühlemann, C., Weber, J. C., Schulte, S., Schulz-Bull, D., & Blanz, T. (2006). Global temperature calibration of the alkenone unsaturation index (UK′37) in surface waters and comparison with surface sediments. *Geochemistry, Geophysics, Geosystems, 7*(2).

Corrège, T. (2006). Sea surface temperature and salinity reconstruction from coral geochemical tracers. *Palaeogeography, Palaeoclimatology, Palaeoecology, 232*(2–4), 408–428.

Cortijo, E., Lehman, S., Keigwin, L., Chapman, M., Paillard, D., & Labeyrie, L. (1999). Changes in meridional temperature and salinity gradients in the North Atlantic Ocean (30–72 N) during the last interglacial period. *Paleoceanography, 14*, 23–33. https://doi.org/10.1029/1998PA900004.

Craig, H., & Gordon, L. I. (1965). Deuterium and oxygen-18 variations in the ocean and marine atmosphere. In E. Tongiorgi (Ed.), *Stable isotopes in oceanographic studies and paleotemperatures* (pp. 9–130). Pisa, Italy: Laboratory of Geology and Nuclear Science.

DeSève, A. M. (1999). Transfer function between surface sediment diatom assemblages and sea-surface temperature and salinity of the Labrador Sea. *Marine Micropaleontology, 36*, 249–267.

De Vernal, A., et al. (2001). Dinoflagellate cyst assemblages as tracers of sea-surface conditions in the Northern Atlantic, Arctic and sub-Arctic seas: The new n = 677 data base and its application for quantitative paleoceanographic reconstruction. *Journal of Quaternary Science, 16*, 681–698.

Duplessy, J. C., Shackleton, N. J., Matthews, R. K., Prell, W. L., Ruddiman, W. F., Caralp, M., et al. (1984). ^{13}C record of benthic foraminifera in the Last Interglacial Ocean: Implications for the carbon cycle and the global deep water circulation. *Quaternary Research, 21*, 225–243.

Duplessy, J. C., Labeyrie, L., Juillet-Leclerc, A., Maitre, F., Duprat, J., & Sarnthein, M. (1991). Surface salinity reconstruction of the North Atlantic Ocean during the last glacial maximum. *Oceanologica Acta, 14*, 311–324.

Duplessy, J. C., Roche, D. M., & Kageyama, M. (2007). The deep ocean during the last interglacial period. *Science, 316*, 89–91.

Dutton, A., Bard, E., Antonioli, F., Esat, T. M., Lambeck, K., & McCulloch, M. T. (2009). Phasing and amplitude of sea-level and climate change during the penultimate interglacial. *Nature Geoscience, 2*, 355–359.

Dutton, A., Carlson, A. E., Long, A. J., Milne, G. A., Clark, P. U., DeConto, R., Horton, B. P., Rahmstorf, S., & Raymo, M. E. (2015). Sea-level rise due to polar ice-sheet mass loss during past warm periods. *Science, 349*. https://doi.org/10.1126/science.aaa4019.

Eiler, J. M., Bergquist, B., Bourg, I., Cartigny, P., Farquhar, J., Gagnon, A., et al. (2014). Frontiers of stable isotope geoscience. *Chemical Geology, 372*, 119–143. https://doi.org/10.1016/j.chemgeo.2014.02.006.

Elderfield, H., Yu, J., Anand, P., Kiefer, T., & Nyland, B. (2006). Calibrations for benthic foraminiferal Mg/Ca paleothermometry and the carbonate ion hypothesis. *Earth and Planetary Science Letters, 250*, 633–649.

Elderfield, H., & Ganssen, G. (2000). Past temperature and d18O of surface ocean waters inferred from foraminiferal Mg/Ca ratios. *Nature, 405*, 442–445.

Elderfield, H., Ferretti, P., Greaves, M., Crowhurst, S., McCave, I. N., Hodell, D. A., et al. (2012). Evolution of ocean temperature and ice volume through the mid-Pleistocene climate transition. *Science, 337*(6095), 704–709.

Englebrecht, A. C., & Sachs, J. P. (2005). Determination of sediment provenance at drift sites using hydrogen isotopes and unsaturation ratios in alkenones. *Geochimica et Cosmochimica Acta, 69*, 4253–4265.

EPICA. (2004). Eight glacial cycles from an Antarctic ice core. *Nature, 429*, 623–628.

Epstein, S., Buchsbaum, R., Lowenstam, H., & Urey, H. C. (1951). Carbonate-water isotopic temperature scale. *Geological Society of America Bulletin, 62*(4), 417–426.

Epstein, S., Buchsbaum, R., Lowenstam, H. A., & Urey, H. C. (1953). Revised carbonate-water isotopic temperature scale. *Geological Society of America Bulletin, 64*(11), 1315–1326.

Emiliani, C. (1955). Pleistocene temperatures. *Journal of Geology, 63*, 538–578.

Galaasen, E. V., Ninnemann, U. S., Irvalı, N., Kleiven, H. K. F., Rosenthal, Y., Kissel, C., et al. (2014). Rapid reductions in North Atlantic deep water during the peak of the last interglacial period. *Science, 343*(6175), 1129–1132.

Gherardi, J. M., Labeyrie, L., Nave, S., Francois, R., McManus, J. F., & Cortijo, E. (2009). Glacial-interglacial circulation changes inferred from 231Pa/230Th sedimentary record in the North Atlantic Region. *Paleoceanography, 24*, PA2204. https://doi.org/10.1029/2008pa001696.

Ghosh, P., Adkins, J., Affek, H., Balta, B., Guo, W., Schauble, E. A., Schrag, D. P., & Eiler, J. M. (2006). ^{13}C-^{18}O bonds in carbonate minerals: A new kind of paleothermometer. *Geochimica et Cosmochimica Acta, 70*, 1 439–1 456.

Gottschalk, J., Vázquez Riveiros, N., Waelbroeck, C., Skinner, L. C., Michel, E., Duplessy, J. C., et al. (2016). Carbon isotope offsets between benthic foraminifer species of the genus Cibicides (*Cibicidoides*) in the glacial sub-Antarctic Atlantic. *Paleoceanography, 31*, 1–20.

Govin, A., Capron, E., Tzedakis, P. C., Verheyden, S., Ghaleb, B., Hillaire-Marcel, C., et al. (2015). Sequence of events from the onset to the demise of the Last Interglacial: Evaluating strengths and limitations of chronologies used in climatic archives. *Quaternary Science Reviews, 129*, 1–36. https://doi.org/10.1016/j.quascirev.2015.09.018.

Gray, W. R., Weldeab, S., Lea, D. W., Rosenthal, Y., Gruber, N., Donner, B., et al. (2018). The effects of temperature, salinity, and the carbonate system on Mg/Ca in *Globigerinoides ruber* (white): A global sediment trap calibration. *Earth and Planetary Science Letters, 482*, 607–620. https://doi.org/10.1016/j.epsl.2017.11.026.

Guihou, A., Pichat, S., Nave, S., Govin, A., Labeyrie, L., Michel, E., et al. (2010). Late slowdown of the Atlantic Meridional overturning circulation during the last glacial inception: New constraint from sedimentary (231 Pa/230Th). *Earth and Planetary Science Letters, 289,* 520–529.

Guihou, A., Pichat, S., Govin, A., Nave, S., Michel, E., Duplessy, J.-C., et al. (2011). Enhanced Atlantic Meridional overturning circulation supports the last glacial inception. *Quaternary Science Reviews, 30,* 1576–1582. https://doi.org/10.1016/j.quascirev.2011.03.017.

Ho, S. L., & Laepple, T. (2016). Flat meridional temperature gradient in the early Eocene in the subsurface rather than surface ocean. *Nature Geoscience, 9*(8), 606.

Hoffman, J. S., Carlson, A. E., Winsor, K., Klinkhammer, G. P., LeGrande, A. N., Andrews, J. T., & Strasser, J. C. (2012). Linking the 8.2 ka event and its freshwater forcing in the Labrador Sea. *Geophysical Research Letters, 39*(18).

Hoffman, J. S., Clark, P. U., Parnell, A. C., & He, F. (2017). Regional and global sea-surface temperatures during the last interglaciation. *Science, 355*(6322), 276–279.

Hönisch, B., Hemming, N. G., Archer, D., Siddall, M., & McManus, J. F. (2009). Atmospheric carbon dioxide concentration across the Mid-Pleistocene transition. *Science, 324,* 1551–1556.

Hönisch, B., Allen, K. A., Russell, A. D., Eggins, S. M., Bijma, J., Spero, H. J., et al. (2011). Planktic foraminifers as recorders of seawater Ba/Ca. *Marine Micropaleontology, 79,* 52–57. https://doi.org/10.1016/j.marmicro.2011.01.003.

Hönisch, B., Allen, K. A., Lea, D. W., Spero, H. J., Eggins, S. M., Arbuszewski, J., et al. (2013). The influence of salinity on Mg/Ca in planktic foraminifers—Evidence from cultures, core-top sediments and complementary δ18O. *Geochimica et Cosmochimica Acta, 121,* 196–213. https://doi.org/10.1016/j.gca.2013.07.028.

Imbrie, J., & Kipp, N. G. (1971). A new micropaleontological method for quantitative paleoclimatology: Application to a late Pleistocene Caribbean Core. In K. K. Turekian (Ed.), *The late Cenozoic glacial ages* (pp. 71–181). Yale University Press.

Jansen, E., & Sjoholm, J. (1991). Reconstruction of glaciation over the 6 Myr from ice-borne deposits in the Norwegian Sea. *Nature, 349,* 600–603.

Kageyama, M., Braconnot, P., Harrison, S. P., Haywood, A. M., Jungclaus, J. H., Otto-Bliesner, B. L., et al. (2018). The PMIP4 contribution to CMIP6—Part 1: Overview and over-arching analysis plan. *Geoscientific Model Development, 11,* 1033–1057. https://doi.org/10.5194/gmd-11-1033-2018.

Kallel, N., Labeyrie, L. D., Juillet-Leclerc, A., & Duplessy, J. C. (1988). A deep hydrological front between intermediate and deep-water masses in the glacial Indian Ocean. *Nature, 333,* 651–655.

Kallel, N., Paterne, M., Duplessy, J. C., Vergnaud-Grazzini, C., Pujol, C., Labeyrie, L., et al. (1997). Enhanced rainfall in the Mediterranean Region during the last sapropel event. *Oceanologica Acta, 20,* 697–712.

Keigwin, L. D., & Swift, S. A. (2017). Carbon isotope evidence for a northern source of deep water in the glacial western North Atlantic. *Proceedings of the National Academy of Sciences, 114*(11), 2831–2835.

Kele, S., Breitenbach, S. F. M., Capezzuoli, E., Meckler, A. N., Ziegler, M., Millan, I. M., et al. (2015). Temperature dependence of oxygen- and clumped isotope fractionation in carbonates: A study of travertines and tufas in the 6–95 °C temperature range. *Geochimica et Cosmochimica Acta, 168,* 172–192. https://doi.org/10.1016/j.gca.2015.06.032.

Khider, D., Huerta, G., Jackson, C., Stott, L. D., & Emile-Geay, J. (2015). A Bayesian, multivariate calibration for *Globigerinoides ruber* Mg/Ca. *Geochemistry, Geophysics, Geosystems, 16,* 2916–2932. https://doi.org/10.1002/2015GC005844.

Kim, J.-H., van der Meer, J., Schouten, S., Helmke, P., Willmott, V., Sangiorgi, F., et al. (2010). New indices and calibrations derived from the distribution of crenarchaeal isoprenoid tetraether lipids: Implications for past sea surface temperature reconstructions. *Geochimica et Cosmochimica Acta, 74,* 4639–4654. https://doi.org/10.1016/j.gca.2010.05.027.

Koç, N., Jansen, E., & Haflidason, H. (1993). Paleoceanographic Reconstructions of surface Ocean conditions in the Greenland, Iceland and Norwegian Seas through the Last 14 Ka based on diatoms. *Quaternary Science Reviews, 12,* 115–140.

Labeyrie, L. D., Duplessy, J. C., & Blanc, P. L. (1987). Variations in mode of formation and temperature of oceanic deep waters over the past 125,000 years. *Nature, 327.*

Labeyrie, L., Duplessy, J. C., Duprat, J., Juillet-Leclerc, A., Moyes, J., Michel, E., et al. (1992). Changes in vertical structure of the North Atlantic Ocean between glacial and modern times. *Quaternary Science Reviews, 11,* 401–413.

Lea, D. W., Mashiotta, T. A., & Spero, H. J. (1999). Controls on magnesium and strontium uptake in planktonic foraminifera determined by live culturing. *Geochimica et Cosmochimica Acta, 63*(16), 2369–2379.

Leduc, G., Sachs, J. P., Kawka, O. E., & Schneider, R. R. (2013). Holocene changes in eastern equatorial Atlantic salinity as estimated by water isotopologues, Earth Planet. *Sci. Lett., 362,* 151–162.

LeGrande, A. N., & Schmidt, G. A. (2006). Global gridded data set of the oxygen isotopic composition in seawater. *Geophysical research letters, 33*(12).

LeGrande, A. N., & Schmidt, G. A. (2011). Water isotopologues as a quantitative paleosalinity proxy. *Paleoceanography, 26,* PA3225. https://doi.org/10.1029/2010pa002043.

Leydet, D. J., Carlson, A. E., Teller, J. T., Breckenridge, A., J. E., Barth, A. M., Ullman, D. J., Sinclair, G., Milne, G. A., Cuzzone, J. K., Caffee, M. W. (2018). Opening of glacial Lake Agassiz's eastern outlets by the start of the Younger Dryas cold period. *Geology, 46,* 155–158. https://doi.org/10.1130/G39501.1.

Lippold, J., Gutjahr, M., Blaser, P., Christner, E., de Carvalho Ferreira, M. L., Mulitza, S., et al. (2016). Deep water provenance and dynamics of the (de) glacial Atlantic meridional overturning circulation. *Earth and Planetary Science Letters, 445,* 68–78.

Lisiecki, L. E. & Raymo, M. E. (2005). A Pleiocene-Pleistocene stack of 57 globally distributed d^{18}O records. *Paleoceanography, 20.* https://doi.org/10.1029/2004PA001071.

Lombard, F., Labeyrie, L., Michel, E., Spero, H. J., & Lea, D. W. (2009). Modelling the temperature dependent growth rates of planktic foraminifera. *Marine Micropaleontology, 70,* 1–7.

Lynch-Stieglitz, J., Curry, W. B., & Slowey, N. (1999). A geostrophic transport estimate for the Florida current from the oxygen isotope composition of benthic foraminifera. *Paleoceanography, 14,* 360–373.

Lynch-Stieglitz, J., Adkins, J. F., Curry, W. B., Dokken, T., Hall, I. R., Herguera, J. C., Hirschi, J.l.J.M., Ivanova, E. V., Kissel, C., Marchal, O., Marchitto, T. M., McCave, I. N., McManus, J. F., Mulitza, S., Ninnemann, U., Peeters, F., Yu, E.-F., & Zahn, R. (2007). Atlantic Meridional overturning circulation during the last glacial maximum. *Science, 316,* 66–69. https://doi.org/10.1126/science.1137127.

Lynch-Stieglitz, J., Schmidt, M. W., Gene Henry, L., Curry, W. B., Skinner, L. C., Mulitza, S., et al. (2014). Muted change in Atlantic overturning circulation over some glacial-aged Heinrich events. *Nature Geoscience, 7,* 144–150. https://doi.org/10.1038/ngeo2045.

Malaizé, B., & Caley, T. (2009). Sea surface salinity reconstruction as seen with foraminifera shells: Methods and cases studies. *The European Physical Journal Conferences, 1,* 177–188. https://doi.org/10.1140/epjconf/e2009-00919-6.

Malmgren, B. A., Kucera, M., Nyberg, J., & Waelbroeck, C. (2001). Comparison of statistical and artificial neuronal network techniques for estimating past sea surface temperatures from planktonic foraminifera census data. *Paleoceanography, 16*, 520–530.

Mann, M. E., Bradley, R. S., & Hughes, M. K. (1998). Global-scale temperature patterns and climate forcing over the past six centuries. *Nature, 392*(6678), 779.

Marchitto, T. M., & Broecker, W. S. (2006). Deep water mass geometry in the glacial Atlantic Ocean: A review of constraints from the paleonutrient proxy Cd/Ca. *Geochemistry, Geophysics, Geosystems, 7*, Q12003.

Marchitto, T. M., Curry, W. B., Lynch-Stieglitz, J., Bryan, S. P., Cobb, K. M., & Lund, D. C. (2014). Improved oxygen isotope temperature calibrations for cosmopolitan benthic foraminifera. *Geochimica et Cosmochimica Acta, 130*, 1–11.

Marcott, S. A., Bauska, T. K., Buizert, C., Steig, E. J., Rosen, J. L., Cuffey, K. M., et al. (2014). Centennial-scale changes in the global carbon cycle during the last deglaciation. *Nature, 514*, 616–619.

MARGO Project Members. (2009). Constraints on the magnitude and patterns of ocean cooling at the last glacial maximum. *Nature Geoscience*. https://doi.org/10.1038/NGEO411.

Mashiotta, T. A., Lea, D. W., & Spero, H. J. (1999). Glacial–interglacial changes in Subantarctic sea surface temperature and δ^{18}O-water using foraminiferal Mg. *Earth and Planetary Science Letters, 170*(4), 417–432.

Mathien-Blard, E., & Bassinot, F. (2009). Salinity bias on the foraminifera Mg/Ca thermometry: Correction procedure and implications for past ocean hydrographic reconstructions. *Geochemistry, Geophysics, Geosystems, 10*, Q12011. https://doi.org/10.1029/2008gc002353.

M'boule, D., Chivall, D., Sinke-Schoen, D., Sinninghe Damsté, J. S., Schouten, S., & van der Meer, M. T. (2014). Salinity dependent hydrogen isotope fractionation in alkenones produced by coastal and open ocean haptophyte algae. *Geochimica et Cosmochimica Acta, 130*, 126–135.

McClymont, E. L., Sosdian, S. M., Rosell-Melé, A., & Rosenthal, Y. (2013). Pleistocene sea-surface temperature evolution: Early cooling, delayed glacial intensification, and implications for the mid-Pleistocene climate transition. *Earth-Science Reviews, 123*, 173–193.

McCorkle, D. C., Martin, P., Lea, D. W., & Klinkhammer, G. (1995). Evidence of a dissolution effect on benthic foraminiferal shell chemistry: δ^{13}C, Cd/Ca, Ba/Ca, and Sr/Ca results from the Ontong Java Plateau. *Paleoceanography, 10*, 699–714.

McGregor, H. V., Evans, M. N., Goosse, H., Leduc, G., Martrat, B., Addison, J. A., et al. (2015). Robust global ocean cooling trend for the pre-industrial Common Era. *Nature Geoscience, 8*, 671–677. https://doi.org/10.1038/ngeo2510.

McManus, J. F., Francois, R., Gherardi, J.-M., Keigwin, L. D., & Brown-Leger, S. (2004). Collapse and rapid resumption of Atlantic meridional circulation linked to deglacial climate changes. *Nature, 428*, 834–837.

Mezger, E. M., Nooijer, L. J., Boer, W., Brummer, G. J. A., & Reichart, G. J. (2016). Salinity controls on Na incorporation in Red Sea planktonic foraminifera. *Paleoceanography and Paleoclimatology, 31*(12), 1562–1582.

Montagna, P., McCulloch, M., Douville, E., López Correa, M., Trotter, J., Rodolfo-Metalpa, R., et al. (2014). Li/Mg systematics in scleractinian corals: Calibration of the thermometer. *Geochimica et Cosmochimica Acta, 132*, 288–310. https://doi.org/10.1016/j.gca.2014.02.005.

Montes, C., Cardona, A., Jaramillo, C., Pardo, A., Silva, J. C., Valencia, V., et al. (2015). Middle Miocene closure of the Central American seaway. *Science, 348*, 226–229.

Mulitza, S., Boltovskoy, D., Donner, B., Meggers, H., Paul, A., & Wefer, G. (2003). Temperature: [delta]18O relationships of planktonic foraminifera collected from surface waters. *Palaeogeography, Palaeoclimatology, Palaeoecology, 202*, 143–152. https://doi.org/10.1016/S0031-0182(03)00633-3.

Müller, P. J., Kirst, G., Ruhland, G., Von Storch, I., & Rosell-Melé, A. (1998). Calibration of the alkenone paleotemperature index U 37 K' based on core-tops from the eastern South Atlantic and the global ocean (60 N-60 S). *Geochimica et Cosmochimica Acta, 62*(10), 1757–1772.

Otto-Bliesner, B. L., Braconnot, P., Harrison, S. P., Lunt, D. J., Abe-Ouchi, A., Albani, S., et al. (2017). The PMIP4 contribution to CMIP6—Part 2: Two interglacials, scientific objective and experimental design for Holocene and Last Interglacial simulations. *Geoscientific Model Development, 10*, 3979–4003. https://doi.org/10.5194/gmd-10-3979-2017.

Ostlund, H. G. (1987). *GEOSECS Atlantic, Pacific, and Indian Ocean Expeditions* (Vol. 7).

PAGES 2 k Consortium. (2013). Continental-scale temperature variability during the past two millennia. *Nature Geoscience, 6*, 339. https://doi.org/10.1038/ngeo1797.

PAGES 2 k Consortium. (2017). A global multiproxy database for temperature reconstructions of the common era. *Scientific Data, 4*, 170088. https://doi.org/10.1038/sdata.2017.88.

Past InterGlacialS Working Group of PAGES. (2016). Interglacials of the last 800,000 years. *Reviews of Geophysics, 54*.

Paul, H. (2002). *Application of novel stable isotope methods to reconstruct paleoenvironments: Compound specific hydrogen isotopes and pore-water oxygen isotopes*. Ph.D. thesis, 149 pp., Swiss Federal Institute of Technology.

Peral, M., Daëron, M., Blamart, D., Bassinot, F., Dewilde, F., Smialkowski, N., et al. (2018). Updated calibration of the clumped isotope thermometer in planktonic and benthic foraminifera. *Geochimica et Cosmochimica Acta, 239*, 1–16. https://doi.org/10.1016/j.gca.2018.07.016.

Petit, J. R., Jouzel, J., Raynaud, D., Barkov, N. I., Barnola, J. M., Basile, I., et al. (1999). Climate and atmospheric history of the past 420,000 years from the Vostok ice core, Antarctica. *Nature, 399*, 429–436.

Prahl, F. G., & Wakeham, S. G. (1987). Calibration of unsaturation patterns in long-chain ketone compositions for paleotemperature assessment. *Nature, 330*, 367–369.

Pisias, N. G., Martinson, D. G., Moore, T. C., Shackleton, N. J., Prell, W. L., Hays, J. D., et al. (1984). High resolution stratigraphic correlation of benthic oxygen isotopic records spanning the last 300,000 years. *Marine Geology, 56*, 119–136.

Railsback, L. B., Gibbard, P. L., Head, M. J., Voarintsoa, N. R. G., & Toucanne, S. (2015). An optimized scheme of lettered marine isotope substages for the last 1.0 million years, and the climatostratigraphic nature of isotope stages and substages. *Quaternary Science Reviews, 111*, 94–106.

Rasmussen, S. O., Bigler, M., Blockley, S. P., Blunier, T., Buchardt, S. L., Clausen, H. B., et al. (2014). A stratigraphic framework for abrupt climatic changes during the last glacial period based on three synchronized Greenland ice-core records: Refining and extending the INTIMATE event stratigraphy. *Quaternary Science Reviews, 106*, 14–28.

Raymo, M. E., Oppo, D. W., & Curry, W. (1997). The mid-Pleistocene climate transition: A deep sea carbon isotopic perspective. *Paleoceanography, 12*(4), 546–559.

Renssen, H., Mairesse, A., Goosse, H., Mathiot, P., Heiri, O., Roche, D. M., et al. (2015). Multiple causes of the Younger Dryas cold period. *Nature Geoscience, 8*(12), 946–949.

Rohling, E., & Bigg, G. (1998). Paleosalinity and $\delta^{18}O$: A critical assessment. *Journal Geophysical Research, 103*, 1307–1318. https://doi.org/10.1029/97JC01047.

Rohling, E. J. (2000). Paleosalinity: Confidence limits and future applications. *Marine Geology, 163*, 1–11.

Rohling, E. J. (2007). Progress in paleosalinity: Overview and presentation of a new approach. *Paleoceanography, 22*, PA3215. https://doi.org/10.1029/2007pa001437.

Rohling, E. J., Foster, G. L., Grant, K., Marino, G., Roberts, A. P., Tamisiea, M. E., et al. (2014). Sea-level and deep-sea-temperature variability over the past 5.3 million years. *Nature, 508*, 477–482.

Rohling, E. J., Marino, G., & Grant, K. M. (2015). Mediterranean climate and oceanography, and the periodic development of anoxic events (sapropels). *Earth-Science Reviews, 143*, 62–97.

Rosell-Melé, A., & Prahl, F. G. (2013). Seasonality of UK′ 37 temperature estimates as inferred from sediment trap data. *Quaternary Science Reviews, 72*, 128–136.

Rosenthal, Y., Boyle, E. A., & Slowey, N. (1997). Temperature control on the incorporation of magnesium, strontium, fluorine, and cadmium into benthic foraminiferal shells from Little Bahama Bank: Prospects for thermocline paleoceanography. *Geochimica et Cosmochimica Acta, 61*(17), 3633–3643.

Rossignol-Strick, M., Nesteroff, W., Olive, P., & Vergnaud-Grazzini, C. (1982). After the deluge: Mediterranean stagnation and sapropel formation. *Nature, 295*(5845), 105.

Saenger, C., Affek, H. P., Felis, T., Thiagarajan, N., Lough, J. M., & Holcomb, M. (2012). Carbonate clumped isotope variability in shallow water corals: Temperature dependence and growth-related vital effects. *Geochimica et Cosmochimica Acta, 99*, 224–242.

Schmidt, G. A., Bigg, G. R., & Rohling, E. J. (1999). Global Seawater oxygen-18 database—v1.21. Available at http://data.giss.nasa.gov/o18data/.

Schmidt, G. A. (1999). Error analysis of paleosalinity calculations. *Paleoceanography, 14*, 422–429.

Schmittner, A., Bostock, H. C., Cartapanis, O., Curry, W. B., Filipsson, H. L., Galbraith, E. D., et al. (2017). Calibration of the carbon isotope composition ($\delta^{13}C$) of benthic foraminifera. *Paleoceanography, 32*, 512–530.

Schauble, E. A., Ghosh, P., & Eiler, J. M. (2006). Preferential formation of 13C–18O bonds in carbonate minerals, estimated using first-principles lattice dynamics. *Geochimica et Cosmochimica Acta, 70*(10), 2510–2529.

Schouten, S., Hopmans, E. C., Schefuß, E., & Damste, J. S. S. (2002). Distributional variations in marine crenarchaeotal membrane lipids: A new tool for reconstructing ancient sea water temperatures? *Earth and Planetary Science Letters, 204*(1–2), 265–274.

Schouten, S., Ossebaar, J., Schreiber, K., Kienhuis, M. V. M., Langer, G., Benthien, A., et al. (2006). The effect of temperature, salinity and growth rate on the stable hydrogen isotopic composition of long chain alkenones produced by Emiliania huxleyi and Gephyrocapsa oceanica. *Biogeosciences, 3*, 113–119.

Schouten, S., Hopmans, E. C., & Damsté, J. S. S. (2013). The organic geochemistry of glycerol dialkyl glycerol tetraether lipids: A review. *Organic Geochemistry, 54*, 19–61.

Schrag, D. P., Adkins, J. F., McIntyre, K., Alexander, J. L., Hodell, D. A., Charles, C. D., et al. (2002). The oxygen isotopic composition of seawater during the last glacial maximum. *Quaternary Science Reviews, 21*(1–3), 331–342.

Shackleton, N. J. (1967). Oxygen isotope analysis and Pleistocene temperature reassessed. *Nature, 2151*, 15–17.

Shackleton, N. J. (1974). Attainment of isotopic equilibrium between ocean water and benthonic foraminifera genus Uvigerina: Isotopic changes in the ocean during the last glacial. dans Les méthodes quantitatives d'étude des variations du climat au cours du Pleistocène CNRS, Gif-sur-Yvette, pp. 203–209.

Shakun, J. D., Clark, P. U., He, F., Marcott, S. A., Mix, A. C., Liu, Z., et al. (2012). Global warming preceded by increasing carbon dioxide concentrations during the last deglaciation. *Nature, 484*, 49–54.

Shanahan, T. M., Hughen, K., McKay, N. P., Overpeck, J., Scholz, C. A., Gosling, W. D., et al. (2016). CO_2 and fire influence tropical ecosystem stability in response to climate change. *Scientific Reports, 6*, 29587.

Sicre, M. A., et al. (2005). Mid-latitude southern Indian ocean response to northern hemisphere Heinrich events. *Earth and Planetary Science Letters, 240*(3–4), 724–731.

Thierstein, H. R., Geitzenauer, K. R., Molfino, B., & Shackleton, N. J. (1977). Global synchroneity of late Quaternary coccolith datum levels validation by oxygen isotopes. *Geology, 5*(7), 400–404.

Thornalley, D. J. R., Oppo, D. W., Ortega, P., Robson, J. I., Brierley, C., Davis, R., et al. (2018). Anomalously weak Labrador Sea convection and Atlantic overturning during the past 150 years. *Nature, 556*, 227–232.

Tierney, J. E., & deMenocal, P. B. (2013). Abrupt shifts in Horn of Africa hydroclimate since the Last Glacial Maximum. *Science, 342*(6160), 843–846.

Tierney, J. E., & Tingley, M. P. (2018). BAYSPLINE: A new calibration for the alkenone paleothermometer. *Paleoceanography and Paleoclimatology, 33*, 281–301. https://doi.org/10.1002/2017PA003201.

Tierney, J. E., & Tingley, M. P. (2014). A Bayesian, spatially-varying calibration model for the TEX_{86} proxy. *Geochimica et Cosmochimica Acta, 127*, 83–106.

Vázquez Riveiros, N., Govin, A., Waelbroeck, C., Mackensen, A., Michel, E., Moreira, S., et al. (2016). Mg/Ca thermometry in planktic foraminifera: Improving paleotemperature estimations for G. bulloides and N. pachyderma left. *Geochemistry, Geophysics, Geosystems, 17*, 1249–1264.

Vázquez Riveiros, N., Waelbroeck, C., Skinner, L. C., Duplessy, J. C., McManus, J. F., Kandiano, E. S., et al. (2013). The 'MIS11 paradox' and ocean circulation: Role of millennial scale events. *Earth and Planetary Science Letters, 371–372*, 258–268.

Vázquez Riveiros, N., Waelbroeck, C., Skinner, L. C., Roche, D. M., Duplessy, J. C., & Michel, E. (2010). Response of South Atlantic deep waters to deglacial warming during terminations V and I. *Earth and Planetary Science Letters, 298*, 323–333.

Waelbroeck, C., Labeyrie, L., Duplessy, J. C., Guiot, J., Labracherie, M., Leclaire, H., et al. (1998). Improving past sea surface temperature estimates based on planktonic fossil faunas. *Paleoceanography, 13*, 272–283.

Waelbroeck, C., Duplessy, J. C., Michel, E., Labeyrie, L., Paillard, D., & Duprat, J. (2001). The timing of the last deglaciation in North Atlantic climate records. *Nature, 412*, 724–727.

Weijers, J. W., Schouten, S., Spaargaren, O. C., & Damsté, J. S. S. (2006). Occurrence and distribution of tetraether membrane lipids in soils: Implications for the use of the TEX86 proxy and the BIT index. *Organic Geochemistry, 37*(12), 1680–1693.

Weldeab, S., Lea, D. W., Schneider, R. R., & Andersen, N. (2007). 155,000 years of West African monsoon and ocean thermal evolution. *Science, 316*, 1303–1307.

Wiersma, A. P., & Renssen, H. (2006). Model–data comparison for the 8.2 ka BP event: Confirmation of a forcing mechanism by catastrophic drainage of Laurentide Lakes. *Quaternary Science Reviews, 25*(1–2), 63–88.

Wit, J. C., De Nooijer, L. J., Wolthers, M., & Reichart, G. J. (2013). A novel salinity proxy based on Na incorporation into foraminiferal calcite. *Biogeosciences, 10*(10), 6375–6387.

Wolhowe, M. D., Prahl, F. G., Probert, I., & Maldonado, M. (2009). Growth phase dependent hydrogen isotopic fractionation in alkenone-producing haptophytes. *Biogeosciences, 6,* 1681–1694.

Yu, E.-F., Francois, R., & Bacon, M. (1996). Similar rates of modern and last-glacial ocean thermohaline circulation inferred from radiochemical data. *Nature, 379,* 689–694.

Zachos, J. C., Shackleton, N. J., Revenaugh, J. S., Pälike, H., & Flower, B. P. (2001). Climate response to orbital forcing across the oligocene-miocene boundary. *Science, 292,* 274–278.

Zachos, J. C., Dickens, G. R., & Zeebe, R. E. (2008). An early Cenozoic perspective on greenhouse warming and carbon-cycle dynamics. *Nature, 451,* 279–283. https://doi.org/10.1038/nature06588.

Zweng, M. M., Reagan, J. R., Antonov, J. J., Locarnini, R. A., Mishonov, A. V., Boyer, T. P., Garcia, H. E., Baranova, O. K., Johnson, D. R., Seidov, D., & Biddle, M. M. (2013). *World Ocean Atlas 2013, Volume 2: Salinity*. In S. Levitus & A. V. Mishonov (Eds.), NOAA Atlas NESDIS 74, p. 39.